Transgenic Animals

Edited by

F. GROSVELD

Laboratory of Gene Structure and Expression, National Institute for Medical Research, London, UK

and

G. KOLLIAS

Molecular Genetics Laboratory, Hellenic Pasteur Institute, Athens, Greece

ACADEMIC PRESS
Harcourt Brace Jovanovich, Publishers
London   San Diego   New York
Boston   Sydney   Tokyo   Toronto

ACADEMIC PRESS LIMITED
24–28 Oval Road
London NW1 7DX

United States Edition published by
ACADEMIC PRESS INC.
San Diego, CA 92101

A catalogue record for this book is available from the British Library

ISBN 0-12-304530-4

Typeset by Colset Private Limited, Singapore
Printed and bound in Great Britain by TJ Press Ltd, Padstow, Cornwall

Contents

Contributors

R. S. P. BEDDINGTON, *AFRC Centre for Animal Genome Research, King's Buildings, West Mains Road, Edinburgh EH9 3JQ, UK*

A. BERNSTEIN, *Department of Medical Genetics, University of Toronto, Toronto, Ontario, Canada*

C. BONIFER, *Institut für Biologie III der Universität Freiburg, Department of Genetics, Schänzlestrasse 1, D-7800 Freiburg/Br., Germany*

M. L. BREITMAN, *Division of Molecular and Developmental Biology, Mount Sinai Hospital Research Institute, 600 University Avenue, Toronto, Ontario, Canada, M5G 1X5*

C. CAMPBELL, *Department of Molecular Genetics, Albert Einstein College of Medicine, 1300 Morris Park Avenue, Bronx, NY 10461, USA*

A. J. CLARK, *AFRC Institute of Animal Physiology and Genetics, Edinburgh Research Station, Roslin, Midlothian EH25 9PS, UK*

S. DOI, *Department of Molecular Genetics, Albert Einstein College of Medicine, 1300 Morris Park Avenue, Bronx, NY 10461, USA*

N. FAUST, *Institut für Biologie III der Universität Freiburg, Department of Genetics, Schänzlestrasse 1, D-7800 Freiburg/Br., Germany*

T. GREWEL, *Institut für Biologie III der Universität Freiburg, Department of Genetics, Schänzlestrasse 1, D-7800 Freiburg/Br., Germany*

F. GROSVELD, *Laboratory of Gene Structure and Expression, National Institute for Medical Research, The Ridgeway, Mill Hill, London NW7 1AA, UK*

A. HECHT, *Institut für Biologie III der Universität Freiburg, Department of Genetics, Schänzlestrasse 1, D-7800 Freiburg/Br., Germany*

D. KIOUSSIS, *Division of Molecular Immunology, National Institute for Medical Research, The Ridgeway, Mill Hill, London NW7 1AA, UK*

G. KOLLIAS, *Molecular Genetics Laboratory, Hellenic Pasteur Institute, 127 Vas Sofias Avenue, GR 115 21 Athens, Greece*

R. KUCHERLAPATI, *Department of Molecular Genetics, Albert Einstein College of Medicine, 1300 Morris Park Avenue, Bronx, NY 10461, USA*

A. L. MELLOR, *Division of Molecular Immunology, National Institute for Medical Research, The Ridgeway, Mill Hill, London NW7 1AA, UK*

W. REIK, *Department of Molecular Embryology, Institute of Animal Physiology and Genetics Research, Babraham, Cambridge CB2 4AT, UK*

W. D. RICHARDSON, *Department of Biology (Medawar Building), University College London, Gower Street, London, WC1E 6BT, UK*

H. SAUERESSIG, *Institut für Biologie III der Universität Freiburg, Department of Genetics, Schänzlestrasse 1, D-7800 Freiburg/Br., Germany*

J. P. SIMONS, *AFRC Institute of Animal Physiology and Genetics, Edinburgh Research Station, Roslin, Midlothian EH25 9PS, UK*

A. E. SIPPEL, *Institut für Biologie III der Universität Freiburg, Department of Genetics, Schänzlestrasse 1, D-7800 Freiburg/Br., Germany*

D. VALERIO, *Gene Therapy Division of the Institute of Applied Radiobiology and Immunology – TNO, PO Box 5815, 2280 HV Rijswijk, The Netherlands*

I. WILMUT, *AFRC Institute of Animal Physiology and Genetics, Edimburgh Research Station, Roslin, Midlothian EH25 9PS, UK*

D. WINTER, *Institut für Biologie III der Universität Freiburg, Department of Genetics, Schänzlestrasse 1, D-7800 Freiburg/Br., Germany*

Preface

The use of transgenic animals has increased dramatically during the last decade and is making a serious impact on our understanding of complex biological systems. It has greatly advanced our knowledge of gene regulation and function, mammalian development, and pathogenesis of disease. We have compiled a book that does not concentrate on techniques or the latest results, but rather a book that would provide an insight into the enormous possibilities that transgenesis provides to address biological questions. We have asked the contributors to address a general area by presenting recent achievements and to discuss, speculate and think about the future. We hope that this will provide an exciting book for investigators who are not familiar with the technology and convince them that the use of transgenesis, despite its substantial cost in equipment, facilities and training, is a very good investment.

We would like to thank all of the authors for their contribution.

F. GROSVELD and G. KOLLIAS

1

The regulatory domain organization of eukaryotic genomes: implications for stable gene transfer

ALBRECHT E. SIPPEL, HARALD SAUERESSIG,
DIANA WINTER, THOMAS GREWAL, NICOLE FAUST,
ANDREAS HECHT and CONSTANZE BONIFER

Institut für Biologie III der Universität Freiburg, Department of Genetics, Schänzlestrasse 1, D-7800 Freiburg/Br., Germany

1. Introduction

The analysis of the molecular mechanisms of regulation of eukaryotic genes is in itself in an evolutionary process. Since the beginning of the recombinant DNA revolution, ever more sophisticated techniques have been developed to describe the structure and function of genes *in vivo*. Initial concentration on DNA coding regions led to the discovery of the exon–intron organization of genes and opened our eyes to the (minigene/protein domain) modular evolution of proteins (Dorit *et al.*, 1990). The analysis of primary transcripts defined the

TRANSGENIC ANIMALS
ISBN 0–12–304530–4

major cornerstones of the basic transcription process, promoter, poly(A) point and termination site. This, in turn, led to the current effort to unravel the mechanism of RNA processing and splicing. In a simplified view, it became clear that a nascent RNA transcript carries the sequence information for nuclear processing steps analogous to the way in which mRNAs carry information for the production of proteins. Linear blueprints code for structures and complex functions and determine individuality by sequence variation.

We are now starting to understand the next dimension of gene function, their regulatory aspect. It is not surprising that similar principles of DNA sequence information present the key for the elucidation of the spatial and temporal aspect of gene activity. A new set of DNA sequence motifs and elements of multiple motifs mostly located in the flanking regions of transcription units are currently discovered and their functional implications are being elucidated. In this chapter on the domain organization of eukaryotic genomes we will outline some of the principles which appear to govern the regulatory DNA information for gene activity. Towards the end we will touch on the implications of these findings for gene transfer into cells and transgenic organisms.

Most current studies on the molecular mechanism of cell type- and cell stage-specific gene activation or inactivation aim to analyse *cis*-active DNA sequences and *trans*-regulatory protein factors involved in promoter, enhancer and silencer function by the use of transient DNA transfer in cells in culture or in *in vitro* transcription systems. However, with the increasing desire to understand chromatin influence on gene regulation it is necessary to include studies which determine the structure of genes in their natural nuclear environment and to determine chromatin function via episomal or chromosomal cellular reinsertion of mutated reporter gene constructs.

Unlike the general chromatin structure, chromatin-mediated processes on gene regulation need to be studied preferentially at the level of individual gene loci. This involves a tremendous downwards shift in concentration to one or two per cell for any structure looked at. Dynamic chromatin analysis demands functional assay systems based on stable transfer of wild-type and mutant DNA constructs into cells and organisms, complementing the analysis of site-specific structures in genomic chromatin.

Higher-order chromatin organization, like nucleosome phasing, nucleosome non-histone protein interaction and torsional and topological problems of genomic DNA, must be considered as a local genomic event before the full impact of chromatin on gene regulation can be understood. In this respect we will mainly present results which we have derived from studies on the chicken lysozyme gene locus and we will consider them in perspective to results derived from one of the most intensively analysed eukaryotic gene regions — the human β-globin gene cluster.

The lysozyme gene is a marker gene for macrophage differentiation (Sippel *et al.*, 1987). It serves a similar role in studies on cell differentiation of the myeloblast–myelomonocyte–macrophage cell lineage of haematopoiesis as globin genes serve for the erythroid cell lineage or immunoglobulin genes serve for the lymphoid cell lineage. The genomic lysozyme locus, however, is organized more simply than these other marker gene regions. In the chicken it contains a single 4 kb gene of four exons and has, as far as is known, no adjacent related gene. The complexity of lysozyme gene regulation, on the other hand, compares well with the regulation of globin genes. Because of its compact chromatin domain the chicken lysozyme locus offers attractive features for the analysis of the role of chromatin in gene regulation.

Lysozyme gene transcription is progressively activated during macrophage development, apparently coupled to the cell differentiation process (Sippel *et al.*, 1988). In the chicken, the same gene is also steroid-inducible in oviduct tubular gland cells (Schütz *et al.*, 1978) and its transcription proceeds from the same promoter as in macrophages (Theisen *et al.*, 1986). The two modes of transcriptional regulation are based on alternative states of chromatin structure which develop in oviduct and myeloid cells (Fritton *et al.*, 1984, 1987). Different sets of regulatory switches in the flanking chromatin of the gene are used to either control transcription by steroid hormones in the oviduct, or activate it constitutively in macrophages (Sippel and Renkawitz, 1989; Sippel *et al.*, 1989). We have found that tissue-specific activation of the gene most likely results from the combined function of, maximally, nine regulatory elements within a chromatin domain of roughly 20 kb (Fritton *et al.*, 1987; Sippel *et al.*, 1988). The regulatory unit ('regulon') appears to be identical with a chromosomal loop, since at its distal ends the chromatin domain is confined by two matrix-attached DNA regions (SARs or MARs), as shown by Phi-Van and Strätling (1988).

Stable DNA transfection experiments have demonstrated that the function of the domain-bordering DNA elements (A-elements for attachment elements) guarantees position-independent reporter gene activity when minidomains were randomly integrated into the genome of tissue culture cells (Stief *et al.*, 1989). In transgenic mice, the DNA of the entire domain mediated consistently high-level, copy number-dependent, position-independent, correctly regulated transcription of the gene (Bonifer *et al.*, 1990). From this we conclude that genomic position effects of randomly integrated transgenes can be overcome if functional chromosomal domains (rather than incomplete subdomain DNA fragments) are stably transferred into the genome of cells and organisms.

2. Chromatin loop domains

Most likely, our early conceptions of the organization of active chromatin in interphase vertebrate nuclei were influenced by pictures of lampbrush chromosomes (Gall, 1956) and of giant polytene chromosomes in cells of the salivary gland of dipteran flies (Beermann, 1952). Both the lateral successive loops of the former and the banding and puffed appearance of highly active chromosomal regions in the latter suggested a looped domain organization for chromosomes of other species as well. Indeed, it was later shown that chromatin of interphase and DNA of metaphase nuclei appeared to be structured in loops constrained by nuclear matrix or chromosomal scaffold material (Benyajati and Worcel, 1976; Cook and Brazell, 1976; Paulson and Laemmli, 1977; Igo-Kemenes and Zachau, 1977).

On lower organizational levels, chromatin is described as a regular structure of DNA wrapped around nucleosomal histone octamers connected by histone H1-covered linker DNA regions (Richmond *et al.*, 1983) and as a 30 nm solenoidal filament in which this 10 nm 'beads on a string' fibre is again wound up into a more compact structure with six nucleosomes per turn (McGhee *et al.*, 1980). Only when structural studies of chromatin are connected to specific gene loci can functional aspects be approached and questions asked on how chromatin is organized in respect to transcriptional active and inactive genes.

2.1. *General DNAaseI sensitivity*

The first bridge from the general chromatin structure to the chromatin structure of a specific vertebrate gene was made by Weintraub and Groudine (1976) when they showed that chicken globin gene DNA in actively transcribing cells is preferentially sensitive towards digestion with DNAaseI and that it is relatively insensitive in non-transcribing fibroblasts. Soon afterwards, Garel and Axel (1976) confirmed these data by showing that the ovalbumin gene is preferentially digested by DNAaseI in chicken oviduct chromatin, but not in chicken liver or erythroid nuclei, in which the gene is never expressed. Later it was found that the general DNAaseI sensitivity is not restricted to the coding region of active genes but also extends to their flanking chromatin (Lawson *et al.*, 1980; Stalder *et al.*, 1980). The first gene locus for which the extent of general DNAase sensitivity was mapped towards its 5′ and 3′ chromosomal borders was the active chicken ovalbumin locus. It includes in a region of 100 kb (approximately 500 nucleosomes) a cluster of three ovalbumin-related genes (Lawson *et al.*, 1982). Other examples for which the extent of the DNAaseI-sensitive domain is known are the chicken GAPDH locus of only 12 kb in size (Alevy *et al.*, 1984), the chicken lysozyme locus of 20–24 kb (Jantzen *et al.*, 1986) and, more recently, the locus of the human apolipoprotein B gene of 45 kb (Levy-Wilson and Fortier, 1989) and the human β-globin gene cluster of more than 150 kb (Forrester *et al.*, 1986).

A schematic diagram of the chicken lysozyme gene domain is shown in Fig. 1.

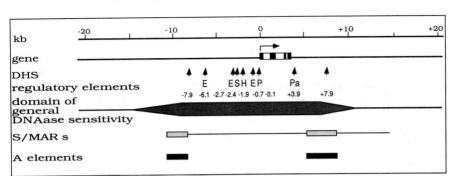

Fig. 1. The active chromatin domain of the chicken lysozyme gene. The diagram shows, from top to bottom: the gene with its four exons (solid bars), three introns (open bars) and the direction of transcription (horizontal arrow); the positions of its nine DNAaseI-hypersensitive sites (DHSs) (vertical arrows); the functional characteristics of the regulatory elements (E, enhancer; S, silencer; H, hormone response element; P, promoter; Pa, poly(A) point) and their location (in kb) with respect to the transcriptional start site; the domain of general DNAaseI sensitivity with its distal regions of transition from open to closed chromatin; the position of scaffold- and matrix-attachment regions (S/MARs) (stippled bars) mapped within the region given by the thin line and the location of the functional A-elements (solid bars).

After cloning and characterizing approximately 50 kb of genomic chicken DNA encompassing the lysozyme gene (Lindenmaier *et al.*, 1979; Baldacci *et al.*, 1981), single-copy fragments along the gene region were used to probe for general DNAaseI sensitivity (Jantzen *et al.*, 1986), for differential chromatin fractionation after micrococcal nuclease digestion and for an analysis of the pattern of nucleosomal repeats (Strätling *et al.*, 1986) in the active locus in oviduct nuclei from the laying hen. All determined parameters mapped the transcribed gene in the centre of a chromatin domain with increased general nuclease sensitivity of DNA and a distinct, more loose pattern of nucleosomes extending over approximately 20 kb. As delineated in Fig. 1, DNAaseI sensitivity at the boundaries of the domain, around −10 and +10 kb with respect to the promoter of the gene, did not drop in one step at defined positions but, rather, decreased gradually over several kilobases to the lower level of DNAaseI sensitivity measured for chromatin of inactive genes. Distal to both transition regions in oviduct nuclei, 'condensed' chromatin of more than 10 kb in length extended in the 5′ and 3′ directions.

2.2. Scaffold- and matrix-attachment regions (S/MARs)

When looped DNA in metaphase chromosomes and interphase nuclei was microscopically examined after extraction of histones, it was noticed that DNA double strands were attached to scaffold or matrix material at the base of the loops (Paulson and Laemmli, 1977). From this it was surmised that specific regions existed along DNA at which attachments to the supporting nuclear structure occur. Two assays were developed to map attachment regions in genomic DNA. The nuclear 'halo' mapping procedure identifies SARs after mild detergent treatment of nuclei (Mirkovitch *et al.*, 1984). The *in vitro* DNA binding assay localizes MARs within cloned DNA (Cockerill and Garrard, 1986). Both assays map the same AT-rich anchorage sequences. In fact, both assays are very much alike because it was shown that exogenously added S/MAR sequences attach to nuclear scaffold material during the lithium 3,5-diiodosalicylate (LIS) preparation of nuclear haloes (Izaurralde *et al.*, 1988). This leaves us in the unfortunate situation that neither of the currently available attachment assays discriminates between true *in vivo* and potential anchorage sites. Potential attachment regions (S/MARs) were found frequently near *cis*-regulatory sequences outside genomic transcription units but not exclusively (for an earlier review see Gasser and Laemmli, 1987; for a more recent list see Mielke *et al.*, 1990).

New evidence supporting the assumption that S/MARs can be *in vivo* anchorage points for genomic loop domains has come from S/MAR mapping in gene loci which also have been used for the mapping of the DNAaseI-sensitive chromatin domain. To our knowledge, results on parallel mapping studies have only been published for two eukaryotic genes, the human apolipoprotein B gene (Levy-Wilson and Fortier, 1989) and the chicken lysozyme gene (Jantzen *et al.*, 1986; Phi-Van and Strätling, 1988). For both genes, attachment regions co-locate with the upstream and the downstream border of the active chromatin domain. As shown in Fig. 1, on the 25 kb DNA of the chicken lysozyme locus used for mapping, S/MARs were found between 11.1 and 8.85 kb upstream of the transcription start site and between 1.3 and 5.0 kb downstream of the poly(A) addition site. The local coincidence of S/MARs with the borders between open and closed chromatin suggests that there might be a functional relationship. The gradual change of general DNAaseI sensitivity could be an expression of the multifocal attachment function observed along the attachment region of several kilobases in length.

2.3. DNAaseI-hypersensitive sites (DHSs)

In addition to the size of the region of general DNAaseI sensitivity and the location of attachment sites there is a third structural feature which has relevance for the looped domain organization of chromatin. Nucleosomal arrays are occasionally interrupted by DHSs. They are more frequently found in the chromatin of active gene loci than in inactive chromatin (for reviews see Elgin, 1988; Gross and Garrard, 1988). Their functional relevance will be discussed later. Here we will just consider their uneven distribution when they are mapped along larger chromosomal regions. When performing a thorough analysis of the presence and absence of DHSs in the chromatin of 50 kb of DNA around the chicken lysozyme gene in a number of expressing and non-expressing cells (Fritton et al., 1983, 1984, 1987) up to nine DHSs were found clustering around the gene (Fig. 1). All are located within the limits set by the active chromatin domain of general DNAaseI sensitivity. The cluster of DHSs is bordered outside of the domain by stretches of more than 15 kb of chromatin in which no such sites can be detected in eight different cell types tested (Sippel, 1988). Since the region in which DHSs occur appears to be constant in size, apparently independent of the transcriptional or regulatory state of the gene, we conclude that the cluster of DHSs is confined within the suggested chromosomal loop. This also indicates that the size of the loop might be stable in different cell types.

3. Regulatory units for gene activity

It is likely that the structural organization of the genome into looped domains facilitates the more than 10 000-fold compaction of DNA in the eukaryotic metaphase nucleus. However, functional aspects might be even more important than structural factors. DNA in mammalian nuclei must have a dynamic conformation favourable for processes like replication and repair, transcription, and local loosening up of chromatin. For theoretical reasons it was predicted that in order for DNA to permit diversity of function it must be topographically partitioned into independent functional units (Cook, 1973; Jackson, 1986; Nelson et al., 1986). In recent years, mainly as a result of DNA transfer studies in combination with studies on chromatin structure, experimental evidence has accumulated which supports the view that the eukaryotic genome is in fact organized in functional domains. We are now collecting data suggesting that chromosomal looped domains are not only structural units but are, in addition, the basic regulatory units for the control of gene activity. Evidence for this hypothesis comes from studies of the function of three principally different types of cis-active DNA signal sequences which are found to be present in chromatin domains: (1) regulatory switches like promoters, enhancers and silencers, (2) regulatory punctuation signals like A-elements and (3) locus activation regions or dominant control regions (these now have a unified nomenclature, and are called locus control regions, LCRs).

3.1. *Regulatory switches* ·

DHSs in eukaryotic chromatin were mapped by the indirect end-labelling procedure (Wu *et al.*, 1979; Nedospasov and Georgiev, 1980), originally to viral regulatory regions (Varshavsky *et al.*, 1978; Scott and Wigmore, 1979), and then to cellular promoters (Wu *et al.*, 1979). Later, it was found in practically every case that DHSs map to DNA regions which contain sequences functioning in transcription, replication, recombination or chromosome segregation and to *cis*-active sites involved in DNA-dependent regulatory processes. As it turned out, DHSs mark positions of short 50–400 bp nucleosome-free chromatin regions in which sequence-specific non-histone DNA-binding proteins have access to their recognition motifs (McGhee *et al.*, 1981; Emerson and Felsenfeld, 1984). For this reason it is not surprising that DHS mapping of chromatin is frequently used to complement cellular transfection studies with DNA constructs to find *cis*-regulatory DNA elements in gene loci. When applied in parallel these independent methods ensure, one with structural and the other with functional information, high biological relevance for any element detected by both strategies.

Our mapping of DHSs in the lysozyme gene region enabled us to compare chromatin structural features in different cell types and to correlate them with the different functional states of the gene. We found that the patterns of DHSs are different in cells in which lysozyme is expressed from those cells in which it is not expressed (Fritton *et al.*, 1983, 1984, 1987). Certain hypersensitive sites appear upon transcriptional activation, either by steroid hormone induction or during cell differentiation. The different modes of transcriptional regulation of the gene correlate with different sets of DHSs in the flanking chromatin, a result which supports the notion that these elements themselves determine the functional state of the gene (Fritton *et al.*, 1984; Reudelhuber, 1984).

Transient DNA-mediated gene transfer in tissue culture cells has proved to be a rapid functional *in vivo* assay for the regulatory specificity of DNA sequences. With this method the first enhancers were mapped in immunoglobulin genes (Banerji *et al.*, 1983; Gillies *et al.*, 1983; Queen and Baltimore, 1983). By transfection of lysozyme DNA-containing reporter gene constructs into different cell lines or primary tissue culture cells of chicken haematopoietic and non-haematopoietic origin we have shown *cis*-regulatory specificities for, currently, five of the seven DHS elements in lysozyme upstream chromatin (Fig. 1):

1. Enhancer at −6.1 kb (−6.1E). The far-upstream transcriptional enhancer is active in all early- and late-type myeloid precursor cells and mature macrophages (Theisen *et al.*, 1986; Grewal *et al.*, 1992). The element can

be mapped as a DHS in chromatin of all cell types in which the lysozyme gene will be active independently of its actual transcriptional state or state of regulation (Fritton *et al.*, 1984; Sippel *et al.*, 1986, 1988).

2. Enhancer at −2.7 kb (−2.7E). This second enhancer is active only in late-type promacrophages (Steiner *et al.*, 1987; Müller *et al.*, 1990); in parallel, the DHS element is only seen in late-type myeloid cells (Fritton *et al.*, 1984, 1987; Sippel *et al.*, 1988).

3. Silencer at −2.4 kb (−2.4S). The *cis* activity of this inhibiting element in different cells coincides with the appearance of a DHS in chromatin (Baniahmed *et al.*, 1987, 1990; Sippel *et al.*, 1988). Results from DNA transfections and chromatin analysis are consistent with the assumption that the −2.4S element is a silencer only in early myeloid differentiation stages and that the full constitutive activation of the lysozyme gene is connected to silencer inactivation in later stages of macrophage maturation (Sippel and Renkawitz, 1989).

4. Steroid hormone response element at −1.9 kb (−1.9H). This inducible enhancer element shows glucocorticoid, progesterone- and oestrogen-dependent *cis*-activation of the homologous as well as heterologous gene promoters in steroid target cells (Hecht *et al.*, 1988, 1990). A DHS at −1.9 kb could only be detected in chromatin of steroid-induced oviduct cells (Fritton *et al.*, 1984) and in activated macrophage-like cells grown from tetradecanoylphorbol acetate (TPA)-induced avian myeloblastosis virus (AMV)-transformed myeloblasts (Müller, 1990; see also Fig. 2).

5. Enhancer–promoter region from −0.7 kb to +0.05 kb (−0.2E/−0.1P). In chromatin the promoter-proximal region of the lysozyme gene is DNAase-hypersensitive only in cell types in which the gene is active or potentially active (Fritton *et al.*, 1984; Sippel *et al.*, 1988).

The combination of structural and functional analysis makes it highly likely that the presense or absence of DNAase hypersensitivity in chromatin signals the actual functioning or non-functioning of individual regulatory elements in the respective cells. Indeed, we could show by genomic footprinting that the absence or presence of the DHS at the −6.1 kb enhancer in different haematopoietic chicken cells is strictly correlated to the simultaneous binding or non-binding of an entire set of six proteins to this DNA region (Borgmeyer, 1987).

The functional specificities, as detected by transient DNA transfection in various transformed haematopoietic chicken cells, show that each element is responsible for a specific subaspect of the global control of lysozyme gene activity. We have to assume a cell type- and cell stage-specific combined action of certain sets of these multiple regulatory switches. The possibility that multiple enhancers act in a combinatorial fashion to generate regulatory diversity was also deduced from transgenic mice experiments with elements from the

developmental step	commitment		maturation	activation	
cell type	bone marrow stem cells	myeloblasts	monocytes macrophages	activated macrophages	
chromatin structure					
transcription	−	+/−	++	++++	

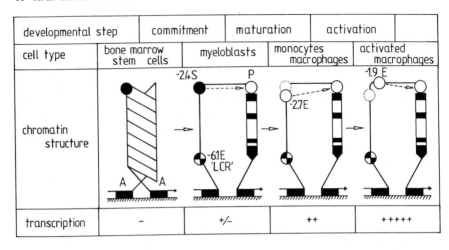

Fig. 2. Schematic diagram of the proposed mechanism of a three-step activation of the chicken lysozyme gene domain as deduced from transformed haematopoietic cell lines representing various stages of myeloid cell differentiation. The structure of the gene locus in bone marrow stem cells cannot be analysed directly but is deduced from the inactive chromatin as seen in total 5-day embryonal cells, adult liver cells and transformed early-type erythroid cells (AEV-transformed erythroblasts, HD3). Diagonal bars indicate the tightly packaged chromatin of the domain, the solid circles the DHS element at −2.4 kb and the solid bars (A) the anchorage regions of the chromosomal loop. In a first step, denoted 'commitment', the domain is proposed to be transformed into open chromatin as seen in transformed early-type myeloid cells (E26 virus-transformed myeloblasts) characterized by the presence of a DHS at the −6.1 kb early enhancer (here tentatively named 'LCR'), a DHS at the −2.4 kb silencer and a DHS at the promoter of the gene (four exons, three introns). In a second step, denoted 'maturation', the chromatin is proposed to change to the structure seen in transformed late-type myeloid cells (MC29 virus-transformed monocytes) and mature blood-derived macrophages. The DHS at the −2.4 kb silencer disappears and is replaced by a DHS at the −2.7 kb late enhancer. Simultaneously, gene transcription is activated (150–300 transcripts per cell). In a third step, denoted activation, the chromatin is proposed to change to the structure seen in macrophage-type cells differentiated from TPA-induced AMV-transformed myeloblasts (BM2 cells). The appearance of a DHS at the −1.9 kb inducible enhancer is simultaneous with a further roughly 100-fold transcriptional activation of the gene.

α-fetoprotein gene locus (Hammer *et al.*, 1987). Similar assumptions were drawn for the regulatory elements mapped in the chicken β-globin locus (Choi and Engel, 1988). Whereas in the lysozyme domain different enhancers cell-specifically activate the same promoter region, in chicken β-globin switching the same central enhancer element activates the different promoter elements of the separate embryonal and adult globin genes. Appropriate stage-specific regulation of the two globin genes is observed only when they are both present on the same plasmid, indicating direct competitive cell type-specific interaction of the enhancer protein complex with the respective promoter protein complexes. A competitive enhancer–promoter interplay can also be seen in the human fetal-to-adult β-globin switch (Stamatoyannopoulos, 1991). Experiments in transgenic mice show that linkage of the superenhancers in the far-upstream locus control region to individual human fetal γ- or adult β-globin genes results in gene expression at all stages of mouse development (Enver *et al.*, 1989; Behringer *et al.*, 1990). Correct developmental control is, however, restored in constructs that contain both the β- and γ-globin genes. This again suggests that the fetal-to-adult gene switch is at least partly mediated by a competitive direct interaction of regulatory complexes in the locus (Behringer *et al.*, 1990; Enver *et al.*, 1990).

The results from several thoroughly analysed gene loci all point in the same direction. Individual regulatory complexes in flanking parts of gene loci are unable to mediate the complete specificity of transcriptional control of nearby genes. Rather, they are responsible for distinct subaspects of global control. Indirect and, recently, direct evidence suggests that the regulatory switch elements are combining their specifities by combinatorial interaction. It appears that, in all cases, interacting elements are located in the same region of generally DNAase sensitive chromatin. In the case of the chicken lysozyme gene locus it can be clearly shown that all regulatory elements are located within the boundaries set for the loop domain of loose chromatin, indicating that structurally defined chromatin domains are equivalent to functional units of regulation ('regulons').

3.2.   *Regulatory 'punctuation marks'*

Chromatin loop domains have distinct borders which become visible when active chromatin abuts inactive chromatin. For a few gene loci the regions in which the chromatin structure changes from the active/open to the inactive/ closed conformation could be shown to co-locate with DNA sequences having increased binding affinity to nuclear scaffold/matrix material (Phi-Van and Strätling, 1988; Levi-Wilson and Fortier, 1989). Sequestration into topological chromatin loops might be the underlying mechanism confining regulatory units and limiting *cis*-acting regulatory influence along the genome. Could this mean that the border region of chromosomal loop structures is a sort of regulatory 'punctuation mark'?

We have tested this possibility for the case of the 5′ border region of the chicken lysozyme gene domain using experiments similar to those applied to other *cis*-acting regulatory elements for transcription, like enhancers or silencers. For the functional analysis in gene transfer experiments we constructed 'minidomains' consisting of the S/MAR element between −12.0 and −9.0 kb (A) of the lysozyme locus on both ends of a DNA construct containing the lysozyme −6.1 kb enhancer (E), the lysozyme promoter region (P) and the chloramphenicol acetyltransferase (CAT) reporter gene (C). When these minidomain constructs were tested in transient transfection experiments in chicken promacrophage cells, in which the endogenous chicken lysozyme gene is active, no difference in reporter gene activity could be seen when compared with DNA constructs lacking the border elements (Stief *et al.*, 1989). The border elements remained neutral with respect to the transcriptional activity of the reporter gene as long as the DNA of the minidomain constructs was not integrated into the cellular genome. To test border element function stably integrated at a random position in cellular genomes we co-transfected mini-domain constructs with plasmids carrying a neomycin resistance gene into promacrophages and isolated neomycin-resistant cell lines that were clonally derived, each representing an individual integration event of the reporter con-struct (Stief *et al.*, 1989). The comparison of reporter gene activity in cell clones with stably integrated constructs lacking border elements clearly showed that, when in a chromosomal position, border elements do not remain neutral. They are functionally relevant for transcription with characteristics indicative for a new type of *cis*-active DNA element.

First, A-elements cause a significant stimulation of transcription of reporter genes randomly inserted into the genome. This stimulation must result from a mechanism quite different from the one used by enhancers, since it cannot be detected in transiently transfected cells. Stimulation of A-elements is additive to enhancer activity and can only be seen if a functional regulatory unit is located between two A-elements. Incomplete functional units cannot be stimulated (Saueressig, 1989).

Secondly, A-elements buffer the position effect normally seen when DNA constructs are randomly inserted into cellular DNA. Whereas all cell lines with constructs lacking A-elements show no correlation between copy number of integrated DNA constructs and reporter gene activity, cell lines with A-element-containing minidomains show position-independent copy number dependency. The clonal distribution clearly demonstrates that A-elements function as regulatory 'punctuation marks' between the inserted reporter gene domain and the unpredictably deregulating influence of neighbouring DNA at the random sites of genomic insertion. This interpretation is supported by the finding that A-elements when placed between enhancers and promoters inhibit enhancer effects in transiently (Stief *et al.*, 1989) and in stably transfected cells (Sippel *et al.*, 1989).

Thirdly, A-elements do not appear to function cell type- or species-specifically. The chicken lysozyme 5′ A element has been shown to act in a stimulatory and position effect buffering way not only in myeloid cells of the chicken but also in a variety of cells of mammalian and non-myeloid origin (Saueressig, 1989; Phi-Van *et al.*, 1990). In addition, the chicken 5′ A-element action is not restricted to cooperation with specific enhancer and promoter elements. It is able to cooperate with heterologous enhancer–promoter elements in the same way as it is seen to cooperate with its homologous *cis*-active transcriptional control elements of the chicken lysozyme gene domain (Saueressig, 1989; Phi-Van *et al.*, 1990).

Because characteristics of A-elements turned out to be different to those of enhancer and silencer elements, we gave these functional DNA elements a new name. We called them A-elements, referring to their scaffold/matrix attachment activity (Stief *et al.*, 1989). The results from functional DNA transfer experiments outlined above define the chicken 5' A-element to be a genomic DNA element mediating position-independent gene expression. Whatever its molecular mechanism of action, it may be only the first example of a border element for regulatory domains frequently to be found in eukaryotic genomes. We have indications that the lysozyme 3' S/MAR too has A-element function (Saueressig *et al.*, 1989). Another DNA element for which *in vitro* scaffold/ matrix attachment coincides with *in vivo* transcriptional activation activity was recently mapped in the 3'-flanking region of the human interferon β gene (Mielke *et al.*, 1990). Domain border elements, functionally homologous to vertebrate A-elements, were also detected in the genome of *Drosophila melanogaster*. Structural domain border elements flanking the two heat shock genes *Hsp70* at the cytogenetic locus 87A7 (Udvardy *et al.*, 1985) are now shown to mediate position-independent correct regulation of randomly integrated reporter gene constructs (Kellum and Schedl, 1991). Blocks inhibiting long-distance regulatory effects have recently been postulated to subdivide the bithorax complex BX-C (Gyurkovics *et al.*, 1990). It is likely that A-elements which confine regulatory units for transcription are only a subset of a much broader family of genomic elements which subdivide chromosomes into functional units of higher-order chromatin organization. As indicated from studies on the phenomenon of position effect variegation, it is likely that block-like DNA elements are located at the borders between sections of euchromatin and heterochromatin (Henikoff, 1990; Paro, 1990). Future work has to show the proposed hierarchical system of chromatin-organizing *cis*-active DNA elements which might overlap with and subdivide clusters of regulatory chromatin domains of transcriptionally active and inactive gene loci.

3.3. The concept of LCR function

Recent efforts to understand dynamic regulatory processes of globin gene clusters in mammals suggest that a third type of DNA element besides regulatory switches (enhancers, silencers) and domain-bordering A elements are involved in the control of chromatin domains during cell differentiation (Townes and

Behringer, 1990; Orkin, 1990). Originally, it was observed that deletions far upstream of the human β-globin gene inactivate its transcriptional activity (Kioussis *et al.*, 1983) and that the deleted chromatin region contains from 6 to 18 kb 5 ' to the ε gene a group of four developmentally stable erythroid-specific DNAase-hypersensitive sites (Tuan *et al.*, 1985; Forrester *et al.*, 1986). A functional assay exploiting human globin expression in transgenic mice was first devised by Grosveld's group (Grosveld *et al.*, 1987) to study the specific activity of this 5 ' upstream region. When the region of clustered DHSs was directly ligated to the β-globin gene, high-level and position-independent gene expression was seen in virtually each transgenic mouse. This result was in stunning contrast to previous unsuccessful efforts to express globin genes in erythroid cells in culture and in transgenic mice with DNA constructs containing only proximal flanking sequences.

It would be inappropriate to describe in this chapter the exciting series of DNA transfection experiments which were initiated by this initial observation. Instead, we will summarize here the three levels of function which might most likely come together in the LCR, previously named DCR for 'dominant control region' by Grosveld's group (Grosveld *et al.*, 1987; Blom van Assendelft *et al.*, 1989) and LAR for 'locus-activating region' by Groudine's group (Forrester *et al.*, 1987).

First, the LCR acts like a superenhancer during the entire switching procedure of β-like globin genes in embryonal yolk sac cells, fetal liver cells and adult erythropoietic bone marrow and peripheral blood reticulocytes. In accordance with the developmentally stable presence of its chromatin structure the LCR turns out to be coresponsible for the correct temporal activation and inactivation of all types of β-like genes of the gene cluster. DNA transfer experiments in erythroid cells and in transgenic mice support a mechanism in which the multifactorial protein–DNA complexes in the LCR region directly interact non-competitively and competitively with the multifactorial protein complexes at the enhancer–promoter regions of the individual globin genes (summarized with appropriate references in Townes and Behringer (1990) and Orkin (1990)).

Secondly, there are a number of indicative observations but no results from direct functional tests for a chromatin-organizing activity of the human β-globin LCR region. In both immature and mature erythroid cells the entire β-globin locus, including active and inactive genes as well as extensive non-transcribed flanking sequences, comprise a continuous domain of general DNAase-sensitive chromatin. The 'open chromatin' structure appears to be dependent on the presence of at least parts of the LCR region, as can be deduced from chromatin analysis of the human β-globin locus in somatic cell hybrids of murine MEL cells with human fibroblasts (Dhar *et al.*, 1989) and lymphoid cells from a patient with the deletional Hispanic (γδβ) thalassaemia (Forrester *et al.*, 1990). In all wild-type, mutated, induced or extinguished cases analysed, a parallel presence has been observed between the LCR 5′ DHS chromatin elements and the far-reaching general DNAase-sensitive chromatin. However, whether LCR elements have a cause-and-effect relationship with the overall chromatin structure of the β-globin gene locus, that is to say, have a direct 'chromatin opener' function, can only be determined with further functional DNA transfection studies.

Thirdly, the same correlation of presence and absence of DHS in the LCR region appears to exist with replication timing of DNA stretches in the entire β-globin locus. Independent of the transcriptional activity of individual genes, active-type chromatin structure and early S phase DNA replication might both be coupled to the same *cis*-active elements in the LCR (Dhar *et al.*, 1989; Forrester *et al.*, 1990), thereby dissociating these functions formally from the transcriptional enhancer effect of the LCR.

It is reasonable to assume that LCR elements are not confined to the human β-globin locus, rather that they are normal and essential master switches of differentially expressed regulatory gene domains. Sequences with LCR-like developmentally stable control function were also found downstream of the human *CD2* gene (Greaves *et al.*, 1989), far upstream of the human α-globin gene cluster (by Higgs, MCR Oxford, mentioned in Orkin, 1990) and by our group in the chicken lysozyme gene domain. The only lysozyme-specific DHS, besides DNAaseI hypersensitivity at the promoter, that is present in all potentially active cell types for lysozyme gene expression, is the DHS at the 6.1 kb early enhancer (Sippel *et al.*, 1988). In addition, the −6.1 DHS is developmentally stable in transformed myeloid cells, representing various developmental stages of macrophage precursor cells. This is in contrast to the DHS at the −2.7 kb late enhancer which is DNAaseI insensitive in early-type myeloblasts. The differential DNAaseI hypersensitivity is reflected by the differential function of the enhancer elements in DNA transfer experiments with various haematopoietic cells in culture (Müller *et al.*, 1990). Our chromatin structural analysis together with our results on the functional specificities of five out of the seven regulatory elements in the 5′-flanking chromatin lead us to propose the stepwise activation model seen in Fig. 2 for the lysozyme gene domain in developing chicken macrophages. The outlined mechanism of activation in three steps is consistent with all our data derived from cellular work. However, it is obvious that the suggested successive steps are merely 'photo-like' derived states of the lysozyme domain in retrovirally transformed myeloid and non-myeloid precursor cells.

The proposed 'closed' chromatin structure of the chromatin domain from the 5′ A-element to the 3′ A-element in haematopoietic stem cells (Fig. 2) can only be deduced from the state of chromatin in different 'never-lysozyme expressing' cells. The early myeloid-specific commitment event in which the LCR function of the −6.1 kb enhancer is involved in a chromatin-opening process is currently merely as correlative as the β-globin LCR function in respect to the suggested dominant function for the activation of its locus in early erythroid cells.

4.   Genomic domain transfer

The introduction of DNA constructs into cells in culture does not offer the full possibility of assaying the regulatory potential of genes during cell differentiation. It lacks the necessary invasive test in the appropriate developing wild-type cells in which a hierarchical order of activation and inactivation events could be studied. Only the possibility of introducing a new DNA into experimental animals or pluripotent stem cells opens the potential for exploration of the effects of DNA sequence modifications on the expression of any gene in developing animals (Brinster and Palmiter, 1982) or developing organ systems, respectively (Doetschman *et al.*, 1985; Shinar *et al.*, 1989).

In the previous sections we have outlined the current evidence from our study of the lysozyme gene locus and from selected examples of the work of others supporting the hypothesis that chromosomal loop domains confine genomic regulatory units for the control of gene activity. It is conceptually difficult to prove that the full complement of *cis*-active DNA sequences for the regulation of a gene or a group of genes is present on a distinct region of genomic DNA. The most rigorous test is its introduction into the genome of a transgenic organism, this providing an opportunity to follow the behaviour of the respective genomic DNA fragment throughout the entire ontogeny, that is to say, in every possible cell type and cell stage of an organism.

Since no appropriate transgenic chicken procedure currently exists, we decided to generate transgenic mice carrying the entire wild-type chicken lysozyme gene domain including DNA sequences from the 5′ end of the 5′ A element at around -12 kb to the 3′ end of the 3′ A element at around $+9.5$ kb. The 21.5 kb chicken lysozyme domain is sufficiently compact and its distal ends are well enough characterized to be transferred end to end without significant DNA deletions. Comparable experiments with other gene loci, e.g. the human β-globin gene domain with its size of, most likely, more than 150 kb and its currently insufficiently characterized distal ends, would create considerably more experimental problems.

Seven founder mice were generated which carried the complete lysozyme locus in copy numbers from two to 70 copies per cell (Bonifer *et al.*, 1990). The qualitative as well as quantitative analysis of specific transcripts from different mouse tissues showed that the chicken transgene, regardless of its random position in the host genome, behaved as an independent regulatory unit in each of the seven mice. Chicken lysozyme RNAs in mouse haematopoietic cells were restricted to macrophage cells, as it is in the donor animal. Transgene transcription in mouse macrophages was consistently high and is, at a gene-to-gene level, comparable to chicken macrophages. Transcription started correctly from the same promoter used in chicken cells, and showed direct correlation with the copy number of the integrated gene domain (Bonifer *et al.*, 1990). When total RNA was analysed in a number of other host tissues, specific transgene transcripts could only be found in brain tissue, again in copy number-dependent levels. Whereas the specificity of transgene expression in haematopoietic cells indicated a high conservation of the cell type-specific apparatus for expression between chicken and mouse, the expression in brain, in which lysozyme is normally not expressed (Cross *et al.*, 1988) showed that this might, however, not hold true for all aspects of the complex regulation of the lysozyme gene. The consistently correct behaviour of the transgene in cell-type specificity and level of expression is most likely the result of the transfer of the gene with its entire regulatory DNA unit, including not only regulatory sequences like enhancer–promoter elements, but also its LCR and the domain-bordering A elements. In contrast to results obtained in DNA transfer experiments with subdomain fragments, our result demonstrates that genomic domain transfer ('regulon transfer') offers a general way to overcome position effects in random DNA transfer.

5.   Conclusions and perspectives

For the solution of problems with increasing complexity, ever more complex experimental systems prove to be the most appropriate choice. For the functional analysis of molecular mechanisms of eukaryotic gene regulation it might turn out that promoter function is most appropriately studied in *in vitro* systems, promoter–enhancer interactions in transient transfection systems, LCR and domain-bordering A-elements in cellular stable transfection systems, and the developmental dynamics of entire regulatory domains in transgenic animal systems. However, there are limits to each experimental system which must be overcome before its full potential can be used. Currently, the true influence of chromatin organization on eukaryotic gene regulation appears to be difficult to approach because any random insertion of a DNA construct into the cellular genome suffers from the unpredictable influence of what we globally call position effect.

'Position effect' is a term used to describe phenomena in which the behaviour of a gene is affected by its location on the chromosome (for a review see Lima-de-Faria, 1983). The change in behaviour can be observed in a variety of ways, such as differences in phenotype, transcriptional efficiency or time of replication. Experimental difficulties caused by the genomic position effect of randomly inserted gene DNA can be circumvented by strategies of altering gene sequences at their natural location through homologous recombination (Capecchi, 1989). Gene targeting by homologous recombination is, however, not applicable in cases for which cellular selection is prevented. For this reason strategies need to be developed to overcome the unpredictable influences of the genomic position effect by other means. By better understanding the mechanisms which cause chromosomal position effects we expect to be able to develop vector systems which allow random genomic integration of transgenes in such a way that they behave independently of their neighbouring DNA sequences.

Electron microscopical observations (Paulson and Laemmli, 1977) and molecular genetic analysis of larger genomic regions (e.g. see Surdej *et al.*, 1990) make it likely that large parts of chromosomes, especially euchromatic regions, consist of arrays of successive loop domains. The congruence of the chromatin loop domain organization with regulatory units for gene activity, as suggested from our studies on the lysozyme gene, might not be equally easy to verify for other gene loci. In the case of the chicken lysozyme locus the active chromatin domain turned out to be particularly small and borders at both ends onto gene domains which do not have the same state of generally DNAase-sensitive chromatin.

Figure 3 shows a schematic diagram outlining the genomic domain transfer problem as viewed from the experimental evidence described in this chapter. Figure 3(a) depicts a section of the euchromatic genome wherein three A-elements confine two adjacent regulatory domains, i.e. regulons, each consisting of a gene and an LCR. In the event of random genomic integration (Fig. 3(b)), transgenes will nearly always enter such a regulatory unit, resulting mostly (but not in every case) in disturbing effects for the host domain and the transgene. Only in the event of integrating directly into essential sites for the host gene will insertional inactivation take place. The statistically more frequent effect will be deregulatory for both host gene and transgene as a result of, for example, interfering interactions between incoming and existing regulatory protein–DNA complexes. Due to the local situation at any random integration site, quite unpredictable effects on the transgene for each independent integration event will occur, providing an explanation for the phenomenon of position effect.

A very different situation arises, however, when transgenes are framed by domain-bordering A-elements (Fig. 3(c) and (d)). According to the absence or presence of essential LCRs, they will be either inactive (Fig. 3(c)) or cell type-specifically active (Fig. 3(d)), respectively. These behavioural differences of transgenes are consistent with recent results from our laboratory (Saueressig, 1989) in which we found that copy number-dependent, position-independent transgene transcription depends entirely on the presence of a functional unit of

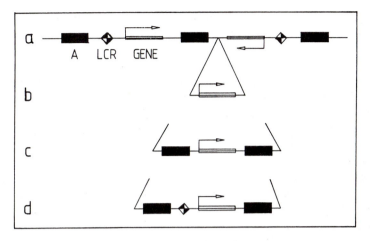

Fig. 3. Schematic diagram of the genomic domain transfer problem. (a) A detail of the genome is shown, with three A-elements (solid bars) confining two chromosomal domains, each domain consisting of a gene (thin open bar with horizontal arrow indicating direction of transcription) and an LCR (diamond). (b) Random insertion of a gene without a regulatory element into one of the domains, resulting in deregulation by the position effect. (c) Insertion of a gene without an LCR or enhancer, but framed by A-elements, resulting in an inactive transgene. (d) Insertion of a complete domain with framing A-elements, an LCR and a gene, resulting in correctly regulated, position-independent, high-level transcriptional activity of the inserted gene.

regulatory elements when A-element-framed reporter genes are transferred into the genome of cells in culture.

The outlined mechanism of transgene behaviour not only suggests a strategy to overcome the genomic position effect but may also explain the recently observed position-independent, highly specific, copy number-dependent transcription of human LCR–β-globin constructs in erythroid cells in culture or in transgenic mice (Grosveld *et al.*, 1987; Blom van Assendelft *et al.*, 1989; Greaves *et al.*, 1989). It is likely that the strong human β-globin LCR by its domain-opening and dominant enhancer activity functionally overrides randomly invaded genomic domains possibly in most, but strictly not all, cases in a position-independent manner (Ryan *et al.*, 1989; Forrester *et al.*, 1990).

Nevertheless, the stable transfection and position effect-considering assay system has the potential to ultimately dissect whichever types of DNA elements are necessary for the construction of improved transgenic vector systems. Stable transfection systems in which position effects of randomly incorporated transgenes are minimized or completely buffered will finally offer the possibility to map chromatin functions to specific DNA sequences by comparison of wild-type and mutant effects, just as they are successfully used for the dissection of enhancer–promoter functions in transient transfection systems. The identification of new types of *cis*-active DNA elements will surely lead to the identification of new protein transfactors and to the discovery of new genes controlling chromosomal functions.

In recent years we have seen an increasing fraction of non-transcribed genomic DNA being recruited as signals for regulatory functions of gene expression and for the logistics of chromosomes in recombination, replication and cell division. Surely the space for 'junk DNA' in intergenic spacers is ever retreating.

Acknowledgements

Many former and present colleagues and collaborators, not appearing as authors, have contributed to our current picture of the chicken lysozyme gene domain. We are greatly indebted to them for practical help, suggestions and discussions. We thank Elke Simpson and Elizabeth Herbrich and Katrina Koehler for secretarial and editorial work, respectively. Ongoing research from our group, covered in this contribution, is supported by grants from the Deutsche Forschungsgemeinschaft (SFB 229/A2), the Fonds der Chemischen Industrie and the German Israeli Foundation for Scientific Research and Development (GIF).

References

Alevy, M. C., Tsai, M. J. and O'Malley, B. W. (1984). DNase I sensitive domain of the gene coding for the glycolytic enzyme glyceraldehyde-3-phosphate dehydrogenase. *Biochemistry* **23**, 2309–2314.

Baldacci, P., Royal, A., Brègègère, F., Abasto, J. P., Cami, B., Daniel, F. and Kourilsky, P. (1981). DNA organisation of the chicken lysozyme gene region. *Nucleic Acids Res.* **9**, 3575–3588.

Banerji, J., Olson, L. and Schaffner, W. (1983). A lymphocyte-specific cellular enhancer is located downstream of the joining region in immunoglobulin heavy chain genes. *Cell* **33**, 729–740.

Baniahmad, A., Muller, M., Steiner, C. and Renkawitz, R. (1987). Activity of two different silencer elements of the chicken lysozyme gene can be compensated by enhancer elements. *EMBO J.* **6**, 2297–2303.

Baniahmad, A., Steiner, C., Köhne, A. C. and Renkawitz, R. (1990). Modular structure of a chicken lysozyme silencer: involvement of an unusual thyroid hormone receptor binding site. *Cell* **61**, 505–514.

Beermann, W. (1952). Chromomerenkonstanz und spezifische Modifikationen der Chromosomenstruktur in der Entwicklung und Organdifferenzierung von Chironomus tentans. *Chromosoma* **5**, 139–198.

Behringer, R. R., Ryan, T. M., Palmiter, R. D., Brinster, R. Z. and Townes, T. M.

(1990). Human γ- to β-globin gene switching in transgenic mice. *Genes Dev.* **4,** 380–389.

Benyajati, C. and Worcel, A. (1976). Isolation, characterization, and structure of the folded interphase genome of Drosophila melanogaster. *Cell* **9,** 393–407.

Blom van Assendelft, G., Hanscombe, O., Grosveld, F. and Greaves, D. R. (1989). The β-globin dominant control region activates homologous and heterologous promotors in a tissue-specific manner. *Cell* **56,** 969–977.

Bonifer, C., Vidal, M., Grosveld, F. and Sippel, A. E. (1990). Tissue specific and position indepent expression of the complete gene domain for chicken lysozyme in transgenic mice. *EMBO J.* **9,** 2843–2848.

Borgmeyer, U. (1987). Das TGGCA-Protein als Strukturbestandteil eines zellspezifischen multifaktoriellen Enhancer–DNA–Proteinkomplexes in vivo. Doctoral thesis, University of Heidelberg.

Brinster, R. L. and Palmiter, R. D. (1982). Induction of foreign genes in animals. *Trends Biochem. Sci.* **7,** 438–440.

Capecchi, M. R. (1989). The new mouse genetics: Altering the genome by gene targeting. *Trends Genet.* **5,** 70–76.

Choi, O.-R. B. and Engel, J. D. (1988). Developmental regulation of β-globin gene switching. *Cell* **55,** 17–26.

Cockerill, P. N. and Garrard, W. T. (1986). Chromosomal loop anchorage of the kappa immunoglobulin gene occurs next to the enhancer in a region containing topoisomerase II sites. *Cell* **44,** 273–282.

Cook, P. R. (1973). Hypothesis on differentiation and the inheritance of gene superstructure. *Nature* **245,** 23–25.

Cook, P. R. and Brazell, J. A. (1976). Conformational constraints in nuclear DNA. *J. Cell Sci.* **22,** 287–302.

Cross, M., Mangledorf, J., Wedel, A. and Renkawitz, R. (1988). Mouse lysozyme M gene: isolation, characterization and expression studies. *Proc. Natl Acad. Sci. USA* **85,** 6232–6236.

Dhar, V., Skoultchi, A. I. and Schildkraut, C. L. (1989). Activation and repression of a β-globin gene in cell hybrids is accompanied by a shift in its temporal replication. *Mol. Cell. Biol.* **9,** 3524–3532.

Doetschman, T. C., Eistetter, H., Katz, M., Schmidt, W. and Kemler, R. (1985). The *in vitro* development of blastocyst-derived embryonic stem cell lines: formation of visceral yolk sac, blood islands and myocardium. *J. Embryol. Exp. Morphol.* **87,** 27–45.

Dorit, R. L., Schoenbach, L. and Gilbert, W. (1990). How big is the universe of exons? *Science* **250,** 1377–1382.

Elgin, S. C. R. (1988). The formation and function of DNase I hypersensitive sites in the process of gene activation. *J. Biol. Chem.* **263,** 19259–19262.

Emerson, B. M. and Felsenfeld, G. (1984). Specific factor conferring nuclease hypersensitivity at the 5' end of the chicken adult β-globin gene. *Proc. Natl Acad. Sci. USA* **81,** 95–99.

Enver, T., Ebens, A. J., Forrester, W. C. and Stamatoyannopoulos, G. (1989). The human β-globin locus activation region alters the developmental rate of a human fetal globin gene in transgenic mice. *Proc. Natl Acad. Sci. USA* **86,** 7033–7037.

Enver, T., Raich, N., Ebens, A. J., Papayannopoulos, T., Constantini, F. and Stamatoyannopoulos, G. (1990). Developmental regulation of human fetal-to-adult globin gene switching in transgenic mice. *Nature* **344**, 309–313.

Forrester, W. C., Thompson, C., Elder, J. T. and Groudine, M. (1986). A developmentally stable chromatin structure in the human β-globin gene cluster. *Proc. Natl Acad. Sci. USA* **83**, 1359–1363.

Forrester, W. C., Takegawa, S., Papayannopoulou, T., Stamatoyannopoulos, G. and Groudine, M. (1987). Evidence for a locus activation region: the formation of developmentally stable hypersensitive sites in globin expressing hybrids. *Nucleic Acids Res.* **15**, 10159–10176.

Forrester, W. C., Epner, E., Driscoll, M. C., Enver, T., Brice, M., Papayannopoulou, T. and Groudine, M. (1990). A deletion of the β-globin locus activation region causes a major alteration in chromatin structure and replication across the entire β-globin locus. *Genes Dev.* **4**, 1637–1649.

Fritton, H. P., Sippel, A. E. and Igo-Kemenes, T. (1983). Nuclease hypersensitive sites in the chromatin domain of the chicken lysozyme gene. *Nucleic Acids Res.* **11**, 3467–3485.

Fritton, H. P., Igo-Kemenes, T., Nowock, J., Strech-Jurk, U., Theisen, M. and Sippel, A. E. (1984). Alternative sets of DNase I-hypersensitive sites characterize the various functional states of the chicken lysozyme gene. *Nature* **311**, 163–165.

Fritton, H. P., Igo-Kemenes, T., Nowock, J., Strech-Jurk, U., Theisen, M. and Sippel, A. E. (1987). DNase I-hypersensitive sites in the chromatin structure of the lysozyme gene in steroid hormone target and non-target cells. *Biol. Chem. Hoppe-Seyler* **368**, 111–119.

Gall, J. G. (1956). On the submicroscopic structure of chromosomes. *Brookhaven Symp. Biol.* **8**, 17–32.

Garel, A. and Axel, R. (1976). Selective digestion of transcriptionally active ovalbumin genes from oviduct nuclei. *Proc. Natl Acad. Sci. USA* **73**, 3966–3970.

Gasser, S. M. and Laemmli, U. K. (1987). A glimpse at chromosomal order. *Trends Genet.* **3**, 16–22.

Gillies, S. D., Morrison, S. L., Oi, V. T. and Tonegawa, S. (1983). A tissue specific transcription enhancer is located in the major intron of a rearranged immunoglobulin heavy chain gene. *Cell* **33**, 717–728.

Greaves, D. R., Wilson, F. D., Lang, G. and Kioussis, D. (1989). Human CD2 3 '-flanking sequences confer high-level, T cell-specific, position-independent gene expression in transgenic mice. *Cell* **56**, 979–986.

Grewal, T., Theisen, M., Borgmeyer, U., Grussenmeyer, T., Rupp, R. A. W., Stief, A., Qian, F., Hecht, A. and Sippel, A. E. (1992). The −6.1 kb chicken lysozyme enhancer is a multifactorial complex composed of several cell-specifically acting subelements. *Mol. Cell Biol.*, in press.

Gross, D. S. and Garrard, W. T. (1988). Nuclease hypersensitive sites in chromatin. *An. Rev. Biochem.* **57**, 159–197.

Grosveld, F., Blom van Assendelft, G., Greaves, D. R. and Kollias, G. (1987). Position-independent, high-level expression of the human β-globin gene in transgenic mice. *Cell* **51**, 975–985.

Gyurkovics, H., Gausz, J., Kummer, J. and Karch, F. (1990). A new homeotic mutation in the Drosophila bithorax complex removes a boundary separating two domains of regulation. *EMBO J.* **9**, 2579–2585.

Hammer, R. E., Krumlauf, R., Camper, S. A., Brinster, R. L. and Tilghman, S. M. (1987). Diversity of alpha-fetoprotein gene expression in mice is generated by a combination of separate enhancer elements. *Science* **235**, 53–58.

Hecht, A., Berkenstam, A., Strömstedt, P.-E., Gustafsson, J. A. and Sippel, A. E. (1988). A progesterone responsive element maps to the far upstream steriod dependent DNase hypersensitive site of chicken lysozyme chromatin. *EMBO J.* **7**, 2063–2073.

Hecht, A. (1990). Strukturelle und funktionelle Charakterisierung des steriodhormonabhängigen Enhancerelements des Lysozymgens und seine Zellspezifität. Doctoral Thesis, University of Heidelberg.

Henikoff, S. (1990). Position-effect variegation after 60 years. *Trends Genet.* **6**, 422–426.

Igo-Kemenes, T. and Zachau, H. G. (1977). Domains in chromatin structure. *Cold Spring Harbor Symp. Quant. Biol.* **42**, 109–118.

Izaurralde, E., Mirkovitch, J. and Laemmli, U. K. (1988). Interaction of DNA with nuclear scaffold *in vitro*. *J. Mol. Biol.* **200**, 111–125.

Jackson, D. A. (1986). Organization beyond the gene. *Trends Biochem. Sci.* **11**, 249–252.

Jantzen, K., Fritton, H. P. and Igo-Kemenes, T. (1986). The DNase I sensitive domain of chicken lysozyme gene spans 24 kb. *Nucleic Acids Res.* **14**, 6085–6099.

Kellum, R. and Schedl, P. (1991). A position effect assay for boundaries of higher order chromosomal domains. *Cell* **64**, 941–950.

Kioussis, D., Vanin, E., de Lange, T., Flavell, R. A. and Grosveld, F. (1983). β-Globin gene inactivation by DNA translocation in γβ-thalassaemia. *Nature* **306**, 662–666.

Lawson, G. M., Tsai, M. J. and O'Malley, B. W. (1980). Deoxyribonuclease I sensitivity of the non-transcribed sequences flanking the 5′ und 3′ ends of the ovomucoid gene and the ovalbumin and its related x and y genes in hen oviduct nuclei. *Biochemistry* **19**, 4403–4411.

Lawson, G. M., Knoll, B. J., March, C. J., Woo, S. L. C., Tsai, M. J. and O'Malley, B. W. (1982). Definition of 5′ and 3′ structural boundaries of the chromatin domain containing the ovalbumin multi-gene family. *J. Biol. Chem.* **257**, 1501–1507.

Levy-Wilson, B. and Fortier, C. (1989). The limits of the DNase I-sensitive domain of the human apolipoprotein gene coincide with the locations of chromosomal anchorage loops and define the 5′ and the 3′ boundaries of the gene. *J. Biol. Chem.* **264**, 21196–21204.

Lima-de-Faria, A. (1983). Processes of directing expression, mutation and rearrangements. *Molecular Evolution and Organization of the Chromosome*, pp. 507–604. Elsevier, Amsterdam.

Lindenmaier, W., Nguyen-Huu, M. C., Lurz, R., Stratmann, M., Blin, N., Wurtz, T., Hauser, H. J., Sippel, A. E. and Schütz, G. (1979). Arrangement of coding and intervening sequences of the chicken lysozyme gene. *Proc. Natl Acad. Sci. USA* **76**, 6196–6200.

McGhee, J. D., Rau, D. C., Charney, E. and Felsenfeld, G. (1980). Orientation of the nucleosome within the higher order structure of chromatin. *Cell* **22**, 87–96.

McGhee, J. D., Wood, W. I., Dolan, M., Engel, J. D. and Felsenfeld, G. (1981). A

200 bp region at the 5' end of the chicken adult β-globin gene is accessible to nuclease digestion. *Cell* **27**, 45–55.

Mielke, C., Kohwi, Y., Kohwi-Shigematsu, T. and Bode, J. (1990). Hierarchical binding of DNA fragments derived from scaffold-attached regions: correlation of properties in vitro and function in vivo. *Biochemistry* **29**, 7475–7485.

Mirkovitch, J., Mirault, M. E. and Laemmli, U. K. (1984). Organization of the higher order chromatin loop: specific DNA attachment sites on nuclear scaffold. *Cell* **39**, 223–232.

Müller, A., Grussenmeyer, T., Strech-Jurk, U., Theisen, M., Stief, A. and Sippel, A. E. (1990). Macrophage differentiation at the genome level: studies of lysozyme gene activation. *Bone Marrow Transplantation* **5**, 13–15.

Müller, A. M. (1990). Die zell- und stadien-spezifische Aktivierung des Hühnerlysozymgens: Funktionsanalyse der −2.7/−2.4 kb Regulationsregion in hämatopoietischen Zellen. Doctoral Thesis, University of Heidelberg.

Nedospasov, S. A. and Georgier, G. P. (1980). Non-random cleavage of SV40 DNA in the compact mini chromosome and free in solution by micrococcal nuclease. *Biochem. Biophys. Res. Commun.* **92**, 532–539.

Nelson, W. G., Pienta, K. J., Barrak, E. R. and Coffey, D. S. (1986). The role of nuclear matrix in the organisation and function of DNA. *An. Rev. Biophys. Chem.* **15**, 457–475.

Orkin, S. H. (1990). Globin gene regulation and switching: circa 1990. *Cell* **63**, 665–672.

Paro, R. (1990). Imprinting a determined state in the chromatin of Drosophila. *Trends Genet.* **6**, 416–421.

Paulson, J. R. and Laemmli, U. K. (1977). The structure of histone depleted metaphase chromosomes. *Cell* **12**, 817–828.

Phi-Van, L. and Strätling, W. H. (1988). The matrix attachment regions of the chicken lysozyme gene co-map with the boundaries of the chromatin domain. *EMBO J.* **7**, 655–664.

Phi-Van, L., von Kries, J. P., Ostertag, W. and Strätling, W. H. (1990). The chicken lysozyme 5' matrix attachment region increases transcription from a heterologous promoter in heterologous cells and dampens postion effects on the expression of transfected genes. *Mol. Cell. Biol.* **10**, 2302–2307.

Queen, C. and Baltimore, D. (1983). Immunoglobulin gene transcription is activated by downstream sequence elements. *Cell* **33**, 741–748.

Reudelhuber, T. (1984). Gene regulation – a step closer to the principles of eucaryotic transcriptional control. *Nature* **311**, 301.

Richmond, T. J., Finch, J. T. and Klug, A. (1983). Studies of nucleosome structure. *Cold Spring Harbor Symp. Quant. Biol.* **47**, 493–501.

Ryan, T. M., Behringer, R. R., Martin, N. C., Townes, T. M., Palmiter, R. D. and Brinster, R. L. (1989). A single erythroid-specific DNase I super-hypersensitive site activates high levels of human β-globin gene expression in transgenic mice. *Genes Dev.* **3**, 314–323.

Saueressig, H. (1989). Die Funktion der A-Elemente des Hühnerlysozymgens: Ihr Zusammenspiel mit anderen cis-Elementen der Transkriptionskontrolle nach stabilem Gentransfer in heterologe Zellen. Diploma Thesis, University of Heidelberg.

REGULATORY ORGANIZATION OF EUKARYOTIC GENOMES 25

Schütz, G., Nguyen-Huu, M. C., Giesecke, K., Hynes, N. E., Groner, B., Wurtz, T. and Sippel, A. E. (1978). Hormonal control of egg-white protein messenger RNA synthesis in the chicken oviduct. *Cold Spring Harbor Symp. Quant. Biol.* **42**, 617–624.

Scott, W. A. and Wigmore, D. J. (1979). Sites in simian virus 40 chromatin which are preferentially cleaved by endonucleases. *Cell* **15**, 1511–1518.

Shinar, D., Yoffe, O., Shani, M. and Yaffe, D. (1989). Regulated expression of muscle-specific genes introduced in mouse embryonal stem cells: inverse correlation with DNA methylation. *Differentiation* **41**, 116–126.

Sippel, A. E. and Renkawitz, R. (1989). The chicken lysozyme gene. In: Renkawitz, R. (ed.) *Tissue Specific Gene Expression*, pp. 185–198. VCH, Weinheim.

Sippel, A. E., Fritton, H. P., Theisen, M., Borgmeyer, U., Strech-Jurk, U. and Igo-Kemenes, T. (1986). The TGGCA protein binds in vitro to DNA contained in a nuclease-hypersensitive region that is present only in active chromatin of the lysozyme gene. In: Grodzicker, T., Sharp, P. ., Botchan, M. R. (eds) *DNA Tumor Viruses: Gene Expression and Replication.* CSH Press (*Cancer Cells* vol. 4, pp. 155–162).

Sippel, A. E., Borgmeyer, U., Püschel, A. W., Rupp, R. A. W., Stief, A., Strech-Jurk, U. and Theisen, M. (1987). Multiple non-histone protein–DNA complexes in chromatin regulate the cell and stage specific activity of an eukaryotic gene. In: Hennig, W. (ed.) *Results and Problems in Cell Differentiation*, vol. 14, pp. 255–269. Springer-Verlag, Berlin.

Sippel, A. E., Theisen, M., Borgmeyer, U., Strech-Jurk, U., Rupp, R. A. W., Püschel, A. W., Müller, A., Hecht, A., Stief, A. and Grussenmeyer, T. (1988). Regulatory function and molecular structure of DNase I-hypersensitive elements in the chromatin domain of a gene. In: Kahl, G. (ed.) *Architecture of Eukaryotic Genes*, pp. 355–369. VCH, Weinheim.

Sippel, A. E., Stief, A., Hecht, A., Müller, A., Theisen, M., Borgmeyer, U., Rupp, R. A. W., Grewal, T. and Grussenmeyer, T. (1989). The structural and functional domain organization of the chicken lysozyme gene locus. In: Eckstein, F. and Lilley, D. M. (eds) *Nucleic Acids and Molecular Biology*, vol. 3, pp. 133–147. Springer-Verlag, Berlin.

Stalder, J., Larsen, A., Engel, J. D., Dolan, M., Groudine, M. and Weintraub, H. (1980). Tissue specific DNA cleavages in the globin chromatin domain introduced by DNase I. *Cell* **20**, 451–460.

Stamatoyannopoulos, G. (1991). Human hemoglobin switching. *Science* **252**, 283.

Steiner, C., Muller, M., Baniahmad, A and Renkawitz, R. (1987). Lysozyme gene activity in chicken macrophages is controlled by positive and negative regulatory elements. *Nucleic Acids Res.* **15**, 4163–4177.

Stief, A., Winter, D. M., Strätling, W. H. and Sippel, A. E. (1989). A nuclear DNA attachment element mediates elevated and position independent gene activity. *Nature* **341**, 343–345.

Strätling, W. H., Dölle, A. and Sippel, A. E. (1986). Chromatin structure of the chicken lysozyme gene domain as determined by chromatin fractionation and micrococcal nuclease digestion. *Biochemistry* **25**, 495–502.

Surdej, P., Got, C., Rosset, R. and Miassod, R. (1990). Supragenic loop organization:

mapping in Drosophila embryos of scaffold-associated regions on a 800 kilo base DNA continuum cloned from the 14B–15B first chromosome region. *Nucleic Acids Res.* **18**, 3713–3722.

Theisen, M., Stief, A. and Sippel, A. E. (1986). The lysozyme enhancer: cell specific activation of the chicken lysozyme gene by a far upstream DNA-element. *EMBO J.* **5**, 719–724.

Townes, T. M. and Behringer, R. R. (1990). Human globin locus activation region (LAR): role in temporal control. *Trends Genet.* **6**, 219–223.

Tuan, D., Solomon, W., Li, Q. and London, M. (1985). The 'β-like globin' gene domain in human erythroid cells. *Proc. Natl Acad. Sci. USA* **82**, 6384–6388.

Udvardy, A., Maine, E. and Schedl, P. (1985). The 87A7 chromomere: identification of novel chromatin structures flanking the heat shock locus that may define the boundaries of higher order domains. *J. Mol. Biol.* **185**, 341–358.

Varshavsky, A. J., Sundin, O. H. and Bohn, M. J. (1978). SV40 viral minichromosome: preferential exposure of the origin of replication as probed by restriction endonuclease. *Nucleic Acids Res.* **5**, 3469–3479.

Weintraub, H. and Groudine, M. (1976). Chromosomal subnuits in active genes have an altered conformation. *Science* **193**, 848–856.

Wu, C., Bingham, P. M., Livak, K. J., Holmgren, R. and Elgin, S. C. R. (1979). The chromatin structure of specific genes: evidence for higher order domains of defined DNA sequence. *Cell* **16**, 797–806.

2
Directed modification of genes by homologous recombination in mammalian cells

SHOICHI DOI, COLIN CAMPBELL and
RAJU KUCHERLAPATI

*Department of Molecular Genetics, Albert Einstein College of Medicine,
1300 Morris Park Avenue, Bronx, NY 10461, USA*

1. Introduction

Purified DNA can be readily introduced into mammalian cells (for a review of methods see Keown *et al.*, 1990). The DNA sequences remain extra-chromosomal for a short period of time (48–72 hours), after which the DNA is lost or is integrated into the cellular genome. If the introduced DNA does not have any significant homology to the cellular DNA, the sites at which it integrates are, for all practical purposes, random (Roth and Wilson, 1988). When the introduced DNA has homology to the cellular sequences, one of two fates await it. Frequently, it integrates into a random location. This integration, of course, is not based on any homology and is referred to as non-homologous recombination (NHR). Occasionally, the introduced DNA undergoes a recombination event based upon homology. This event is referred to as homologous recombination (HR).

Gene transfer resulting in transient expression of extrachromosomal sequences or integration at random sites has proved to be very useful in identifying *cis*-acting elements that are important for gene expression, elements which are important for tissue specificity and developmental stage-specific expression (for a review see Kucherlapati and Skoultchi, 1984). Gene transfer systems into somatic mammalian cells of various mammalian species as well as microinjection into fertilized mouse embryos have served, and continue to, as important molecular biological tools (for a review see Jaenisch, 1988). However, the random integration of the exogenously integrated DNA sequences has at least two major adverse consequences: (1) the site at which the DNA becomes integrated has profound effects on the expression of the introduced gene (position effect) (see e.g. Nandi *et al.*, 1988); (2) it is now known that endogenous genes, or the introduced genes, can be activated or inactivated, depending on the location at which the sequences are integrated (e.g. see Gossler *et al.*, 1989). Although such unpredicted gene activation and inactivation events have proved to be very useful in identification of important genes and genetic elements, the inability to predetermine the sites of integration has proved to be an obstacle for many experiments. The ability to target genes to specific locations in the mammalian genome would facilitate defined gene inactivation, gene mutation or gene correction (see Figs 1 and 2). Each of these three types of manipulations, in turn, has significance in addressing a number of biological problems.

Through the efforts of several investigators, it has now been shown that it is possible to modify mammalian genes by HR. Several of the early experiments to study HR utilized established cell lines. The availability of proliferating mouse embryonic stem (ES) cells from which mice could be produced has become a powerful adjunct to homologous recombination.

Evans and Kaufman (1981) have shown that cells derived from the inner cell mass of mouse blastocysts can be adapted to grow in culture and that, under appropriate circumstances, the cells can be maintained in continuous culture and stimulated to continue their developmental programme by reintroduction into the blastocoel of blastocysts followed by implantation into the uterus of a pseudopregnant female. The fact that such cells can continue to divide in culture makes it possible to modify them by HR and to use the modified cells to generate chimeric mice. If the germ line of the chimeric mice is colonized by the modified cells, the modification introduced into these cells *in vitro* can be transmitted genetically to all the succeeding generations. The combination of HR and ES cell technologies is becoming very important in the genetic analysis of mammals. In this chapter we summarize the methods for accomplishing HR, their relative merits, and the future of HR technology.

2. Gene targeting in yeast cells

The initial impetus to study HR in mammalian cells came from the observation that random integration of genes resulted in unpredictable, if not abnormal, behaviour in terms of gene expression and that integration of DNA sequences in the yeast *Saccharomyces cerevisiae* is almost exclusively through HR (Hinnen *et al.*, 1978). Although the frequency with which the homologous integration events could be obtained (one colony per microgram of input DNA per 10^8 spheroplasts (Szostak *et al.*, 1983)) is relatively low, the ease with which yeast cells can be cultured and the availability of a few selectable markers resulted in the widespread use of this method. In initial experiments, a circular plasmid was used for transformation, which resulted in the integration of the introduced DNA into the chromosomal site by an integration mechanism analogous to that proposed for bacteriophage λ by Campbell (1971). Later, it was shown that double-strand breaks in the input DNA within the regions of homology results in a 10–1000-fold increase in HR and the site of the break directs the site of integration of the plasmid (for a review of these results see Szostak *et al.* (1983)). Thus, if the plasmid contained two genes A and B and was linearized by cutting in A, the integration preferentially occurred into gene A in the chromosome. If the cut was introduced into B, integration occurred into site B. These and additional observations about the effects of double-strand cuts lead to the double-strand break repair model of recombination (Szostak *et al.*, 1983). Although very useful in achieving disruption of the target gene (also referred to as gene knock-out), the initial circular or O-type recombination substrates in yeast resulted in duplication of the segment of the gene located in the input plasmid and a secondary intrachromosomal recombination was necessary to achieve true gene replacement.

Rothstein (1983) described a strategy to achieve a one-step gene replacement by utilizing a replacement or Ω-type vector. This strategy is perhaps the most widely employed to target, inactivate or correct genes. A simplistic view of how HR is achieved with the different substrates indicates that a single cross-over event is needed for integration of an O-type vector and a double-crossover or gene conversion is needed for integration of an Ω-type vector. Since conventional transmission genetics indicates that single crossovers within a given region will be more frequent than double crossovers, it may be expected that O-type vectors will be more efficient than Ω-type vectors. In practice, the efficiency of HR using the O-type or Ω-type vector is almost identical. All of these and several other observations in regard to gene targeting in yeast cells influenced the design of vectors and experiments to attempt gene modification in mammalian cells.

3.   Extrachromosomal recombination in mammalian cells

For HR to be of widespread use as a genetic tool, it is necessary that gene modification is achieved in somatic or mitotic cells. Although HR is a normal cellular process and occurs in meiosis, at the outset it was not known whether mammalian somatic cells have the ability to mediate this reaction. It was long felt that mitotic crossing over will be an indication of the presence of functional HR enzymes in somatic cells. Although early experiments to detect mitotic crossing over in mammalian cells did not succeed (Tarrant and Holliday, 1977; Rosenstraus and Chasin, 1978), later experiments did succeed in demonstrating mitotic recombination in mammalian cells (Cavanee *et al.*, 1983; Wasmuth and Vockhall, 1984).

The availability of plasmid vectors carrying genes whose expression could be selected for in mammalian cells made it possible to design experiments to test somatic cells for their ability to recombine DNA sequences. One of the early experiments of this nature involved the use of mutant herpes simplex virus thymidine kinase (*tk*) genes. When two plasmids, each of which carried a different mutation in the *tk* gene, were introduced into mouse *Ltk*⁻ cells, it was observed that some proportion of the cells grew in medium containing hypoxanthine, aminopterin and thymidine (HAT), indicating that the cells have a functional *tk* gene (Miller and Temin, 1983; Shapira *et al.*, 1983; Small and Scangos, 1983). A careful molecular analysis of the DNA of these cells indicated that a functional *tk* gene was reconstructed by HR. Similar experiments were conducted by several investigators using different selectable genes (DeSaint Vincent and Wahl, 1983). One of these selectable genes which proved to be very useful was the bacterial neomycin phosphotransferase (*neo*) gene. This gene was inserted into a pSV2 vector such that it could be expressed both in bacterial and mammalian cells (Southern and Berg, 1982). When this gene is expressed in mammalian cells, they acquire resistance to an amino-glycoside analogue (G418). Since mammalian cells do not have a *neo* gene, they are naturally sensitive to G418 and *neo* expression can be used as a dominant selectable marker. When two different deletion mutations of pSV2 *neo* were introduced into mouse or human cells, Kucherlapati *et al.* (1984) observed that a functional gene is generated by homologous recombination. They also showed that introduction of double-strand breaks in one or both of the input plasmids within the *neo* gene resulted in an increase in recombination. Examination of the recombination products revealed that double-strand break repair played an important role in generating the functional *neo* gene (Song *et al.*, 1985). The use of these and similar plasmids provided valuable information about the nature of the recombination machinery in mammalian cells.

4.   Artificial chromosomal targets

The success of homologous recombination observed with plasmids provided the basis for designing experiments to target chromosomal sequences. Since the efficiency of HR involving chromosomal sequences was not known, it was felt that the availability of a selection system to isolate the HR events would be most desirable. The starting material for a set of experiments was a plasmid which had two dominant selectable genes, *neo* and the xanthine phosphoribosyl transferase (*gpt*) gene, each of which is independently expressed (Southern and Berg, 1982). We constructed a deletion derivative of this plasmid, pSV2neoDL SV2gpt. This plasmid carried a 248 bp deletion in the 5′ end of the *neo* gene, rendering it non-functional in bacterial as well as in mammalian cells. This plasmid was introduced into human EJ cells by calcium phosphate precipitation and *gpt*$^+$ colonies were isolated in mycophenolic acid, aminopterin and xanthine (MAX) medium (Mulligan and Berg, 1981). Examination of these cell lines revealed that each contained integrated copies of *neoDL* and *gpt*. The integrated *neoDL* gene, which for most purposes behaves as a cellular gene, was used as the target for gene modification. Several independently derived cells containing *neoDL* were transfected with pSV2/neoDR which carried a 283 bp deletion at the 3′ end of the *neo* gene. Cells which contained a functional *neo* gene were selected by using G418. The frequency with which such colonies were obtained was compared to that obtained by using pSV2neo.

Since pSV2neo can integrate anywhere in the genome to give G418^R colonies, the frequency with which such colonies arise can be used as a measure of NHR. Since the G418^R colonies obtained from transfection with *neoDR* must involve interaction with the chromosomal *neoDL* target, that frequency can be considered to reflect HR. The ratio of the HR:NHR frequencies provides a measure of the efficiency of HR. In these experiments, Song *et al.* (1987) observed that different cell lines gave different HR:NHR ratios with ranges between 1:74 and 1:500. Similar experiments by other groups using a number of different targets gave ratios as divergent as 1:100–10 000 (Smith and Berg, 1984; Lin *et al.*, 1985; Thomas *et al.*, 1986).

Molecular analysis of the G418^R colonies revealed that they are indeed the result of HR between the chromosomal locus and the introduced plasmid. Studies from several different laboratories revealed that three different types of events can generate a functional *neo* gene. In some instances, the input plasmid underwent a reciprocal recombination event resulting in the integration of the input plasmid into the chromosomal locus and yielding a functional *neo* gene in the process. A second class of cells had their functional *neo* gene regenerated by correction of the target gene by gene conversion or double reciprocal recombination. In the third class, the *neo* gene in the input plasmid was corrected by a similar mechanism and the corrected plasmid integrated at a different site.

5. Chromosomal gene targeting

Although experiments involving modification of an integrated selectable gene, such as the *neo* gene described above, can be considered strong evidence for HR of cellular genes, it was felt that what is true for artificial targets may not hold true for cellular genes. The first definitive evidence for modification of

GENE INACTIVATION

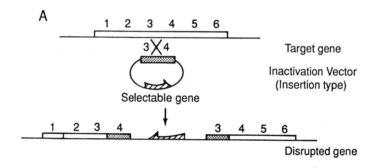

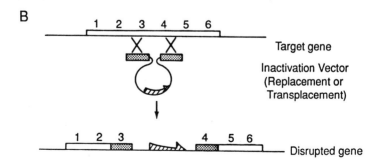

Fig. 1. Strategies for gene inactivation by homologous recombination.

a cellular gene was provided by Smithies *et al.* (1985). In these experiments, the target gene was the human β-globin. Initially, several targeting vectors were constructed to test recombination at the globin locus (Smithies *et al.*, 1984). However, the vector that proved most useful contained the following elements: an incomplete β-globin gene and an appropriate unique restriction enzyme recognition site within, which could be used to linearize the plasmid, a bacterial *SupF* gene capable of suppressing amber termination mutants in *E. Coli*, and SV2neo, whose expression can be selected in mammalian cells. This plasmid was introduced into human cells or rodent × human somatic cell hybrids containing a single copy of human chromosome 11 bearing the globin locus. Stable transfectants were isolated by selection in G418. Each of these G418-resistant colonies results from either a random integration of the plasmid or HR. HR would result in disruption of an 11.1 kb *XbaI* fragment which contains the β-globin gene, and the creation of two new *XbaI* fragments, one of which is 16.7 kb and the other 7.7 kb. The 7.7 kb fragment is expected to

GENE CORRECTION

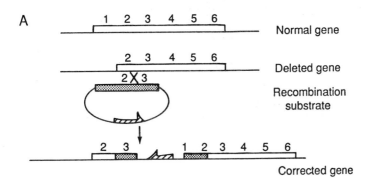

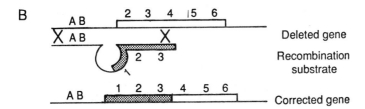

Fig. 2. Strategies for gene correction by homologous recombination.

contain the complete β-globin gene flanked at its 5' end by the bacterial *SupF* gene. This unique structure is not present in the input plasmid, or the target locus, and cannot be generated by NHR. The presence of the novel recombination product was detected by rescuing the fragment from pools of transfectants and testing this rescued DNA for the desired structure. After successfully rescuing the appropriate fragment, a cell line containing the desired modification was isolated.

The globin gene targeting experiments revealed that (1) normal cellular targets can be modified by HR, (2) assays based on screening for novel junction fragments created by recombination are an appropriate method to detect HR events, (3) transcriptional activity of the target gene is not required for gene targeting, and (4) the ratio of HR:NHR events involving cellular genes could be greater than 1:1000.

6.   ES cells

6.1.   Use of ES cells for HR

HR has became a powerful genetic tool in yeast because it is not only relatively easy to obtain cells which have undergone HR, but it is then relatively straight-forward to obtain homozygosity at the modified locus. Since mammalian cells are diploid, it is necessary to have methods to modify both copies of an autosomal gene or to have an alternative method for achieving homozygosity of the modified locus. Embryonic stem cells paved the way to achieve this goal. Based on the work of several investigators, Evans and Kaufman (1981) have shown that it is possible to establish cells derived from the inner cell mass of mouse blastocysts in culture. They also demonstrated that the cells are capable of resuming their full developmental potential when they are reintroduced into a mouse blastocyst and the blastocyst is introduced into the uterus of a pseudopregnant female. Under appropriate circumstances, the introduced embryonic cells could con-tribute to all of the embryonic tissues, including the germ line. This observation raised the possibility that mice carrying a modified gene in a heterozygous or homozygous state can be obtained by modifying one of the two copies of the target gene in the ES cells by HR and introducing the resulting cells into appropriate blastocysts.

6.2.   Gene modification of ES cells

The first attempts at gene modification by HR in mouse ES cells were aimed at the selectable hypoxanthine phosphoribosyl transferase (HPRT) gene. In one set of experiments a gene correction was attempted. The cells (E14TG2a) were derived from normal ES cells and are HPRT$^-$. This cell line was obtained as a spontaneous mutant and is thioguanine (TG)-resistant by virtue of a deletion of part of HPRT. The deletion removed exons 1, 2 and (part of) 3, and an unknown portion at the 5' end (Hooper et al., 1987). Doetschman et al. (1987) constructed an O-type plasmid substrate containing a part of the human HPRT gene (exons 1 and 2) and a part of the mouse gene. This plasmid was introduced into E14TG2a cells by electroporation and HPRT$^+$ cells were isolated in HAT. These investigators were able to show that the HPRT$^+$ cells have resulted from HR, and the frequency with which these events occurred was about one per million cells electroporated.

A second series of experiments which utilized the mouse HPRT gene as the target were designed to inactivate the gene. In one series of experiments, Thomas and Capecchi (1987) made constructs of both the O type (insertion) and Ω type (replacement or transplacement). In these cases the introduced DNA contained sequences homologous to the eighth exon of the gene and a dominant selectable marker *neo*. Thomas and Capecchi isolated TG^R G418^R colonies and showed that many of the colonies are the result of HR. They observed that the nature of the recombination substrate (of the insertion type or replacement type) had little effect on HR efficiency. In another set of experiments, Doetschman et al. (1988) utilized an Ω type vector which carried a promoterless *neo* gene imbedded into the HPRT coding sequences. They introduced this plasmid into HPRT$^+$ ES cells by electroporation and isolated TG^R colonies. Although the introduced DNA contained a *neo* gene, Doetschman et al. did not select for *neo* expression. They examined the TG^R colonies for HR events by screening by the polymerase chain reaction (Saiki et al., 1985) to detect a novel junction fragment which can be created only through an HR event. Positive cell lines were demonstrated to have resulted from HR by Southern blot hybridization of the cellular DNA.

Although the number of cells in the starting population which were modified by HR in the two sets of HPRT inactivation experiments was comparable (approximately one per million cells), the HR:NHR ratios in the two experiments were quite different. Although Doetschman et al. did not have a functional *neo* gene in the input construct to directly test transfection efficiencies reflecting NHR events, they estimate that the HR:NHR ratio is about 1:15. Thomas et al. (1986) obtained different ratios when they used different amounts of homology (see below). In a construct with the least amount of homology, the ratio was 1:40 000. Part of the reason for this discrepancy lies in the fact that Thomas et al. used a version of *neo* which would be expressed most efficiently in ES cells. This results in a greater sampling of integration sites and, consequently, a greater number of NHR events.

7.   Strategies to detect HR events involving non-selectable genes

Part of the reason for the successes in modification of the mouse HPRT gene is that the expression of the gene can be selected for in both a positive and a negative way. Unfortunately, such selective schemes are not available for most of the genes in the genome. As a result, alternative strategies to identify and isolate HR events are needed.

In cases where the target cell number is limiting it is desirable to have methods which result in a significant increase in absolute HR efficiency. Methods which effectively change the HR:NHR event ratios either by decreasing the efficiency of NHR events or by eliminating NHR events are also useful. If alteration of HR or NHR efficiencies is not possible, it would be desirable to have efficient screening strategies to detect the rare HR events. Approaches covering all of these possibilities have been described.

7.1.   Increasing HR efficiencies

Thomas *et al.* (1987) examined the effect of introducing different numbers of copies of the targeting plasmid on HR. They reported that they did not observe any significant differences when they microinjected 5–100 copies of input DNA into cells. Since most of the targeting experiments are now conducted by electroporation and as it is difficult to quantify the amount of DNA that goes into each cell and is available to undergo HR, this approach of changing DNA concentrations to increase HR does not hold much promise. Zhang and Wilson (1990) examined the effect of varying the target number on HR. For these experiments, the target was an amplifiable dihydrofolate reductase (DHFR) gene. They attempted targeting to this locus when it was present in one copy or at 200 copies. They observed that the targeting frequencies in these two cases are indistinguishable. Based upon these data, they conclude that the target number is not the limiting feature for recombination.

A striking observation that was made by Thomas *et al.* is that a two-fold increase in the amount of homology results in a 10-fold increase in HR. Capecchi (1989) reported that this logarithmic increase in HR holds true for a homology range of 2–15 kb. A similar result which tests the effect of homology on recombination has been reported at the immunoglobulin gene locus by Shulman *et al.* (1990). In this report, a two-fold increase in homology stimulated HR by nearly 10-fold. Based on these observations, we can conclude that a large amount of homology between the input and target sequences is desirable. The ability to clone large fragments of DNA into yeast artificial chromosomes (YACs) (Burke *et al.*, 1987) and recent demonstrations of the ability to introduce these large DNA fragments into mammalian cells (Pachnis *et al.*, 1990; Pavan *et al.*,

1990) present the possibility of testing the effect of homologies of up to 1 Mb on HR.

7.2. *Effect of heterology on HR*

Both O-type and Ω-type recombination substrates contain sequences which are not homologous to the target. It is conceivable that the frequency of HR may depend on the amount of homology as well as the size of the heterologous sequences present. Two experiments support the view that low amounts of heterology are beneficial for HR. In one report Zimmer and Gruss (1989) presented results of microinjection of a *hox-1.1* gene into which 20 nucleotides of heterology was introduced. When this disruption vector was used, they observed that approximately 1:600 cells injected with the DNA had undergone an HR event. Since microinjection yields a stable transfection efficiency of 10–20%, in these experiments the HR:NHR ratio was estimated to be 1:30–60. Similarly, Steeg *et al.* (1990) introduced a 5.8 kb DNA containing two base pair differences from the target into ES cells. The target was the RNA *pol* II gene. In these experiments, they observed that the HR:NHR ratio was also about 1:30. These results, although suggestive, do not provide definitive evidence that low levels of heterology increase HR. While no direct evidence exists, it is possible that there is a threshold level of heterology beyond which the cell will not be able to discriminate the substrates. Thus, it is advisable to keep the amount of heterology to a minimum during the design and construction of recombination substrates.

7.3. *Altering or reducing NHR efficiencies*

The observations by Thomas *et al.* (1987) provide a strategy for altering the levels of effective NHR efficiencies. They observed that when the *neo* gene, located in the targeting vector, was modified to have an enhancer–promoter combination and translation start signals which are ideally suited for the cell type of interest, the transfection efficiency can be improved by as much as 100–200-fold. Although similar differences have not been observed by other investigators, it is clear that such modifications could alter the transfection efficiencies. Based upon these results, it is possible to argue that the use of inefficient transcriptional signals for the selectable gene, yielding lower levels of NHR events, might be desirable. This simplistic view is complicated by the fact that the chromatin structure of different target genes may permit different levels of expression of the introduced gene. If the target gene structure does not permit expression of the selectable gene unless it has a strong promoter, the selection system employed may be selecting against the types of desired events. On the other hand, the use of highly effective enhancer–promoter combinations may unnecessarily increase the NHR efficiency. Alternative strategies to circumvent this issue are now available.

7.4. Use of fusion genes

In one strategy, an incomplete selectable gene is incorporated into the targeting vector. The selectable gene lacks appropriate 5′ and/or 3′ sequences which would be needed for its expression. When the vector undergoes HR the helper or selectable gene now acquires the signals needed for its correct expression and the cell that harbours that event will survive in the selective medium. When the targeting vector integrates at a random location the probability that such a location fortuitously contains the needed signals for the expression of the helper gene would be low, resulting in death of that cell in the selective medium. Enrichment by 50–100-fold in HR events has been achieved in this way. This strategy, originally applied to artificial targets (Jasin and Berg, 1988; Sedivy and Sharp, 1989) has now been extended to several genes. Schwartzberg *et al.* (1989) and Charron *et al.* (1990) have utilized this approach in constructing their targeting vectors for c-*abl* and n-*myc*, respectively. Both groups report very favourable HR:NHR event ratios in their experiments. Thomas and Capecchi (1987) and Joyner *et al.* (1989) eliminated the poly(A) addition signals for the helper *neo* gene in their constructs to inactivate HPRT, and *en-2* in mouse ES cells. Joyner *et al.* report a modest two-fold enrichment in homologous recombinants using this strategy. It must be noted that for this approach to be successful the target gene must be transcriptionally active or be activated by the HR event itself.

7.5. Positive–negative selection

Mansour and colleagues (1988) developed a method which permits positive selection for the correctly targeted events while negatively selecting for non-homologous events. The general strategy is to employ an Ω-type vector for gene inactivation. The non-homologous portion in the Ω vector will contain the gene whose expression can be selected for in a positive manner. A negatively selectable gene, in this case the herpes simplex virus (HSV) *tk* gene, was incorporated at one end of the homologous arm in the Ω vector and the construct introduced into cells. Since NHR is mediated by the free ends (Roth and Wilson, 1988), cells in which such an event has occurred will have a functional HSV *tk* gene which can be selected against by using the nucleotide analogues acyclovir or gancyclovir (GANC). Since HR has to be mediated by the homology present in the arms of the Ω vector, such an event will result in retention of the positive selectable gene in the loop and loss of the negative selectable gene. Cells which survive G418 and GANC selection are expected to have a substantial enrichment for HR events. Mansour *et al.* (1988) and Thomas and Capecchi (1990) have shown that they can obtain as much as a 1000-fold or greater enrichment for HR events using this method. Although similar levels of enrichment have not been observed by other workers, it is clear that the positive–negative selection (PNS) system significantly enhances the ability to recover the targeted events (Johnson *et al.*, 1989; Dechiara *et al.*, 1990).

8. Screening for HR events

In cases where the target gene is not selectable or the targeting vector does not contain a selectable gene, it is necessary to have an appropriate and rapid screening system to isolate the desired targeted cell lines. Several different screening strategies are available. We have already described a screening approach that was used by Smithies *et al.* (1985) to identify and isolate HR events at the human β-globin locus. The general strategy included generation of pools of colonies and sib selection of pools which tested positive for the HR event. The strategy of sib selection is very useful in conjunction with the polymerase chain reaction. Experiments of this nature were described by two groups (Zimmer and Gruss, 1989; Joyner *et al.*, 1989) and other experiments using this approach are underway in several laboratories.

A second screening strategy was employed by Jasin *et al.* (1990) to isolate a targeted event at the CD4 locus in human T cells. They designed the targeting vector in such a way that an HR event would result in disruption of the CD4 gene with the simultaneous activation of a *Thy-1* epitope included in the targeted vector. They were able to screen the transfected cells by cell-sorting methods and reported a 100-fold enrichment of homologous recombinants. This method, called epitope addition, or variations of it should prove to be very valuable additions to the repertory of gene targeters.

A number of different groups have reported success at gene modification by HR. A list of the genes involved is presented in Table 1.

9. Applications of HR

Perhaps the most widely used application of HR, at this time, is to understand gene function. In most of the experiments, the target cell type is mouse ES cells. The general strategy of these experiments is to inactivate or modify one copy of a mouse gene of choice and obtain germ line transmission of the modified gene. If successful, mice which are heterozygous or homozygous for

Table 1. Summary of gene targeting experiments

Targeted gene	Species (cells)	Method of introduction	Approach	Homology (kb)/heterology	HR:NHR (actual, effective)[a]	Reference
β-Globin	Human (Hu11 hybrid)	Electroporation	Selection (G418)	4.6/8.5	1:1100, 1:1100	Smithies et al. (1985)
HPRT	Mouse (ES)	Electroporation	Selection (G418, 6-TG)	4.0–9.1/1.0	1:950–40 000, 1:1	Thomas et al. (1987)
HPRT	Mouse (ES)	Electroporation	Selection (G418, 6-TG)	1.2 + 0.132/2.0	1:5000, 6:9	Doetschman et al. (1988)
HPRT	Mouse (ES)	Electroporation	Positive–negative selection (G418, GANC)	9.1/1.0	1:4000, 19:24	Mansour et al. (1988)
int-2	Mouse (ES)	Electroporation	Positive–negative selection (G418, GANC)	10.0/1.0	1:40000, 1:20	Mansour et al. (1988)
IgH (μ chain)	Mouse (Sp6)	Electroporation	Selection (G418) Screening (PCF)	4.3/5.7	1:1000, 4:5	Baker et al. (1988)
β₂-Microglobulin	Mouse (ES)	Electroporation	Selection (G418) Screening (PCR)	4.4/1.2	1:100, 1:100	Koller et al. (1989)
β₂-Microglobulin	Mouse (ES)	Electroporation	Selection (G418, 6-TG) Screening (PCR)	10.0/1.2	1:25, 1:25	Zijlstra et al. (1989)
APRT	Chinese hamster (CHO mutant)	Calcium phosphate	Selection (ALASA)	2.6/4.5	1:8000, 1:2	Adair et al. (1989)
MHC (IIEα)	Mouse eggs	Microinjection	Screening (Southern)	2.6–8.9	1:500, 1:500	Brinster et al. (1989)
c-fos, Adipsin, aP2	Mouse (ES)	Electroporation	Positive–negative selection (G418, GANC)	?/1.2, ?/1.2, 4.5/1.2	1:20 000, 1:16; 1:25 000, 1:10; 1:3300, 1:12	Johnson et al. (1989)

Gene	Cell type	Method	Selection/Screening	HR:NHR actual	HR:NHR effective	Reference
En-2	Mouse (ES)	Electroporation	Selection (G418) Screening (PCR)	?/1.0	1:260, 1:260	Joyner et al. (1989)
Hox-1.1	Mouse (ES)	Microinjection	Screening (PCR)	1.57/0.02	1:640, 5:12	Zimmer and Gruss (1989)
c-abl	Mouse (ES)	Electroporation	Promoterless neo (G418)	7.5/0.6	1:3400, 1:34	Schwartzberg et al. (1989)
n-myc	Mouse (ES, pre-B cell)	Electroporation	Promoterless neo (G418)	?/1.0	1:500, 1:5	Charron et al. (1990)
CD4	Human (T cell)	Electroporation	Epitope addition	0.4 + 2.9/?	1:900, 8:11	Jasin et al. (1990)
int-1	Mouse (ES)	Electroporation	Positive–negative selection	13.5/1.0	$1:5 \times 10^6$, 1:400	Thomas and Capecchi (1990)
DHFR	Chinese hamster (CHO, CHO400)	Electroporation	Positive–negative selection (G418, FIAU)	4.6/1.2	1:100, 1:2	Zhang and Wilson (1990)
IGF-II	Mouse (ES)	Electroporation	Positive–negative selection (G418, GANC)	1.6 + 8.3/1.2	1:500, 1:40	DeChiara et al. (1990)
RNA polII	Mouse (ES)	Electroporation	Selection (amanitin)	5.8/2 bp	1:30, 1:1	Steeg et al. (1990)
Hox-3.1	Mouse (ES)	Electroporation	Selection (G418)	6.8 + 1.5/7.2	1:40, 1:40	Le Mouellie et al. (1990)
HPRT	Human	Electroporation	Selection (G418, 6-TG)	3.4 + 1.0/1.2	1:40 000, 2:72	C. Campbell and R. Kucherlapati (unpublished)

[a] HR:NHR actual, estimated ratio of HR events to random integration events; HR:NHR effective, the ratio of number of colonies with HR to the number of tested colonies.

the modification can be obtained and the effect of the modification examined. The most extensively studied modification of this nature is the inactivation of the mouse β_2-microglobulin. In two separate experiments, Koller and Smithies (1989) and Zijlstra *et al.* (1989) inactivated one copy of the mouse β_2-microglobulin gene in ES cells. The resulting cells were injected into mouse blastocysts and germ line transmission was obtained. β_2-Microglobulin is a single-copy gene which is necessary for the expression of the MHC class I antigens on the cell surface. It would be possible to study the effects of lack of MHC class I proteins on the development of the immune system and other effects on the normal development and viability of the organism. Results from both groups indicate that mice homozygous for β_2-microglobulin inactivation are viable but they lack a particular class of mature T cells, the $CD8^+$ cells. The effects of the absence of this particular class of cells on the immune response and the role of class I gene expansion in phenomena such as tissue rejection are currently being studied.

Several other modifications achieved by HR have been successfully transmitted through the germ line. These include *en-2*, a homeobox-containing gene, c-*abl*, the cellular homologue for the oncogene *abl*, and insulin-like growth factor. The nature of the modifications and the effects of these modifications have been described (Joyner *et al.*, 1989; Schwartzberg *et al.*, 1989; DeChiara *et al.*, 1990).

In addition to studies of gene modifications in ES cells, it is expected that modifications in established cell lines would be important areas for study. Although several genes have been modified by HR in several established cell lines, studies of gene structure–function relationships require homozygosity. One concern in this respect is that unless the efficiency of HR is significantly enhanced it may not be possible to obtain cell lines in which both copies of the gene are inactivated. It is interesting to note that Jasin *et al.* (1990) observed that one in nine cell lines which had an inactivated CD4 gene was homozygous for the modification. Although such homozygosity could result from simultaneous inactivation of both gene copies, it is felt that this result is probably due to non-disjunctional loss of the chromosome carrying the unmodified gene. Whatever the case may be, these results indicate that it may be relatively easy to obtain cells containing homozygous modifications.

One of the potential applications of HR is in the area of human gene therapy. The general strategy for gene therapy is as follows: isolate the population of affected cells or their precursors from a patient suffering from the genetic disease, modify the mutant gene by HR and replace the cells carrying the normal gene into the patient. Modification of the mutant gene by HR is ideal because such genes are regulated properly (Nandi *et al.*, 1988) and one can be assured that there are no adverse effects that may result from random integration of introduced DNA sequences. Even though gene correction (Doetschman *et al.*, 1987) and targeted integration (Smithies *et al.*, 1985) have been demonstrated, these methods have not yet been applied to gene therapy. Part of the reason for the lack of attempts in this direction is the difficulty in obtaining appropriate pluripotential stem cells in sufficiently large quantities to perform the desired modifications. Methods to grow large numbers of certain precursor cells such as keratinocytes, and certain other cell types, are currently available and it is anticipated that intensive efforts in this area will be forthcoming.

10. Conclusion

When DNA is introduced into yeast cells, integration of the DNA into the cellular chromosome is mediated predominantly by HR. Gene modification in yeast through HR has become a powerful genetic tool. It was expected that the ability to modify genes in mammalian cells in a similar fashion would prove to be equally powerful. There is now ample evidence that mammalian genes can be modified by HR. In some instances the efficiencies of recombination are quite favourable and in others it is less efficient. Efforts at understanding the mechanisms of recombination and the parameters that influence its efficiency are underway. It is expected that HR in somatic cells opens up new possibilities for genetic investigation of mammalian genes.

References

Adair, G. M., Nairn, R. S., Wilson, J. H., Seidman, M. M., Brotherman, K. A., MacKinnon, C. and Scheerer, J. B. (1989). Targeted homologous recombination at the endogenous adenine phosphoribosyltransferase locus in Chinese hamster cells. *Proc. Natl Acad. Sci. USA* **86**, 4574–4578.

Baker, M. D., Pennell, N., Bosnoyan, L. and Shulman, M. J. (1988). Homologous recombination can restore normal immunoglobulin production in a mutant hybridoma cell line. *Proc. Natl Acad. Sci. USA* **85**, 6432–6436.

Brinster, R. L., Braun, R. E., Lo, D., Avarbock, M. R., Oram, F. and Palmiter, R. D. (1989). Targeted correction of a major histocompatibility class IIEα gene by DNA microinjected into mouse cells. *Proc. Natl Acad. Sci. USA* **86**, 7087–7091.

Burke, D. T., Carle, G. F. and Olson, M. V. (1987). Cloning of large segments of exogeneous DNA into yeast by means of artificial chromosomal vectors. *Science* **236**, 806–812.

Campbell, A. (1971). In: Hershey, A. R. (ed) *The Bacteriophage Lambda*, pp. 13–44, Cold Spring Harbor Laboratory, New York.

Capecchi, M. R. (1989). Altering the genome by homologous recombination. *Science* **244**, 1288–1292.

Cavanee, W. K., Dryza, T. P., Phillips, R. A., Benedict, W. F., Godbout, R., Gallie, B. L., Strong, L., Murphee, A. L. and White, R. L. (1983). Expression of recessive

44 S. DOI *ET AL.*

alleles by chromosomal mechanisms in retinoblastoma. *Nature (London)* **305**, 779–784.

Charron, J., Malynn, B. A., Robertson, E. J., Goff, S. P. and Alt, F. W. (1990). High frequency disruption of the n-myc gene in embryonic stem and pre-B cell lines by homologous recombination. *Mol. Cell. Biol.* **10**, 1799–1804.

Dechiara, T. M., Efstratiadis, A. and Robertson, E. J. (1990). A growth-deficiency phenotype in heterozygous mice carrying an insulin-like growth factor II gene disrupted by targeting. *Nature (London)* **345**, 78–80.

deSaint Vincent, B. R. and Whal, G. M. (1983). Homologous recombination in mammalian cells mediates formation of a functional gene from two overlapping gene fragments. *Proc. Natl Acad. Sci. USA* **180**, 2002–2006.

Doetschman, T., Gregg, R. G., Maeda, N., Hooper, M. L., Melton, D. W., Thompson, S. and Smithies, O. (1987). Targeted corrections of a mutant HPRT gene in mouse embryonic stem cells. *Nature* **330**, 576–578.

Doetschman, T., Maeda, N. and Smithies, O. (1988). Targeted mutation of the *Hprt* gene in mouse embryonic stem cells. *Proc. Natl Acad. Sci. USA* **85**, 8583–8587.

Evans, M. J. and Kaufman, M. H. (1981). Establishment in culture of pluripotential cells from mouse embryos. *Nature* **292**, 154–156.

Gossler, A., Joyner, A. L., Rossant, J. and Skarnes, W. C. (1989). Mouse embryonic stem cells and reporter constructs to detect developmentally regulated genes. *Science* **244**, 463–465.

Hinnen, A., Hicks, J. B. and Fink, G. R. (1978). Transformation of yeast. *Proc. Natl Acad. Sci. USA* **75**, 1929–1933.

Hooper, M. L., Hardy, K., Handyside, A., Hunter, S. and Monk, M. (1987). HPRT deficient (Lesch–Nyhan) mouse embryos derived from germ-line colonization by cultured cells. *Nature* **326**, 292–295.

Jaenisch, R. (1988). Transgenic animals. *Science* **240**, 1468–1474.

Jasin, M. and Berg, P. (1988). Homologous integration in mammalian cells without target gene selection. *Genes Dev.* **2**, 1353–1363.

Jasin, M., Elledge, S. J., Davis, R. W. and Berg, P. (1990). Gene targeting at the human CD4 locus by epitope addition. *Genes Dev.* **4**, 157–166.

Johnson, R. S., Sheng, M., Greenberg, M. E., Kolodner, R. D., Papiannou, V. E. and Spiegelman, B. M. (1989). Targeting of non-expressed genes in embryonic stem cells via homologous recombination. *Science* **245**, 1234–1236.

Joyner, A. L., Skarnes, W. L. and Rossant, J. (1989). Production of a mutation in mouse *En-2* gene by homologous recombination in embryonic stem cells. *Nature (London)* **338**, 153–156.

Keown, W., Campbell, C. R. and Kucherlapati, R. (1990). Methods for introducing DNA into mammalian cells. *Meth. Enzymol.* **185**, 527–537.

Koller, B. H. and Smithies, O. (1989). Inactivating the β_2 microglobulin locus in mouse embryonic stem cells by homologous recombination. *Proc. Natl Acad. Sci. USA* **86**, 8932–8935.

Kucherlapati, R. and Skoultchi, A. I. (1984). Introduction of purified genes into mammalian cells. *CRC Crit. Rev. Biochem.* **16**, 349–379.

Kucherlapati, R. S., Eves, E. M., Song, K. Y., Morse, B. S. and Smithies, O. (1984).

Homologous recombination between plasmids in mammalian cells can be enhanced by treatment of input DNA. *Proc. Natl Acad. Sci. USA* **81**, 3153–3157.

LeMouellie, H., Lallomand, Y. and Brulet, P. (1990). Targeted replacement of the homeobox gene *Hox-3* by the *Escherichia coli lacZ* in Mouse Chimeric Embryos. *Proc. Natl Acad. Sci. USA* **87**, 4712–4716.

Lin, F. L., Sperle, K. and Sternberg, N. (1985). Recombination in mouse L-cells between DNA introduced into cells and homologous chromosomal sequences. *Proc. Natl Acad. Sci. USA* **82**, 1391–1395.

Mansour, S. L., Thomas, K. R. and Capecchi, M. R. (1988). Disruption of the proto-oncogene *int-2* in mouse embryo-derived stem cells: a general strategy for targeting mutations to non-selectable genes. *Nature (London)* **336**, 348–352.

Miller, C. K. and Temin, H. M. (1983). High efficiency ligation and recombination of DNA fragments by vertebrate cells. *Science* **220**, 606–609.

Mulligan, R. and Berg, P. (1981). Selection for animal cells that express the *Escherichia coli* gene coding for xanthine-guanine phosphoribosyltransferase. *Proc. Natl Acad. Sci. USA* **78**, 2072–2076.

Nandi, A. K., Roginski, R. S., Gregg, R. G., Smithies, O. and Skoultchi, A. I. (1988). Regulated expression of genes inserted at the human chromosome β-globin locus by homologous recombination. *Proc. Natl Acad. Sci. USA* **85**, 3845–3849.

Pachnis, V., Pevny, L., Rothstein, R. and Constantini, F. (1990). Transfer of a yeast artificial chromosome carrying human DNA from *Saccharomyces cerevisiae* into mammalian cells. *Proc. Natl Acad. Sci. USA* **87**, 5109–5113.

Pavan, W. J., Hieter, P. and Reeves, R. H. (1990). Modification and transfer into an embryonal carcinoma cell line of a 360 kb human-derived artificial chromosome. *Mol. Cell. Biol.* **10**, 4163–4169.

Rosenstraus, M. J. and Chasin, A. (1978). Separation of linked markers in Chinese hamster cell hybrids – mitotic recombination is not involved. *Genetics* **90**, 735–760.

Roth, D. and Wilson, J. (1988). Illegitimate recombination in mammalian cells. In: Kucherlapati R. and Smith, G. R. (eds) *Genetic Recombination*, pp. 621–654, ASM, Washington, DC.

Rothstein (1983). One step gene disruption in yeast. *Meth. Enzymol.* **101**, 202–211.

Saiki, R. K., Scharf, S., Faloona, F., Mullis, K. B., Horn, G. and Erlich, H. A. (1985). Enzymatic amplification of β-globin sequences and restriction site analysis for diagnosis of sickle cell anemia. *Science* **230**, 1350–1354.

Schwartzberg, P. O., Goff, S. P. and Robertson, E. J. (1989). Germ-line transmission of a c-abl mutation produced by targeted gene disruption of ES cells. *Science* **246**, 799–803.

Sedivy, J. and Sharp, P. (1989). Positive genetic selection for gene disruption in mammalian cells by homologous recombination. *Proc. Natl Acad. Sci. USA* **86**, 227–231.

Shapira, G., Stachelek, J. L., Letson, A., Soodak, L. and Liskay, R. M. (1983). Novel use of synthetic oligonucleotide insertion mutants for the study of homologous recombination in mammalian cells. *Proc. Natl Acad. Sci. USA* **80**, 4827–4831.

Shulman, M. D., Nissen, L. and Collins, C. (1990). Homologous recombination in hybridoma cells: dependence on time and fragment length. *Mol. Cell. Biol.* **10**, 4466–4472.

Small, J. and Scangos, G. (1983). Recombination during gene transfer into mouse cells can restore the function of deleted genes. *Science* **219**, 174–176.

Smith, A. J. H. and Burg, P. (1984). Homologous recombination between defective *neo* genes in mouse 3T6 cells. *Cold Spring Harbor Symp. Quant. Biol.* **49**, 171–181.

Smithies, O., Koralewski, M. A., Song, E. Y. and Kucherlapati, R. S. (1984). Homologous recombination with DNA introduced into mammalian cells. *Cold Spring Harbor Symp. Quant. Biol.* **49**, 161–170.

Smithies, O., Gregg, R. G., Boggs, M. A., Koralewski, M. A. and Kucherlapati, R. (1985). Insertion of DNA sequences into the human chromosome β-globin locus by homologous recombination. *Nature (London)* **317**, 230–234.

Song, K. Y., Chekuri, L., Rauth, S., Erlich, S, and Kucherlapati, R. S. (1985). Effect of double-strand breaks on homologous recombination in mammalian cells and extracts. *Mol. Cell. Biol.* **5**, 3331–3336.

Song, K. Y., Schwartz, F., Maeda, N., Smithies, O. and Kucherlapati, R. (1987). Accurate modification of a chromosomal plasmid by homologous recombination in mammalian cells. *Proc. Natl Acad. Sci. USA* **84**, 6820–6824.

Southern, P. J. and Berg, P. (1982). Transformation of mammalian cells to antibiotic resistance with a bacterial gene under control of the SV40 early region promoter. *J. Mol. Appl. Genet.* **1**, 327–341.

Steeg, C. M., Ellis, J. and Bernstein, A. (1990). Introduction of specific point mutations into RNA polymerase II by gene targeting in mouse embryonic stem cells: evidence for a DNA mismatch repair mechanism. *Proc. Natl Acad. Sci. USA* **87**, 4680–4684.

Szostak, J. W., Orr-Weaver, T. L., Rothstein, R. J. and Stahl, F. H. (1983). The double strand break repair model for recombination. *Cell* **33**, 25–35.

Tarrant, G. M. and Holliday, R. (1977). A search for allelic recombination in Chinese hamster cell hybrids. *Mol. Gen. Genet.* **156**, 273–279.

Thomas, K. R. and Capecchi, M. R. (1987). Site-directed mutagenesis by gene targeting in mouse embryo-derived stem cells. *Cell* **51**, 503–512.

Thomas, K. R. and Capecchi, M. R. (1990). Targeted disruption of the mouse *Int-1* proto-oncogene results in severe abnormalities in mid-brain and cerebellar development. *Nature (London)* **346**, 847–850.

Thomas, K. R., Folger, K. R. and Capecchi, M. R. (1986). High frequency targeting of genes to specific sites in the mammalian genome. *Cell* **44**, 419–428.

Wasmuth, J. J. and Vock-Hall, L. (1984). Genetic demonstration of mitotic recombination in cultured Chinese hamster cell hybrids. *Cell* **36**, 697–707.

Zhang, H. and Wilson, J. H. (1990). Gene targeting in normal and amplified cell lines. *Nature (London)* **344**, 170–173.

Zijlstra, M., Li, E., Sajjadi, F., Subramani, S. and Jaenisch, R. (1989). Germ-line transmission of a disrupted β₂ microglobulin gene produced by homologous recombination embryonic stem cells. *Nature* **342**, 435–438.

Zimmer, A. and Gruss, P. (1989). Production of chimeric mice containing embryonic stem (ES) cells carrying a homeobox *Hox1. 1* allele mutated by homologous recombination. *Nature (London)* **338**, 150–153.

3
Transgenic strategies in mouse embryology and development

ROSA S. P. BEDDINGTON

AFRC Centre for Animal Genome Research, King's Buildings,
West Mains Road, Edinburgh EH9 3JQ, UK

1. Introduction

The combination of classical genetics, cell lineage studies and molecular biology has proved the most penetrating and rewarding alliance in defining and resolving key developmental issues. The power of such a three-pronged attack is, of course, exemplified by the current pace of discovery in *Drosophila*. In the past decade the persuasive elegance of precise molecular description and manipulation has assumed supremacy and dominates most contemporary developmental work. However, one should not forget that, without the painstaking accumulation

TRANSGENIC ANIMALS
ISBN 0-12-304530-4

and description of developmental mutations and the experimental analysis of cell behaviour during development, the prettiest molecular picture provides little more than a slice of detailed phenomenology. The molecular map of an embryo is as important as any meticulous histological description, but neither morphology nor molecules alone can elucidate developmental mechanism. The context for molecular function must, in the first instance, be supplied by experimental embryology and genetic examination or alteration of phenotype.

The gradient generated by the caudal (cad) gene product in Drosophila, both the distribution of the mRNA (Levine et al., 1985; Mlodzik et al., 1985) and the protein (MacDonald and Struhl, 1986), provides an excellent cautionary lesson regarding interpretation of molecular pattern. That a gradient or gradients must be the primary determinant of initial body pattern in insects had been conclusively established by surgical ligation experiments conducted more than 30 years ago (see Sander, 1976; Schubiger et al., 1977). The gradient generated by the cad gene has all the features expected of such a primary morphogenetic determinant. It is present during the stages at which the body plan is being established and the graded distribution of the protein is independent of zygotic gene expression. However, genetic experiments demonstrate unequivocally that cad cannot be the primary posterior determinant because it is not a true maternal effect mutation in that zygotic expression can largely rescue the cad⁻ phenotype (MacDonald and Struhl, 1986). On the other hand, genetic evidence supports and confirms the molecular indication that, for example, the bicoid gene product is indeed a primary pattern determinant in Drosophila (Driever and Nusslein-Volhard, 1988).

In discussing the application of transgenic technology to the study of mammalian development the multidisciplinary approach should not be overlooked. There is no ready short cut, afforded by molecular genetics, that will obviate the need for classical genetics, lineage information and micromanipulative experiments. However, even if transgenesis cannot make the mouse reproduce more quickly or guarantee an unequivocal clonal lineage analysis, transgenics promise to supply invaluable tools for all aspects of mammalian embryology. In this chapter their most promising contributions to the study of cell behaviour and determination during development, to advancing developmental genetics and for engineering specific molecular perturbations in the embryo will be discussed in turn. The advances possible are enormous but, as will be emphasized in this chapter and illustrated where possible by examples from other organisms, there is also a need for circumspection and the enduring difficulties confronting mammalian developmental biologists should not be underestimated.

2. Determination in development

If one is to assign causative roles to particular molecules during development, one must first know the sequence of determinative events at the cellular level. In other words, one must chart the time and location of apparently stable changes in the behaviour of individual cells, or populations of cells, with respect to their commitment to particular differentiation pathways. This provides an essential framework of developmental events from which can be derived specific questions amenable to molecular analysis. If one turns again to *Drosophila*, it was clonal lineage analysis which defined when cells become committed to particular parasegments or compartments and, thereby, established some of the principal cellular rules for development of body pattern (see Lawrence, 1981). A chart of the predictable and progressive restrictions that occur in well-defined groups of cells supplies both a timetable of when and where one might expect instrumental molecules to be present and also a precise blueprint against which to test the effects of molecular perturbation.

Mammalian embryology suffers from a distinct lack of such a timetable or blueprint. The principal reason for this is the difficulty in following the development of cells in viviparous animals. Adequate experimental access to the mammalian embryo is restricted to preimplantation stages of development and, therefore, to resolving the allocation and segregation of the extraembryonic tissue lineages. The body pattern emerges and the organ primordia begin to form only after implantation, when the embryo is relatively inaccessible. It is salutory to realize that, unlike any of the other major vertebrates or lower organisms popular in developmental biology, no determinative event, or inductive interaction, has been identified directly in any mammalian embryo during the onset of fetal organization. Even in the mouse, the best studied mammalian embryo, we have no direct knowledge of the cellular rules for generating the different embryonic tissues in the correct temporal and spatial sequence or for establishing axial pattern during gastrulation.

These are the details one needs to know and cannot simply borrow from other species. If important boundaries, delineating differently determined populations of cells, exist during the onset of fetal organization these cannot simply be triangulated onto the mouse embryo using information gained in other organisms of quite different three-dimensional design and porportion. More importantly, the mammalian embryo may employ certain developmental strategies not common in lower organisms. For example, the impressive regulative properties of the preimplantation mouse embryo extend into postimplantation development (e.g. Snow and Tam, 1979), and there is extensive cell mixing within most nascent tissues during early organogenesis (Beddington *et al.*, 1989). This indicates that commitment to particular axial or geographical domains and the segregation of specific tissue types may be a less synchronous and more labile process than in many lower organisms. Presumably, individual tissue primordia do become self-contained populations at some point, no longer giving rise to other kinds of tissue and no longer recruiting additional cells from unrelated neighbours. However, the process may differ from tissue to tissue. The most precocious of fetal organs, the heart mesoderm, probably arises early in gastrulation and quickly becomes a segregated tissue lineage. Similarly, primordial germ cells are probably set aside early in gastrulation (see Eddy, 1975; Snow, 1981). On the other hand, it is clear that neuroectoderm and mesoderm continue to form at the posterior end of the embryo for several days (see Rugh, 1968). It is not known in the mouse whether all emergent neuroectoderm is equivalent or if it arises with a specific axial identity. Can nascent caudal neuroectoderm form anterior central nervous system (CNS) structures if transplanted into an appropriate location in earlier-stage embryos? Similarly, do mammalian somites have a distinct and stable axial character corresponding to their rostrocaudal position? If so, at what stage of somite formation is this axial identity tag acquired and is it shared by all somitic derivatives? These are fairly elementary questions and yet they are fundamental to making sense of the role of any specific regionalized gene expression. No molecule can be a determinant if there is no corresponding determination.

What then can transgenics offer in helping to resolve mammalian cell lineages? Undoubtedly, the most straightforward and least controversial contribution is the potential supply of new single-cell markers to facilitate chimeric studies. In addition, retroviruses promise to be a useful method for creating marked clones of cells at any stage of development, although at present the validity of clonal lineage analyses using available vectors and detection systems is somewhat equivocal. More suspect claims of applying transgenesis to study cell lineages have been made. These include the retrospective analysis of adult retroviral mosaics derived from infection of preimplantation embryos (Soriano and Jaenisch, 1986), and using genetic ablation to determine lineage relationships in specific tissues.

2.1. Cell markers

In order to follow cell fate or to assess developmental potential in the embryo it is necessary to be able to distinguish between the progeny of the cell, or cells, whose behaviour is to be studied and the rest of the embryo. For some purposes exogenous markers such as vital dyes can be used but, in general, a genetic marker, in conjunction with transplantation and chimeric studies, offers a superior means of ensuring faithful inheritance by all progeny over any length of time. Unfortunately, in the mouse there are few suitable polymorphic loci whose products can be distinguished, either immunologically or histochemically, in individual cells in a tissue section or whole mount thus preserving spatial information. Most gene products which exhibit such readily detectable polymorphisms, such a β-glucuronidase or lectin-binding surface molecules, are neither ubiquitously nor constitutively expressed (see Ponder, 1987) and, therefore, are of only limited use in studying the provenance of particular tissues where expression is high. Transgenics, however, provide a ready source of new markers in the form of novel genes or DNA sequences. Two kinds of marker have been generated in stable transgenic strains: those that rely on gene expression and those that do not. Each has its merits and drawbacks.

2.1.1. *Nuclear transgenic markers*

Pronuclear injection of DNA often results in multiple copies of the transgene arranged as a head-to-tail concatamer inserted at a single integration site (see Palmiter and Brinster, 1986). In one transgenic strain, approximately 1000 copies of the human β-globin gene are inserted in this way. The transgene is not expressed but can be detected in sections by DNA *in situ* hybridization (Lo, 1986). This marker has been used successfully in chimera studies (Clarke *et al.*, 1988; Thomson and Solter, 1988a, b) to assess the tissue-specific selection against androgenetic, parthenogenetic or gynogenetic cells in chimeras (see Section 3.2.4). In many respects this marker is similar to the quail nucleolar marker used for avian lineage analysis (Le Douarin, 1973). However, the β-globin transgenic marker has the advantage that it can be used in intraspecific chimeras, or eventually congenic chimeras, and that detection is not restricted to interphase nuclei. At present, detection is restricted to sectioned material, which poses some problems in that nuclei of different size require different optimal section widths for maximal detection ($>90\%$ of transgenic cells, Thomson and Solter, 1988). Also, laborious reconstruction has to be used in order to regain three-dimensional information regarding cell position and it is more difficult to determine cell phenotype and position using a nuclear marker where the cell outline is not defined by the marker. In particular, nuclear markers are inadequate for analysing the distribution of cells, such as neurons, with tortuous cytoplasmic processes projecting long distances from the nucleus. Similarly, they are not suitable for resolving the arrangement of intimately juxtaposed cells in multilayered tissues, such as the labyrinth of the placenta (Duval, 1892). Despite the drawbacks, a nuclear non-expression marker is wholly reliable, in the sense that its limitations can be unequivocally defined and are not subject to poorly understood transcriptional and translational influences.

2.1.2. Expression markers

A transgene providing a constitutive and ubiquitous cytoplasmic expression marker would provide the ideal marker for lineage studies. In theory, this would seem a straightforward undertaking, requiring only a reporter gene, whose product is cell-autonomous and has no effect on development, controlled by a promoter derived from any universally expressed housekeeping or viral gene. Although the repertoire of available reporter genes is still quite small, a variety of suitable exogenous genes, such as the *Escherichia coli lacZ* gene, or the firefly (Ow *et al.*, 1986; de Wet *et al.*, 1987) or marine bacterial (Engebrecht *et al.*, 1985) luciferase gene, do exist whose products can be easily detected in mammalian tissue. However, so far no combination of promoter and enhancer sequences has been found which is capable of effecting truly constitutive and ubiquitous gene expression. This is not particularly surprising considering both the influence of integration site on gene expression and the incomplete characterization of most potentially ubiquitous promoters in terms of their full coterie of enhancer elements. Some ubiquitous promoters, such as herpes simplex thymidine kinase, appear to be too weak to overcome position effects (Allen *et al.*, 1988). Other ostensibly strong promoters produce unexpectedly tissue-specific or stage-specific gene expression in transgenics: for example, the rat β-actin promoter while generating ubiquitous *lacZ* expression in the midgestation embryo does not maintain ubiquitous expression in any tissue in the latter half of gestation or postnatally (Beddington *et al.*, 1989). It is possible that, using 'gene trap' constructs (Gossler *et al.*, 1989) consisting of a reporter gene without a promoter, recombination into embryonic stem (ES) cells may fortuitously produce insertion into a host gene which is active in all cells. The potential of this strategy remains to be proved by the production of germ line chimeras generating hemizygous offspring expressing the reporter gene ubiquitously and constitutively.

So far the only lineage analysis utilizing a reporter gene in chimeras has used the transgenic strain in which a rat β-actin promoter is coupled to *lacZ* and results in ubiquitous and constitutive expression of bacterial β-galactosidase in all embryonic tissues of the midgestation mouse conceptus (Beddington *et al.*, 1989). Even though ubiquitous expression is not maintained later, this strain has proved very useful for studying early organogenesis, allowing for the first time direct visualization of the distribution of inner cell mass clones in nascent fetal primordia (Beddington *et al.*, 1989). The opportunity to inspect the distribution of clonal descendants in whole mount, prior to sectioning, provides immediate three-dimensional information. This is invaluable for studying growth patterns in an embryo which is curved and therefore impossible to section in the same plane relative to the embryonic axis along its length. Extensive cell mixing, similar to that described in zebrafish (Kimmel and Warga,

1986), is evident during gastrulation, transgenic and wild-type cells being intimately intermingled in the epiblast and nascent mesoderm, and, although there is an indication of coherent growth for limited periods in certain epithelial primordia, some mixing appears to continue in all tissues throughout early organogenesis.

2.2. Retroviral mosaics

Unfortunately in the mouse, mitotic recombination is an extremely rare event and cannot be used to generate mosaics analogous to those used in *Drosophila* (West, 1984). However, it has been shown that infection of preimplantation embryos with Moloney murine leukaemia virus (Mo-MuLV) (Jaenisch et al., 1975) or injection of it into early postimplantation embryos (Jaenisch, 1976) results in mosaic offspring. Retroviruses should integrate rarely and at random into accessible cells and, whereas wild-type virus can spread infectiously from cell to cell, replication-defective viruses cannot spread horizontally and thus become a heritable marker for all clonal descendants of the target cell. The first attempt at lineage analysis using retroviruses was a retrospective analysis of mosaic offspring following infection of preimplantation embryos with replication-competent retrovirus (Soriano and Jaenisch, 1986). Both the distribution of independent retroviral integration sites and the relative contribution of cells containing a particular integration site to a variety of tissues in live-born mosaic animals were analysed. However, the conclusions that the fetus arises from no more than eight cells (because the extent of mosaicism in any tissue was always 12% or a multiple of 12%) and that the germ line is segregated earlier than fetal primordia and from fewer cells (because it transmitted unique integration sites not found in somatic tissues) must be viewed with some caution. The small number of cells available at the time of infection may be responsible for the apparent quantal contribution of retrovirally marked cells to adult tissues and the use of replication-competent virus, and thus the formal possibility of new integration sites appearing due to infectious spread, confuses any interpretation of the significance of unique integration sites peculiar to particular tissues.

The advent of replication-defective retroviral vectors promised a more elegant method for generating clonal mosaics at any stage of development and in any tissue. Recently, direct analyses of lineage have been undertaken using this strategy. Vectors containing *lacZ* were used to study lineage in the early postimplantation mouse embryo (Sanes et al., 1986) and in the neonatal rat retina and developing cerebral cortex (Price et al., 1987; Price and Thurlow, 1988) although, perhaps, the most compelling lineage analysis using this approach was that performed on the chick tectum, which revealed dramatic and convincing radial growth (Gray *et al.*, 1988). In all these studies a patch of stained cells was assumed to be a clone and the distribution of the clone and its cell type constituents was taken as a reflection of the potency and growth pattern of the cell into which the virus initially integrated.

The theoretical defence for using retroviral vectors for studying lineage is persuasive. However, in practice the approach is dogged by problems associated with the selective efficiency of viral integration and expression of reporter genes in different kinds of tissue. Some of the early vectors relied on the Mo-MuLV long terminal repeat (LTR) as the promoter for *lacZ* expression (Price *et al.*, 1987; Turner and Cepko, 1987; Price and Thurlow, 1988) and clearly there are developmental stage- and tissue-specific effects which influence the efficiency of this promoter. It is well known that the LTR is repressed during preimplantation development (Jaenisch *et al.*, 1975; Jahner *et al.*, 1982) and that it is an inefficient promoter in haemopoeitic stem cells (Williams *et al.*, 1986). Furthermore, the LTR does not become an efficient promoter in all postimplantation embryonic tissues simultaneously but shows a precise and tissue-specific order of activation during early organogenesis (Savatier *et al.*, 1990). This raises the possibility that the LTR may be particularly susceptible to repression in a variety of primary progenitor cells and that lineage information in these tissues may be biased towards downstream events in the differentiation pathway. Use of internal promoters provides an obvious solution, but these in turn can be subject to transcriptional interference (e.g. Emerman and Temin, 1984) and the site of retroviral integration may influence expression (e.g. Kratchowil *et al.*, 1989). If retroviruses have preferred integration sites (Shih *et al.*, 1988), normally integrating into regions of active chromatin (Barklis *et al.*, 1986; Peckham *et al.*, 1989), one may envisage a disturbing scenario where the apparent distribution of clonal descendants reflects not lineage restrictions but constraints on integration site in different cell types, these sites subsequently exhibiting differential transcriptional activity in different differentiated progeny. It would appear that reliable use of this technique to study cell lineages must await a means of identifying single retroviral inserts in individual cells irrespective of their expression.

A separate problem is the unequivocal recognition of clones when there is no control over which cell the virus integrates into at the time of infection. For example, in the early postimplantation embryo, where there is considerable cell mixing, the clonal descendants of a cell may become widely distributed throughout the embryo (Lawson *et al.*, 1987), making the distinction between different experimental clones marked with the same reporter gene extremely difficult. Finally, there are limitations imposed by the accessibility of certain tissues to infection and their susceptibility to infection. For example, visceral and gut endoderm, which can be readily exposed to viral particles, are particularly refractory to infection as judged by their failure to express a reporter gene unless integration occurs via infection first of the epiblast and its subsequent differentiation into endoderm (Savatier *et al.*, 1990).

Despite these limitations which, at present, demand cautious interpretation of retroviral lineage analyses, the potential uses of retroviruses in studying development should not be underestimated. They have already proved to be efficient insertional mutagens with the added benefit of providing a suitable tag for cloning the disrupted endogenous gene (Schnieke *et al.*, 1983; Harbers *et al.*, 1984). They also provide an invaluable tool for producing inappropriate expression of any gene which can be cloned into the vector and for observing its effect in a small population of cells surrounded by otherwise normal tissue (e.g. see Thompson *et al.*, 1989). This may be far more revealing of gene function than studying the consequences of aberrant expression in all cells of a tissue or an embryo.

2.3. *Genetic ablation*

The principal of genetic ablation is to destroy all cells expressing a particular gene (Breitman *et al.*, 1987; Palmiter *et al.*, 1987; Evans, 1989). This is achieved by linking the sequence coding for the 'poisonous' element of a toxin (such as the diptheria toxin subunit A which kills cells by modifying elongation factor 2 and inhibiting protein synthesis (Collier, 1975)) to a tissue-specific promoter. When this transgenic promoter becomes active at the appropriate stage of tissue differentiation, toxin is produced and causes cell death. It has been claimed that such directed lethality to specific cell types can contribute to the study of cell lineages. Despite some problems in the efficiency and penetrance of reproducible ablation, it is clearly an elegant method for eliminating certain cell types and perhaps evaluating the function of cells that express specific genes. However, it is inappropriate for studying cell lineages (Beddington, 1988). Cells expressing a particular tissue-specific product are not necessarily ancestrally related and, therefore, genetic ablation cannot provide information on the lineal relationship of the cells it kills. An obvious, if extreme, example of such 'convergent' differentiation is the formation of muscle in the nematode where five of the primary lineages can give rise independently to differentiated muscle (Sulston *et al.*, 1983) Genetic ablation is also a treacherous tool for the study of cell–cell interactions and regulation where adjacent tissues might transdifferentiate to compensate for the loss of a particular cell type. Such regulation, which would have important implications regarding the relationship between cell lineage and differentiation, would be obscured in transgenics designed for genetic ablation since any cell that did transdifferentiate would, in effect, commit suicide.

2.4.   Lineage analysis and gene function

Lineage analysis has an important role in assessing the significance of certain patterns of gene expression. One important question to be answered in whether mixing occurs across boundaries of gene expression. For example, most *Hox* genes show rather precise domains of expression, particularly in the CNS (see Holland and Hogan, 1988), and it is important to know whether cells can mix across these boundaries or are confined to the domains defined by *Hox* gene expression. If *Hox* genes serve as determinants there should not be mixing as this would degrade the spatial pattern of determination. Unfortunately, chimeras created during preimplantation development are unlikely to answer this question because the progeny of even a single inner cell mass cell are dispersed throughout the fetus and therefore uninformative regarding mixing across boundaries of gene expression. The construction of postimplantation chimeras or production of retroviral mosaics are more likely to demonstrate constraints on mixing during organogenesis.

The fate of postimplantation tissue in *in vitro* chimeras (Beddington, 1981) has been tested using the β-actin *lacZ* transgenic strain. Transplantation of a single transgenic somite to replace one, either in the same or a different

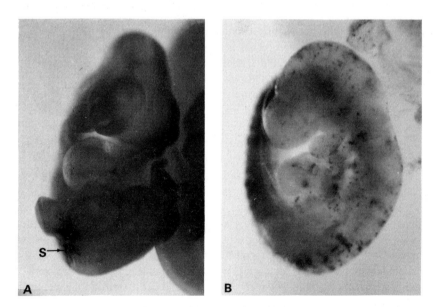

Fig. 1. (A) Transgenic somite (S) expressing *lacZ* transplanted to an early somite stage embryo subsequently developed *in vitro* for 48 hours. (B) Blastocyst injection chimera containing ES cells expressing *lacZ*.

position, in a wild-type host followed by 2 or 3 days development *in vitro* (Beddington and Martin, 1989) has been used to study the fate of somitic cells (Fig. 1(A)). Preliminary experiments from such experiments indicate that mammalian somites may give rise to endothelial cells as well as the classic sclerotome, myotome and dermatome derivatives, and also suggest that the sequence of limb bud colonization may differ from that described in avian embryos (Beddington and Martin, 1989).

The ability to transplant marked tissue in the postimplantation mouse embryo and study its development also provides a powerful method for testing the state of cell determination with respect to the expression of a given gene. In other words, does the expression of a given gene always coincide with a given developmental fate? In *Drosophila* the cell-autonomous role of homeotic genes in development was convincingly demonstrated using mosaic analysis. Clones of mutant Ultrabithorax (*Ubx*) cells, marked with a suitably linked gene whose phenotype was easily visible in the cuticle, develop abnormally regardless of the genetic status of their neighbours (see Lawrence and Morata, 1983). In other words *Ubx* must be a key gene in determining segment identity because mutant cells consistently formed segmental structures characteristic of one segment despite being located in another. Surrounding tissue had no effect on this aberrant development nor did the *Ubx* cells affect their neighbours. Conversely, the segment polarity gene wingless (*wg*) was shown not to act cell-autonomously because mutant cells could be rescued in mosaic flies (Morata and Lawerence, 1977; Baker, 1987). This indicates that the influence of *wg* is mediated by cell interaction. A more general example of non-autonomy is that of the apterous (*ap⁴*) mutation, where mosaic analysis suggests that many of the abnormalities caused by this mutation correlate with mutant Malpighian tubules and may, therefore, stem from the production of defective haemolymph (Wilson, 1981).

Although mosaic analysis of this kind is not possible in the mouse, one can relocate tissue in order to move it from one region, expressing a particular repertoire of, for example, *Hox* genes and characterized by a particular developmental fate, into a different region characterized by a different fate and a different profile of *Hox* expression. Using transgenically marked cells in such relocation experiments should demonstrate whether *Hox* genes act cell-autonomously and, therefore, can be viewed as potential primary determinants of axial identity.

3. Developmental molecular genetics

One of the most important tools for any developmental biologist trying to determine the underlying molecular mechanisms of form and pattern in the embryo is a means of studying the effects of perturbed gene expression during development. There can be little doubt that the bank of developmental mutants obtained from deliberately saturating the *Drosophila* genome with point mutations (Nusslein-Volhard and Wieschaus, 1980) provided the impetus for a concerted molecular dissection of *Drosophila* development. In principle, transgenics offer a similar opportunity to mammalian developmental biologists, the insertion of exogenous DNA acting as a covenient mutagen (which provides a contiguous tag for recovering the endogenous gene) or as a means of altering the pattern of normal gene expression.

Generation of distinct embryonic phenotypes by mutation may be used either to identify candidate developmental genes or to assess the developmental function of known genes. Random mutagenesis can be used to identify previously unknown genes involved in embryogenesis by searching for suggestive phenotypes resulting from gene malfunction. Alternatively, it is anticipated that mutation (or ectopic expression) may reveal the normal embryonic function of cloned genes known to be expressed in development. These include genes identified from their effects in adult or tissue culture experimental systems, such as oncogenes or growth factors (see Adamson, 1987), and genes isolated by virtue of their homology with proven developmental genes in lower organisms, such as the *Hox* genes (see Holland and Hogan, 1988). In both cases correct interpretation of the mutant phenotype is central to the argument that mutation can help to reveal function.

Figure 2 summarizes the various ways in which transgene insertion has been used to identify and mutate potentially important developmental genes. This includes random insertional mutagenesis (Schnieke *et al.*, 1983; Mark *et al.*, 1985; Robertson, 1986), selective screening for insertion into transcribed genes via enhancer (Allen *et al.*, 1988) or gene trap (Gossler *et al.*, 1989) strategies and specific targeting of insertion into known genes by homologous recombination (Schwartzberg *et al.*, 1989; S. Thompson *et al.*, 1989; Zijlstra *et al.*, 1989). The principal methods for generating such mutations and implementing gene traps are dealt with elsewhere in this book and will not be considered further in this chapter. Instead, the following sections will concentrate on more general aspects of identifying and exploiting mammalian developmental mutants, many of which are familiar problems to mammalian developmental geneticists. Transgenesis may be able to create mutants but can it also help in their analysis?

First, how does one screen for informative developmental mutants and ensure that one does not follow false trails leading to genes far downstream or only peripherally involved in developmental processes? Secondly, how does one ascertain the primary function of a gene from studying the effects of its aberrant expression in an abnormal embryo? Thirdly, will redundancy within gene families in the mammalian genome serve to minimize the frequency of recognizably abnormal phenotypes, the function of one gene being executed when necessary by another? Fourthly, evolution may have selected for those genes

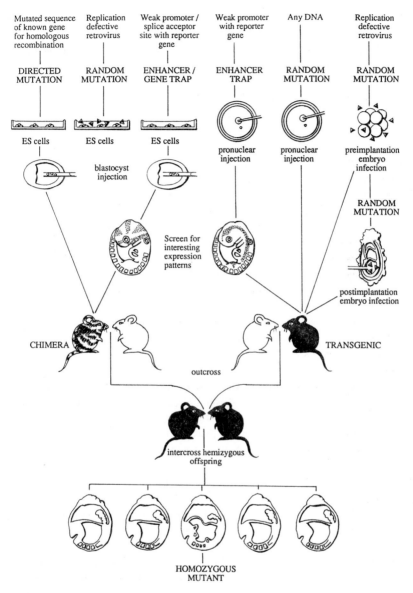

Fig. 2. Diagram of the various methods used to identify and mutate endogenous developmental genes using transgenes.

which are particularly efficient in coordinating specific events within cell populations, such as growth factors, to be used repeatedly in embryogenesis for different purposes. If such genes are required for embryonic survival, some of their later functions may be inaccessible to analysis by mutation (*intralocus epistasis*) unless conditional mutants can be made.

3.1.   Recognition of developmental genes from phenotype

In its broadest sense a developmental mutation is any mutation which disrupts development. However, there are many genes, e.g. housekeeping genes, whose products are likely to be indispensable to most cells and when disrupted will result in cell, and eventually embryonic, death. These are not the genes likely to reveal the underlying genetic control of tissue diversification. Instead, the developmental biologist is interested in mutations which are not, in themselves, cell-lethal but which cause inappropriate development. Such disrupted development may or may not be compatible with continued viability of the conceptus.

Undoubtedly the most straightforward mutations to recognize are those which do not kill the embryo but which produce specific tissue deficiencies or morphological disruption. Here there is no complication regarding general cell lethality. However, it should be remembered that reproducible defects, or phenocopies, can also arise due to the effects of quite non-specific teratogenic agents applied at particular stages of development. For example, many teratogens which kill cells during gastrulation and therefore disrupt growth equilibrium can generate vertebral abnormalities (e.g. see Russell and Russell, 1954; Gregg and Snow, 1983). In addition, mutation in a relatively ubiquitous gene involved in pyrimidine synthesis in *Drosophila* (the gene locus rudimentary) results in surprisingly precise abnormalities of the wings (Falk, 1977). It should also be emphasized that it is important to isolate multiple alleles at a locus since these may have widely differing effects. This is illustrated by the decapentaplegic gene complex in *Drosophila* where different mutations in this complex produce effects ranging from quite trivial alterations to wing morphology to extensive ventralization of the embryo (Gelbart *et al.*, 1985)

In a few cases, transgene insertion has produced replica phenotypes for pre-existing mutations and complementation studies have indicated that insertion has indeed produced new alleles at known loci. For example, insertions causing mutations in limb deformity (Woychik *et al.*, 1985), dystonia muscularis (Kothary *et al.*, 1988), Purkinje cell degeneration (Kruwlewski *et al.*, 1989), downless (Shawlot *et al.*, 1989,) pygmy (Xiang *et al.*, 1990) and hotfoot (Gordon *et al.*, 1990) genes have been reported. In addition, transgene insertion has produced a new autosomal recessive mutant (called legless), which exhibits pleiotropic effects in newborn mice, including abnormal limb and craniofacial development and a high frequency of visceral inversion (McNeish *et al.*, 1988, 1990) These fortuitous insertions should now permit cloning of the disrupted genes and characterization of their products.

In most cases it is likely that mutation in genes that play a critical role early in tissue diversification or pattern formation, some of which may also act later in development, will prove to be embryonic lethals when homozygous. However, most of mammalian embryology takes place within the uterus and, therefore, the embryos are not readily accessible at all developmental stages. This makes a systematic screen for postimplantation abnormalities extremely difficult and laborious and it is no accident that most of the original mouse embryonic lethals were found due to dominant heterozygous effects on pigmentation or gross anatomy (e.g. *Ve*, velvet coat; A^y, yellow; *T*, brachyury (see Lyon and Searle, 1989)). This, in fact, is also true in *Drosophila* where the dominant homeotic mutants (e.g. *Antp* and *Ubx*) were among the first developmental mutations to be recognized and studied.

Homozygous lethals generated by insertional mutagenesis have two advantages in such screens. Despite the relatively long reproductive cycle of the mouse, which places any saturation mutagenesis programme in mammals on a very different timescale from that required for insects, the absence of homozygous transgenics can be detected relatively quickly by appropriate breeding programmes. Ascertaining the stage at which they die and characterizing the defects is still tedious, but at least it is possible, using probes for the inserted transgene or flanking sequences, to produce a convincing correlation between an abnormal phenotype and homozygosity for the mutation. In the past, most embryonic lethals could only be identified unequivocally once they were clearly abnormal and, therefore, it was difficult to study the inception or cell autonomy of the malformation.

Technical considerations aside, the problem remains of how to sift the peripheral genetic effects from those of the mainstream genes governing development. Can one rely on phenotype to distinguish between the two? Indeed, can one assume in the first place that there will be sufficient parallels between mammalian and arthropod development that critical developmental genes in the mouse can be revealed by a somewhat simplistic search for cell-autonomous defects? It is probably too early in mammalian developmental genetics to give a satisfactory answer to these questions. In practice the relevance to mammals of the hierarchical models of gene interaction devised to explain early *Drosphila* development (Ingham, 1988) will only be known when the function of many genes which affect mammalian development have been fully characterized. Simple considerations of developmental biochemistry highlight some of the difficulties ahead.

There seem to be certain hot spots of embryonic lethality in early mouse development such as the peri-implantation period. Although zygotic gene expression starts very early in mouse development, at the two-cell stage (see Schultz, 1986), it is possible that some maternal products, in the absence of the relevant zygotic genes, can support development until accelerated growth or changes in the metabolic demands of the conceptus make zygotic protein synthesis essential. For some gene products this may well coincide with implantation when the embryo starts to grow faster (Snow, 1976) and there are indications of general changes in gene expression associated with early postimplantation development. For example, the embryo becomes permissive to infection by certain retroviruses, X inactivation occurs and a number of cell surface antigens change (see Gardner and Beddington, 1988). Carbohydrate metabolism has also been shown to change during early development (see Kaye, 1986). During preimplantation development there is a shift from reliance on lactate and pyruvate to glucose as a source of energy, thus requiring a different repertoire of enzymes. Immediately after implantation the pentose phosphate pathway seems to predominate and it is not until organogenesis is well underway that oxidative phosphorylation becomes a significant source of energy. Therefore, one might expect that many of the homozygous mutations which kill embryos during preimplantation or early postimplantation development, often at exactly the stages when important lineage segregations and pattern formation are occurring, may be due to the absence of quite humdrum metabolic constituents. Although metabolic shifts are undoubtedly relevant to these developmental changes they are not likely to be instrumental. Consequently, many early postimplantation lethals may have to be painstakingly analysed in order to find mutations effecting developmental as opposed to general biochemical deficiencies. Of the transgenic insertional mutants that have been described which die soon after implantation (e.g. see Mark et al., 1985; Soriano et al., 1987) the nature of the disrupted gene has yet to be characterized.

3.2.   Determining developmental function

There are now many cloned genes which have been shown to be expressed during mouse development but whose developmental function remains unknown. Much speculation regarding gene function has accompanied in situ hybridization and antibody studies which show intriguing stage- or tissue-specific expression patterns of particular genes. However, function cannot be deduced from expression pattern. Eventually, developmental function can only be determined by combining an intimate knowledge of the molecular and cell biological properties of the molecule with a precise understanding of its relevant source and site of action in the embryo and the effects it has on a particular developmental sequence.

Consequently, studying mutant phenotypes is only a start in trying to identify the primary effects of a gene in the embryo. Null mutants may reveal telling disturbances in pattern or form but, for many genes, simple absence of gene function may be too crude a tool to reveal specific function or the exact site of action of a gene. The effects of expressing a gene at the wrong time or in the wrong tissue during development could be informative, but there are instances in *Drosophila* where ectopic expression has proved remarkably benign (see below). One of the more fruitful strategies may be to study the effects of mutant cells developing alongside wild-type ones in mosaic or chimeric embryos.

3.2.1. *Plural strategies*

Before discussing the relative merits of these approaches in the mouse a brief consideration of certain aspects of the development of the compound eye in *Drosophila* will serve to illustrate some elementary points demonstrating the importance of combining molecular, genetic and mosaic approaches. Figure 3 illustrates the fixed sequence of differentiation of the eight photoreceptors in an ommatidium. This pattern of developmental fate is not due to ancestral directives because there is no fixed lineage relationship within an ommatidium (Ready *et al.*, 1976; Lawrence and Green, 1979). Instead, it must be determined by the environment, or sequential inductive interactions (Tomlinson and Ready, 1987a). Two mutants, sevenless and bride of sevenless (boss) cause

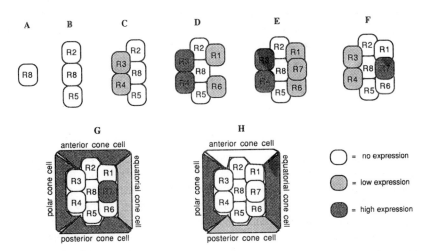

Fig. 3. Diagram of the developmental sequence of photoreceptor differentiation in the ommatidium of *Drosophila melanogaster*. R1–R8 are photoreceptors. Cone cells are lens-secreting cells. Shaded areas show the level of expression of the sevenless gene product. (After Rubin (1989) and Tomlinson *et al.* (1987)).

the cell that would normally differentiate into the last photoreceptor (R7) to form instead a lens-secreting cell (see Rubin, 1989). These genes complement each other and, therefore, must have different functions, and yet they produce the same phenotype. Thus, phenotype alone cannot resolve function where cell interactions are at work. The sevenless protein is expressed in all photoreceptors except R2, R5 and R8 (Fig. 3). In particular, it is abundant in R3, R4 and R7 at their junctional contacts with R8, although showing rather more sustained expression in R7. In addition, it is also expressed strongly in surrounding lens-secreting cone cells (Tomlinson et al., 1987). Clearly, this pattern of sevenless expression cannot immediately explain the very specific phenotype of the mutant. Ubiquitous ectopic expression of sevenless, achieved by linking it to the heat shock promoter hsp70, in mutant larvae lacking the wild-type sevenless gene, results in rescue of a subset of ommatidia which are at a particular stage of differentiation when the sevenless product is induced by heat shock. However, this abnormally widespread expression of sevenless does not result in any detectable abnormality of the retina or any other structure, regardless of when or for how long ectopic expression is induced (Basler and Hafen, 1989; Bowtell et al., 1989). Again, this argues that the distribution of the sevenless product cannot explain its precise effect on the R7 cell, and that ectopic expression reveals little except that sevenless is required only during a limited sensitive phase of ommatidia development. Mosaic analysis of sevenless and boss has added important information because it transpires that sevenless acts cell-autonomously (Harris et al., 1976; Campos-Ortega et al., 1979; Tomlinson and Ready, 1987b), R7 having an absolute requirement for this gene product, whereas boss is non-autonomous within a single ommatidium, and the inappropriate cone differentiation of R7 occurs only when the adjacent R8 cell is lacking the boss gene (Reinke and Zipursky, 1988). Therefore, mosaic analysis suggests a mechanism whereby sevenless codes for a receptor, as its DNA sequence predicts, which must interact with a ligand in order for R7 to differentiate into a photoreceptor. The indifference to expression of sevenless elsewhere in the ommatidium, especially in cells (R3 and R4) also adjacent to R8 and, therefore, likely to be exposed to the ligand, suggests either that sevenless may be a redundant component of other aspects of pattern formation in the Drosophila retina or that it is expressed too late in the differentiation of other retinal cells to transform them into R7 cells. In other words, it may be the precise time and place of ligand production, which may be related to boss expression, which determines the final pattern of recruitment and differentiation of photoreceptor cells. This extremely elegant series of experiments, on a remarkably stereotyped developmental sequence, demonstrates how important it is to use complementary molecular and genetic approaches in order to uncover the relevant aspects of gene expression and function (Rubin, 1989).

3.2.2. *Null mutants*

Pioneering work earlier this century, describing the abnormal morphological consequences of those developmental mutations that had been identified, illustrates the difficulty of disentangling primary and secondary effects in abnormal embryos (see Lyon and Searle, 1989). It is relatively easy to describe the final abnormal phenotype but very much harder to detect the initial deviation from normal which causes it. Again, this is not dissimilar to teratological studies where a number of diverse agents may result in a common defect sometimes quite remote from their primary site of action (e.g. Gregg and Snow, 1983).

In order to characterize precisely mutant development one needs to have stringent criteria for assessing normal development. This is less straightforward than it might seem. For a start, the range of developmental stages represented within a single litter, from both outbred and inbred strains of mice, which can sometimes span almost a day's worth of development, indicates that that there can be a certain amount of leeway in gestational time-keeping (Kaufman, 1990). It is not clear to what extent this stems from differences in the time of fertilization, implantation or actual growth rates. Unfortunately, it is also impossible to know what degree of retardation is compatible with normal development and birth because one cannot return a retarded postimplantation embryo to the uterus and let it develop to term (Beddington, 1985). Therefore, unlike animals with a more precise developmental timetable where age is tightly correlated with developmental stage, this is probably a poor measure of normality for mouse development. Instead, one needs precise physiological or morphological features to assess normality. However, even quite elaborate scoring systems for assessing normality rely on fairly gross features (Brown, 1990). Consequently, subtle deviations from normal are extremely difficult to identify even when tracing the provenance of very specific defects.

Any discussion of the usefulness of null mutants, particularly where directed mutation of known genes by homologous recombination is being countenanced, has to consider the problem of redundancy in the genome. The phenotype of the *Mov-13* mutant embryos illustrates the potential for one gene to take the place of another during mouse development. In these mutants a retrovirus has interrupted the $\alpha1(I)$ collagen gene (Schnieke *et al.*, 1983; Lohler *et al.*, 1984) and led to the complete absence of type 1 collagen synthesis (except in odontoblasts and possibly osteoblasts, where the gene may be reactivated (Kratchowil *et al.*, 1989)). In normal embryos, type I collagen is first detected just after gastrulation (9th day) and is present in the cranial, cardiac and somitic mesoderm as well as some extraembryonic membranes, including the amnion, chorion and visceral yolk sac (Leivo *et al.*, 1983). Later in development it is produced by most mesodermal tissues and its abundance in various epithelial–mesenchymal organs, and the consequences of chemically inhibiting collagen synthesis, has meant that type I collagen has been invoked as a critical stabilizing element during morphogenesis (Bernfield *et al.*, 1973). However, in the *Mov-13* mutants the epithelial branching morphogenesis in organs such as the salivary gland is entirely normal despite the absence of type I collagen (Kratchowil *et al.*, 1986). The probable explanation for this is that other collagens, particularly type III, can substitute for the type I deficiency. In fact, homozygous *Mov-13* embryos die surprisingly late (between the 12th and 14th day of gestation), primarily because the major blood vessels are weak and prone to haemorrhage. One may argue that the collagen family is a special case and that there is no absolute requirement for a specific type of collagen fibrillar protein in the matrix so long as such proteins are present.

It is known that the absence of other gene products is also compatible with normal development. For example, null mutants for hypoxanthine–guanine phosphoribosyltransferase (Hooper *et al.*, 1987; Kuehn *et al.*, 1987), cytoplasmic malic enzyme (*Mod-1* locus (Johnson *et al.*, 1981)) and several of the major histocompatibility Q region products (O'Neill *et al.*, 1986) result in apparently normal mice. The recent demonstration that failure to express MHC class I antigens on the cell surface, due to the absence of β_2-microglobulin, does not compromise normal development or fertility (Zijlstra *et al.*, 1990) indicates that these antigens, although present during normal development, are dispensable. In addition, various deletions affecting the albino (Ruppert *et al.*, 1988) and short ear loci (Russell, 1971) and deletions ablating the IB1075 gene, which is expressed in the brain (Kingsley *et al.*, 1990), do not appear to compromise development. Although the list is still very small it does serve as a reminder that functional substitution by related genes may be a confusing element in the study of null mutants. This may be particularly true for those genes, such as *Hox*, which are members of quite large gene families. In the next few years, when null mutations have been made for a wide variety of candidate developmental genes, the extent of redundancy, itself an intriguing facet of eukaryotic genomes, may be more realistically evaluated.

3.2.3. *Ectopic expression*

Inappropriate expression of genes, either in the wrong tissue or at the wrong time, can shed light on their function. A good example of this comes from the study of oncogenes where their expression following gene transfer into cell lines *in vitro* results in predictable disturbances in cell growth and/or differentiation (see Bishop, 1987). Such gain-of-function effects may be more readily interpreted in moderately homogenous tissue culture cell lines than among the heterogenous cell types present in an embryo. For example, the production of transgenic animals expressing oncogenes may have confirmed their role in neoplasia but the stochastic appearance of tumours in these animals only serves to emphasize the plurality of causes leading to growth imbalance. The primary effects of oncogene expression, which lead eventually to pathological conditions in the animal, remain elusive (see Hanahan, 1989).

The ectopic expression of the sevenless gene during *Drosophila* development (see above) illustrates that, for genes involved in cell–cell interactions, ectopic expression may be too blunt an instrument to reveal normal function. Even ectopic expression of putative transcription factors such as the Deformed (*Dfd*) homeotic gene in *Drosophila* produces a very complicated phenotype only partially intelligible in the light of a considerable knowledge of the role of *Dfd* itself and other homeotic genes (Kuziora and McGinnis, 1988).

In the mouse, ectopic expression of the *Hox-1.1* gene results in craniofacial abnormalities including malformation of the palate (Balling *et al.*, 1989; Kessel *et al.*, 1990). It is thought that any disturbance which reduces the number of neural crest cells entering the facial primordia will produce cleft palate (Ross and Johnston, 1972), and it is known that a variety of agents, including phenytoin, cortisone (see Johnston, 1986), methotrexate, thalidomide, hyperthermia and retinoids (Sulik *et al.*, 1988), can result in similar craniofacial abnormalities. It has been argued that the similarity in defects arising either from administering exogenous retinoic acid or from ectopic *Hox-1.1* expression may reflect the normal induction of *Hox* genes by retinoic acid during embryogenesis (Balling *et al.*, 1989). However, bearing in mind the diverse agents which can cause comparable craniofacial defects, this requires independent substantiation.

The full potential of ectopic expression as a strategy to reveal function will only properly be realized when ectopic expression can be carefully controlled. This may be achieved either by designing efficient inducible promoters which will allow the time and duration of expression to be controlled, or by generating both stage- and tissue-specific expression by judicious use of available promoters. For example, the significance of the elaborate, complementary expression patterns of the various *Hox-5* genes active in the limb bud (Dollé *et al.*, 1989; Lewis and Martin, 1989) may be resolved by expressing the different members of the *Hox-5* gene complex in inappropriate domains. A precise way of doing this might be to swap promoters between the different *Hox-5* genes and study the effects on limb development in first-generation transgenics.

3.2.4. Chimeric analysis

It is now possible to generate specific mutations by targeting transgenes to recombine homologously with particular endogenous genes in ES cells. At present, emphasis is placed on obtaining germ line transmission of the ES cell genotype after making chimeras in order to establish mouse lines heterozygous for the mutation. From interbreeding these, homozygotes can be generated in which the effect of any recessive mutation can be studied. However, examining the effect of a mutation in an embryo or adult in which all the cells are mutant may be less revealing than observing the behaviour of mutant cells in chimeras where they are intermixed with wild-type ones. This can be instructive for a number of reasons. First of all, it may reveal whether a mutation which results in early embryonic lethality is a general cell-lethal or has a tissue-specific effect. If the mutation is cell-lethal for all cell types it will not be possible to rescue mutant cells by forming chimeras whereas, if it primarily affects only certain tissues, it should be possible to form viable chimeras. If the mutant cells can be rescued in chimeras, it may be easier to identify which tissue is primarily affected by the mutation since chimeric embryos will survive beyond the normal stage of lethality. For example, mutant cells in competition with wild-type ones may be specifically selected against in those tissues where they are deleterious. Alternatively, abnormalities in particular tissues may correlate precisely with particular mutant population patterns. Thirdly, it should be possible to determine whether or not the gene product acts cell-autonomously or can be rescued by neighbouring wild-type cells. This may help to resolve gene function at the cellular level.

Specific selection against mutant cells in particular tissues of chimeras has been observed. For example, cells trisomic for chromosome 16 are excluded from the blood, spleen, thymus and bone marrow of adult chimeras (Cox et al., 1984). Probably, the most detailed studies of this kind have been those following the development of parthenogenetic, androgenetic or gynogenetic cells in chimeras. A systematic elimination of parthenogentic cells from skeletal muscle, pancreas and liver in adult chimeras has been found using electrophoretic allozyme markers to distinguish between parthenogenetic and normal cells (Fundele et al., 1989). A more exact study using the transgenic β-globin nuclear marker to distinguish between the different genotypes (see Section 2.1) showed that *individual* androgenetic blastomeres aggregated with eight-cell embryos only rarely contributed to the embryo proper or to extraembryonic mesoderm and endoderm in the early postimplantation conceptus, whereas single parthenogenetic and gynogenetic blastomeres showed the reciprocal colonization pattern, apart from the fact that they too were seldom seen in extraembryonic endoderm (Thomson and Solter, 1988). Using the same marker, aggregation of *whole* parthenogenetic eight-cell embryos with normal ones produced the same results, except that there was no apparent exclusion from extraembryonic endoderm derivatives (Clarke et al., 1988). This discrepancy concerning extraembryonic endoderm colonization may reflect the different methods used for producing chimeras and highlight the importance of studying cell fate in normal-sized embryos rather than double ones. However, the principle of using chimeras to monitor developmental preferences or specific exclusions is well illustrated by these studies.

ES cells themselves provide the ideal vehicle for conducting chimeric analyses of mutations since they have been shown to colonize all tissues of the conceptus (Beddington and Robertson, 1989). These may be ES cells in which specific mutations have been introduced *in vitro* or they may be cell lines which have been isolated from mutant embryos (e.g. t^{w5}/t^{w5}; Magnuson et al., 1983). The advantage of using ES cells rather than embryonic cells is that the genotype can be confirmed before creating chimeras. Equally important is the opportunity provided for marking the mutant ES cells with an independent lineage marker (see Section 2.1) so that mutant development can be followed at the single-cell level (Fig. 1(B)). Obviously, any dominant mutation can be examined using ES cells containing only one altered allele. For recessive mutations it will be necessary to obtain germ line transmission and to derive ES cell lines from homozygous mutant embryos, providing the mutation does not compromise the ability of blastocysts to generate cell lines. However, there may also be an argument for generating homozygosity directly in ES cells by selecting for homologous recombination into each allele of a particular gene. This would not obviate the need for germ line transmission, as it will still be necessary to characterize the phenotype of intact homozygotes, but it may be invaluable in studying the effects of mutations causing early embryonic lethality.

One drawback of chimeric analysis in the mouse is that effectively one is restricted to creating the chimeras during preimplantation development. This means that cells of the genotype to be studied are present in every fetal tissue, whereas it might be easier to assess gene function if only certain lineages were populated with mutant cells. There is no obvious solution to this problem unless ES cells can be made to incorporate into the postimplantation embryo where they may colonize only a subset of tissues. That cells can get into the post-implantation embryos and form tissue-specific chimeras has been demonstrated with both early haemopoietic cells (Weissman et al., 1978) and neural crest cells (Jaenisch, 1985). In addition, haemopoietic chimeras can be made late in gestation by injection of cells into the placenta (Fleischman and Mintz, 1979; Toles et al., 1989). It should also be remembered that mixing genotypes in organ culture can provide useful information regarding cell autonomy of gene function, as demonstrated by the ability of normal skin to rescue homozygous Steel (*Sl/Sl*) melanoblasts (Mayer, 1973). It remains to be seen whether ES cells can differentiate directly into specific tissues following introduction into the postimplantation embryo or if they can be made to participate in tissue-specific differentiation using certain organ culture combinations. If so, this may allow for a more refined 'mosaic' analysis of genetic malfunction.

4.   Conclusion

Clearly, transgenesis has much to offer the developmental biologist. It promises to provide a much broader genetic basis for embryological studies and to supply some of the tools required for analysing new mutants. The current euphoria among mammalian developmental biologists is not misplaced for it is now possible to contemplate experiments in the mouse which were impossible 10 years ago. However, transgenics can only contribute to certain facets of development. It does not do away with the need for experimental embryology or classical genetics, and as yet has contributed little, conceptually or practically, to the problems of morphogenesis and acquisition of form. In this chapter some of the basic difficulties in studying mammalian cell lineages and developmental mutants have been highlighted, not to undermine the importance of transgenics but to emphasize the necessity for retaining a multidisciplinary approach.

Acknowledgements

I thank Karen Downs, Phil Ingham and Julian Lewis for their useful, constructive criticisms of the manuscript.

References

Adamson, E. D. (1987). Oncogenes in development. *Development* **99**, 449–471.

Allen, N. D., Cran, D. G., Barton, S. C. *et al.* (1988). Transgenes as probes for active chromosomal domains in mouse development. *Nature* **333**, 852–855.

Baker, N. E. (1987). Molecular cloning of sequences from wingless, a segment polarity gene in Drosophila: the spatial distribution of a transcript in embryos. *EMBO J.* **6**, 1765–1773.

Balling, R., Mutter, G., Gruss, P. and Kessel, M. (1989). Craniofacial abnormalities induced by ectopic expression of the homeobox gene Hox-1.1 in transgenic mice. *Cell* **58**, 337–347.

Barklis, E., Mulligan, R. C. and Jaenisch, R. (1986). Chromosomal position of virus mutation permits retrovirus expression in embryonal carcinoma cells. *Cell* **47**, 391–399.

Basler, K. and Hafen, E. (1989). Ubiquitous expression of sevenless: position dependent specification of cell fate. *Science* **243**, 931–934.

Beddington, R. S. P. (1981). An autoradiographic analysis of the potency of embryonic ectoderm in the 8th day postimplantation mouse embryo. *J. Embryol. Exp. Morph.* **64**, 87–104.

Beddington, R. S. P. (1985). The development of 12th to 14th day fetuses following re-implantation of pre- and early primitive streak stage mouse embryos. *J. Embryol. Exp. Morph.* **88**, 281–291.

Beddington, R. S. P. (1988). Toxigenics: strategic cell death in the embryo. *Trends Genet.* **4**, 1–2.

Beddington, R. S. P. and Martin, P. (1989). An in situ transgenic enzyme marker to monitor migration of cells in the mid-gestation mouse embryo. *Mol. Biol. Med.* **6**, 263–274.

Beddington, R. S. P. and Robertson, E. J. (1989). An assessment of the development potential of embryonic stem cells in the midgestation mouse embryo. *Development* **105**, 733–737.

Beddington, R. S. P., Morgenstern, J., Land, H. and Hogan, A. (1989). An in situ transgenic enzyme marker for the midgestation mouse embryo and the visualization of inner cell mass clones during early organogenesis. *Development* **106**, 37–46.

Bernfield, M. R., Cohn, R. H. and Bannerjee, S. D. (1973). Glycosaminoglycans and epithelial organ formation. *Amer. Zool.* **13**, 1067–1084.

Bishop, J. M. (1987). The molecular genetics of cancer. *Science* **235**, 305–311.

Bowtell, D. D. L., Simon, M. A. and Rubin, G. M. (1989). Ommatidia in the developing Drosophila eye require and can respond to sevenless for only a restricted period. *Cell* **56**, 931–936.

Breitman, M. L., Clapoff, S., Rossant, J. *et al.* (1987). Genetic ablation: targeted expression of a toxin gene causes microphthalmia in transgenic mice. *Science* **238**, 1563–1565.

Brown, N. A. (1990). Routine assessment of morphology and growth: scoring systems and measurements of size. In: Copp, A. J. and Cockroft, D. L. (eds) *Postimplantation Mammalian Embryos: A Practical Approach*, pp. 93–108. IRL, Oxford.

Campos-Ortega, J. A., Jurgens, G. and Hofbauer, A. (1979). Cell clones and pattern formation: studies on sevenless, a mutant of Drosophila melanogaster. *Wilhelm Roux's Arch. Dev. Biol.* **186**, 27–50.

Clarke, H., Varmuza, S., Prideaux, V. and Rossant, J. (1988). The developmental potential of parthenogenetically derived cells in chimeric mouse embryos: implications for action of imprinted genes. *Development* **104**, 175–182.

Collier, R. J. (1975) Diptheria toxin: mode of action and structure. *Bacteriol. Rev.* **39**, 54–85.

Cox, D. R., Smith, S. A., Epstein, L. B. and Epstein, C. J. (1984). Mouse trisomy 16 as an animal model for human trisomy 21 (Down syndrome): formation of viable trisomy 16–diploid mouse chimeras. *Dev. Biol.* **101**, 416–424.

de Wet, J. R., Wood, K. V., DeLuca, M., Helsinki, D. R. and Subramani, S. (1987). Firefly luciferase gene: structure and expression in mammalian cells. *Mol. Cell. Biol.* **7**, 725–737.

Dollé, P., Izpisúa-Belmonte, J.-C. Falkenstein, H. Renucci, A. and Duboule, D. (1989). Coordinate expression of the murine Hox-5 complex homoeobox-containing genes during limb pattern formation. *Nature (London)* **342**, 767–772.

Driever and Nusslein-Volhard, C. (1988). The bicoid protein determines position in the Drosophila embryo in a concentration dependent manner. *Cell* **54**, 95–104.

Duval, M. (1892). *Le Placenta des Rongeurs.* Paris.

Eddy, E. M. (1975). Germ plasm and the differentiation of the germ cell line. *Int. Rev. Cytol.* **43**, 229–280.

Emerman, M. and Temin, H. M. (1984). Genes with promoters in retrovirus vectors can be independently suppressed by an epigenetic mechanism. *Cell* **39**, 459–467.

Engebrecht, J., Simon, M. and Silverman, M. (1985). Measuring gene expression with light. *Science* **227**, 1345–1347.

Epstein, C. J. (1986). *The Consequences of Chromosome Imbalance*, Cambridge University Press, New York.

Evans, G. A. (1989). Dissecting mouse development with toxigenics. *Genes Dev.* **3**, 259–263.

Falk, D. (1977). Genetic mosaics of the rudimentary locus of Drosophila melanogaster: a genetical investigation into the physiology of pyrimidine synthesis. *Dev. Biol.* **58**, 134–147.

Fleischman, R. A. and Mintz, B. (1979). Prevention of genetic anaemia in mice by microinjection of normal haematopoietic stem cells into the fetal placenta. *Proc. Natl Acad. Sci. USA* **76**, 5736–5740.

Fundele, R., Norris, M. L., Barton, S. C., Reik, W. and Surani, M. A. (1989) Systematic elimination of parthenogenetic cells in mouse chimaeras. *Development* **106**, 29–35.

Gardner, R. L. and Beddington, R. S. P. (1988). Multi-lineage 'stem cells' in the mammalian embryo. *J. Cell Sci.* **10**(Suppl.), 11–27.

Gelbart, W. M. *et al.* (1985). The decapentaplegic gene complex in Drosophila melanogaster. *Cold Spring Harbor Symp. Quant. Biol.* **50**, 119–125.

Gordon, J. W., Vehlinger, J., Dayani, N. *et al.* (1990). Analysis of the Hotfoot (ho) locus by creation of an insertional mutation in a transgenic mouse. *Dev. Biol.* **137**, 349–359.

Gossler, A., Joyner, A., Rossant, J. and Skarnes, W. (1989). Mouse embryonic stem cells and reporter constructs to detect developmentally regulated genes. *Science* **244**, 463–465.

Gray, G. E., Glover, J. C., Majors, J. and Sanes, J. R. (1988). Radial arrangement of clonally related cells in the chicken optic tectum: lineage analysis with a recombinant retrovirus. *Proc. Natl Acad. Sci. USA* **85**, 7356–7360.

Gregg, B. C. and Snow, M. H. L. (1983). Axial abnormalities following disturbed growth in mitomycin C-treated mouse embryos. *J. Embryol. Exp Morph.* **73**, 135–149.

Hanahan, D., (1989). Transgenic mice as probes into complex systems. *Science* **246**, 1265–1275.

Harbers, K., Kuehn, M., Delius, H. and Jaenisch, R. (1984). Insertion of retrovirus into the first intron of alpha1(I) collagen gene leads to embryonic lethal mutation in mice. *Proc. Natl Acad. Sci. USA* **81**, 1504–1508.

Harris, W. A., Stark, W. S. and Walker, J. A. (1976). Genetic dissection of the photoreceptor system in the compound eye of Drosophila melanogaster. *J. Physiol. (London)* **256**, 415–439.

Holland, P. W. H. and Hogan, B. L. M. (1988). Expression of homeobox genes during mouse development: a review. *Genes Dev.* **2**, 773–782.

Hooper, M., Hardy, K., Handyside, A., Hunter, S. and Monk, M. (1987). HPRT-deficient (Lesch–Nyhan) mouse embryos derived from germline colonization by cultured cells. *Nature* **326**, 292–295.

Ingham, P. (1988). The molecular genetics of embryonic pattern formation in Drosophila. *Nature* **335**, 25–34.

Jaenisch, R. (1976). Germ line integration and Mendelian transmission of the exogenous Moloney leukaemia virus. *Proc. Natl Acad. Sci. USA* **73**, 1260–1264.

Jaenisch, R. (1985). Mammalian neural crest cells participate in normal development when microinjected into postimplantation mouse embryos. *Nature* **318**, 181–183.

Jaenisch, R., Fan, H. and Croker, B. (1975). Infection of preimplantation mouse embryos and of newborn mice with leukaemia virus: tissue distribution of viral DNA and RNA and leukomogenesis in the adult animal. *Proc. Natl Acad. Sci. USA* **72**, 4008–4012.

Jahner, D., Stuhlmann, H., Stewart, C. *et al.* (1982). De novo methylation and expression of retroviral genomes during mouse embryogenesis. *Nature* **298**, 623–628.

Johnson, F. M., Chasalow, F., Lewis, S. E., Barnett, L. and Lee, C.-Y. (1981). A null allele at the Mod-1 locus of the mouse. *J. Hered.* **72**, 134–136.

Johnston, D. R. (1986). *The Genetics of the Skeleton.* Clarendon Press, Oxford.

Kaufman, M. H. (1990). Morphological stages of postimplantation embryonic development. In: Copp, A. J. and Cockroft, D. L. (eds) *Postimplantation Mammalian Embryos: A Practical Approach.* IRL, Oxford.

Kaye, P. L. (1986). Metabolic aspects of the physiology of the preimplantation embryo. In: Rossant, J. and Pedersen, R. A. (eds) *Experimental Approaches to Mammalian Embryonic Development*, pp. 267–292. Cambridge University Press, New York.

Kessel, M., Balling, R. and Gruss, P. (1990). Variations of cervical vertebrae after expression of a Hox1.1 transgene in mice. *Cell* **61**, 301–308.

Kimmel, C. B. and Warga, R. M. (1986). Tissue-specific cell lineages orginate in the gastrula of zebrafish. *Science* **231**, 365–368.

Kingsley, D. M., Rinchik, E. M., Russell, L. B. *et al.* (1990). Genetic ablation of a mouse gene expressed specifically in brain. *EMBO J.* **9**, 395–399.

Kothary, R. *et al.* (1988). A transgene containing *lacZ* inserted into the dystonia locus is expressed in neural tube. *Nature* **335**, 435–437.

Kratchowil, K., Dziadek, M., Löhler, J., Harbers, K. and Jaenisch, R. (1986). Normal epithelial branching in the absence of collagen I. *Dev. Biol.* **117**, 596–606.

Kratchowil, K., von der Mark, K., Kollar, E. J. *et al.* (1989). Retrovirus-induced insertional mutation in Mov13 mice affects collagen I expression in a tissue-specific manner. *Cell* **57**, 807–816.

Kruwlewski, T. F., Neuman, P. E. and Gordon, J. W. (1989). Insertional mutation in a transgenic mouse allelic with Purkinje cell degeneration. *Proc. Natl Acad. Sci. USA* **86**, 3709–3712.

Kuehn, M. R., Bradley, A., Robertson, E. J. and Evans, M. J. (1987). A potential animal model for Lesch–Nyhan syndrome through introduction of HPRT mutations into mice. *Nature* **326**, 295–298.

Kuziora, M. A. and McGinnis, W. (1988). Autoregulation of a Drosophila homeotic selector gene. *Cell* **55**, 477–485.

Lawrence, P. A. (1981). The cellular basis of segmentation in insects. *Cell* **26**, 3–10.

Lawrence, P. A. and Green, S. M. (1979). Cell lineage in the developing retina of Drosophila. *Dev. Biol.* **71**, 142–152.

Lawrence, P. A. and Morata, G. (1983). The elements of the bithorax complex. *Cell* **35**, 595–601.

Lawson, K. A., Pedersen, R. A. and van der Geer, S. (1987). Cell fate, morphogenetic movement and population kinetics of embryonic endoderm at the time of germ layer formation in the mouse. *Development* **101**, 627–652.

Le Douarin, N. M. (1973). A biological cell labelling technique and its use in experimental embryology. *Dev. Biol.* **3**, 217–222.

Leivo, I., Vaheri, A., Timpl, R. and Wartiovaara, J. (1983). Appearance and distribution of collagens and laminin in the early mouse embryo. *Dev. Biol.* **76**, 100–114.

Levine, M., Harding, K., Wedeen, C. *et al.* (1985). Expression of the homeo box gene family in Drosophila. *Cold Spring Harbor Symp. Quant. Biol.* **50**, 209–222.

Lewis, J. and Martin, P. (1989). Limbs: a pattern emerges. *Nature (London)* **342**, 734–735.

Lo, C. (1986). Localization of low abundance DNA sequences in tissue sections by in situ hybridization. *J. Cell Sci.* **8**, 143–162.

Lohler, J. R. Timple, R. and Jaenisch, R. (1984). Embryonic lethal mutation in mouse collagen I gene causes rupture of blood vessels and is associated with erythropoietic and mesenchymal cell death. *Cell* **38**, 597–605.

Lyon, M. E. and Searle, A. G. (1989). *Genetic Variants and Strains of the Laboratory Mouse*, 2nd edn. Oxford University Press, Oxford.

MacDonald, P. M. and Struhl, G. (1986). A molecular gradient in early Drosophila and its role in specifiying body pattern. *Nature* **324**, 537–545.

McNeish, J. D., Scott, W. J. and Potter, S. S. (1988). Legless, a novel mutation found in PHT1-1 transgenic mice. *Science* **241**, 837–839.

McNeish, J. D., Thayer, J., Walling, K. *et al.* (1990). Phenotypic characterization of the transgenic mouse insertional mutatation, *legless*. *J. Exp. Zool.* **253**, 151–162.

Magnuson, T., Martin, G. R., Silver, L. M. and Epstein, C. J. (1983). Studies on the viability of t^{w5}/t^{w5} embryonic cells *in vitro* and *in vivo*. In: Silver, L. M., Martin, G. R. Strickland, S. (eds) *Cold Spring Harbor Conferences on Cell Proliferation*, pp. 671–681. Cold Spring Harbor Laboratory, Cold Spring Harbor.

Mark, W. H., Signorelli, K. and Lacy, E. (1985). An insertional mutation in a transgenic mouse line results in developmental arrest at day 5 of gestation. *Cold Spring Harbor Symp. Quant. Biol.* **50**, 453–463.

Mayer, T. (1973). Site of gene action in steel mice: analysis of the pigment defect by ectoderm–mesoderm recombinations. *J. Exp. Zool.* **184**, 345–352.

Mlodzik, M., Fjose, A. and Gehring, W. J. (1985). Isolation of caudal, a Drosophila homeo box-containing gene with maternal expression, whose transcripts form a concentration gradient at the pre-blastoderm stage. *EMBO J.* **4**, 2961–2969.

Morata, G. and Lawrence, P. A. (1977). The development of wingless, a homeotic mutation of Drosophila. *Dev. Biol.* **56**, 227–240.

Nusslein-Volhard, C. and Wieschaus, E. (1980). Mutations affecting segment number and polarity in Drosophila. *Nature* **287**, 795.

O'Neill, A. E., Reid, K., Garberi, J. C., Karl, M. and Flaherty, L. (1986). *Immunogenetics* **24**, 368–373.

Ow, D. W., Wood, K. V., DeLuca, M. *et al.* (1986). Transient and stable expression of the firefly luciferase gene in plant cells and transgenic plants. *Science* **234**, 856–859.

Palmiter, R. D. and Brinster, R. L. (1986). Germ-line transformation of mice. *An. Rev. Genet.* **20**, 465–499.

Palmiter, R. D., Behringer, R. R., Quaife, C. J. *et al.* (1987). Cell lineage ablation in transgenic mice by cell-specific expression of a toxin gene. *Cell* **50**, 435-443.

Peckham, I., Sobel, S., Comer, J., Jaenisch, R. and Barklis, E. (1989). Retrovirus activation in embryonal carcinoma cells by cellular promoters. *Genes Dev.* **3**, 2062-2071.

Ponder, B. (1987). Cell marking techniques and their application. In: Monk, M. (ed.) *Mammalian Development: A Practical Approach*, pp. 115-138. IRL Press, Oxford.

Price, J. and Thurlow, L. (1988). Cell lineage in the rat cerebral cortex: a study using retroviral mediated gene transfer. *Development* **104**, 473-482.

Price, J., Turner, D. and Cepko, C. (1987). Lineage analysis in the vertebrate nervous system by retrovirus mediated gene transfer. *Proc. Natl Acad. Sci. USA* **84**, 158-160.

Ready, D. F., Hanson, T. E. and Benzer, S. (1976). Development of the Drosophila retina, a neurocrystalline lattice. *Dev. Biol.* **53**, 217-240.

Reinke, R. and Zipursky, S. L. (1988). Cell–cell interaction in the Drosophila retina: the bride of sevenless gene is required in photoreceptor cell R8 for R7 cell development. *Cell* **55**, 321-330.

Robertson, E. J. (1986). Pluripotential stem cell lines as a route into the mouse germ line. *Trends Genet* **2**, 9-13.

Ross, R. B. and Johnston, M. C. (1972). *Cleft Lip and Palate*. Williams and Wilkins, Baltimore.

Rubin, G. (1989). Development of the Drosophila retina: inductive events studied at single cell resolution. *Cell* **57**, 519-520.

Rugh, R. (1968). *The Mouse*. Burgess, Minneapolis.

Ruppert, S., Müller, G., Kwon, B. and Schütz, G. (1988). Multiple transcripts of the mouse tyrosinase gene are generated by alternative splicing. *EMBO J.* **7**, 2715-2722.

Russell, L. B. (1971). Definition of functional units in a small chromosomal segment of the mouse and its use in interpreting the nature of radiation-induced mutations. *Mutat. Res.* **11**, 107-123.

Russell, L. B. and Russell, W. L. (1954). Analysis of the changing radiation response of the developing mouse embryo. *J. Cell. Comp. Physiol.* **43** (Suppl. 1), 103-147.

Sander, K. (1976). Specification of the basic body pattern in insect embryogenesis. *Adv. Insect Physiol.* **12**, 125-238.

Sanes, J. R., Rubinstein, J. L. R. and Nicolas, J.-F. (1986). Use of recombinant retrovirus to study postimplantation cell lineage in mouse embryos. *EMBO J.* **2**, 3133-3142.

Savatier, P., Morgenstern, J. and Beddington, R. S. P. (1990). Permissiveness to murine leukaemia virus expression during preimplantation and early post-implantation mouse development. *Development* **109**, 655-665.

Schnieke, A., Harbers, K. and Jaenisch, R. (1983). Embryonic lethal mutation in mice induced by retrovirus insertion into the alpha1(I) collagen gene. *Nature* **304**, 315-320.

Schubiger, G., Mosely, R. C. and Wood, W. J. (1977). Interaction of different egg parts in determination of various body regions in Drosophila melanogaster. *Proc. Natl Acad. Sci. USA* **74**, 2050-2054.

Schwartzberg, P., Goff, S. P. and Robertson, E. J. (1989). Germ-line transmission of a c-abl mutation produced by targeted gene disruption in ES cells. *Science* **246**, 799–802.

Schultz, G. A. (1986). Utilization of genetic information in the preimplantation mouse embryo. In: Rossant, J. and Pedersen, R. A. (eds) *Experimental Approaches to Embryonic Development*, pp. 239–265. Cambridge University Press, New York.

Shawlot, W., Siciliano, M. J., Stallings, R. L. and Overbeek, P. A. (1989). Insertional inactivation of the downless gene in a family of transgenic mice. *Mol. Biol. Med.* **6**, 299–308.

Shih, C.-C., Stoye, J. P. and Coffin, J. M. (1988). Highly preferred targets for retrovirus integration. *Cell* **53**, 531–537.

Snow, M. H. L. (1976). Embryo growth during the immediate postimplantation period. *Embryogenesis in Mammals. CIBA Foundation Symposium*, vol. 40 (new series), pp. 53–70. Elsevier, Amsterdam.

Snow, M. H. L. (1981). Autonomous development of parts isolated from primitive-streak-stage mouse embryos. Is developmental clonal? *J. Embryol. Exp. Morph.* **Suppl. 65**, 269–287.

Snow, M. H. L. and Tam, P. P. L. (1979). Is compensatory growth a complicating factor in mouse teratology? *Nature* **279**, 554–557.

Soriano, P. and Jaenisch, R. (1986). Retroviruses as probes for mammalian development: allocation of cells to the somatic and germ cell lineages. *Cell* **46**, 19–29.

Soriano, P., Gridley, T. and Jaenisch, R. (1987). Retroviruses and insertional mutagenesis in mice: proviral integration at the Mov34 locus leads to early embryonic death. *Genes Dev.* **1**, 366–375.

Sulik, K. K., Cook, C. S. and Webster, W. S. (1988). Teratogens and craniofacial malformations: relationships to cell death. *Development* **103** (Suppl.), 213–232.

Sulston, J. E., Schierenberg, E., White, J. G. and Thomson, J. N. (1983). The embryonic cell lineage of the nematode *Caenorhabditis elegans. Dev. Biol.* **100**, 64–119.

Thompson, S., Clarke, A. R., Pow, A. M., Hooper, M. L. and Melton, D. W. (1989). Germline transmission and expression of a corrected HPRT gene produced by gene targeting in embryonic stem cells. *Cell* **56**, 313–321.

Thompson, T. C., Southgate, J., Kitchener, G. and Land, H. (1989). Multistage carcinogenesis induced by ras and myc oncogenes in a reconstituted organ. *Cell* **56**, 917–930.

Thomson, J. A. and Solter, D. (1988a). The developmental fates of androgenetic, parthenogenetic and gynogenetic cells in chimeric gastrulating mouse embryos. *Genes Dev.* **2**, 1344–1351.

Thomson, J. A. and Solter, D. (1988b). Transgenic markers for mammalian chimeras. *Roux's Arch. Dev. Biol.* **197**, 63–65.

Tomlinson, A. and Ready, D. F. (1987a). Neuronal differentiation in the Drosophila ommatidium. *Dev. Biol.* **120**, 366–376.

Tomlinson, A. and Ready, D. F. (1987b). Cell fate in the Drosophila ommatidium. *Dev. Biol.* **123**, 264–275.

Tomlinson, A., Bowtell, D. D. L., Hafen, E. and Rubin, G. M. (1987). Localization of the sevenless protein, a putative receptor for positional information, in the eye imaginal disc of Drosophila. *Cell* **51**, 143–150.

Toles, J. F., Chui, D. H. K., Belbeck, L. W., Starr, E. and Barker, J. E. (1989).

Hemopoietic stem cells in murine embryonic yolk sac and peripheral blood. *Proc. Natl Acad. Sci. USA* **86**, 7456–7459.

Turner, D. L. and Cepko, C. (1987). Cell lineage in the rat retina: a common progenitor for neurons and glia persists late in development. *Nature* **328**, 131–136.

Weissman, I. L., Papaioannou, V. E. and Gardner, R. L. (1978). Fetal hematopoietic origin of the hematolymphoid system. In: Clarkson, B., Marks, T. and Till, J. (eds) *Differentiation of Normal and Neoplastic Hematopoietic Cells*, pp. 33–47. Cold Spring Harbor Laboratory, Cold Spring Harbor.

West, J. D. (1984). Cell markers. In: Douarin, N. L. and McLaren, A. (eds) *Chimaeras in Developmental Biology* pp. 39–63. Academic Press, London.

Williams, D. A., Orkin, S. H. and Mulligan, R. C. (1986). Retrovirus-mediated transfer of human adenosine deaminase gene sequences into cells in culture and into murine hematopoietic cells in vivo. *Proc. Natl Acad. Sci. USA* **83**, 2566–2570.

Wilson, T. G. (1981). A mosaic analysis of the *apterous* mutation in *Drosophila melanogaster*. *Dev. Biol.* **85**, 434–445.

Woychik, R. P., Stewart, T. A., Davis, L. G., D'Eustachio, P. and Leder, P. (1985). An inherited limb deformity created by insertional mutagenesis in a transgenic mouse. *Nature* **318**, 36–40.

Xiang, X., Benson, K. F. and Chada, K. (1990). Mini-mouse: disruption of the pygmy locus in a transgenic insertional mutant. *Science* **247**, 967–969.

Zijlstra, M., Li, E., Saijadi, F., Subramani, S. and Jaenisch, R. (1989). Germline transmission of a disrupted β2-microglobulin gene produced by homologous recombination in embryonic stem cells. *Nature* **342**, 435–438.

Zijlstra, M., Bix, M., Simister, M. E. *et al.* (1990). Beta 2-microglobulin deficient mice lack CD4-8+ cytolytic T cells. *Nature* **344**, 742–746.

4
The study of gene regulation in transgenic mice

GEORGE KOLLIAS* and FRANK GROSVELD[†]

*Molecular Genetics Laboratory, Hellenic Pasteur Institute, 127 Vas Sofias Avenue, GR115 21 Athens, Greece
[†] Laboratory of Gene Structure and Expression, National Institute for Medical Research, The Ridgeway, Mill Hill, London NW7 1AA, UK

1. Introduction

The ability to introduce and express exogenous genetic information in the context of a developing organism allows the investigation of the mechanisms regulating specific gene expression and the role of such expression in normal development and pathological states. Since the first success at producing transgenic mice using microinjection of eggs with cloned genes (Gordon et al., 1980), numerous investigators have employed this technique to study various developmental processes. The use of transgenic mice to study gene regulation has steadily increased, because it offers particular advantages over cell culture systems:

1. The gene is present in every cell type from the F_1 generation. This allows the study of gene expression in many cell types, including those which cannot easily be transfected or maintained in cell culture.
2. The gene can be studied throughout the entire developmental programme,

which is particularly important when studying developmentally regulated genes (see below).

3. The gene is injected as a fragment without a second gene coding for a selection marker as is normally used in transfection protocols in cultured cells. Selection results in integration into 'active' regions of the host cell chromatin and this may influence the behaviour of the gene under study. In addition the transgene will already be present in a normal chromatin configuration before differentiation to the cell type which expresses the gene. In contrast, in transfection experiments the gene is introduced as naked DNA into the expressing cell.

The main disadvantages of transgenic experiments are the cost of the equipment, training time, maintenance of mice and, when breeding is important, the time to complete a full round of experiments (for details see Hogan et al., 1986; Dillon and Grosveld, 1991).

The identification of the regulatory regions that determine the tissue and developmentally specific expression of a gene forms the basis of many other more biologically oriented experiments such as determining the function of a transgene product in the context of the entire animal or the manipulation of certain tissues and biological processes as described in several of the other chapters.

2.   Experimental design for optimal gene expression

There are several methods for introducing foreign DNA into the germ cells of mice (reviewed in Palmiter and Brinster, 1986). Choosing the right technique depends on the purpose of the experiment. For example, infection of preimplantation embryos with natural or recombinant retroviruses (Jaenisch, 1976; Jahner et al., 1985; van der Putten et al., 1985) can be employed when high-efficiency single-copy integration is required, provided DNA construct size limitations or interference of retroviral DNA sequences with expression of the transgene are not a problem. Introducing genes into mouse embryonic stem (ES) cells and then incorporating these cells into the blastocyst of developing embryos represents another route to transgenesis (Bradley et al., 1984; see below). Although still the most demanding technique, manipulation of ES cells in culture is currently the only means for perturbing endogenous gene expression by homologous recombination in transgenic mice. However, in experiments where gene expression regulatory mechanisms are studied or gene addition is intended, the most rewarding techique has been the direct injection of DNA into one of the pronuclei of mouse zygotes (Gordon and Ruddle, 1983; Brinster et al., 1985; Hogan et al., 1986; Fig. 1).

Several considerations should be taken into account when recombinant gene

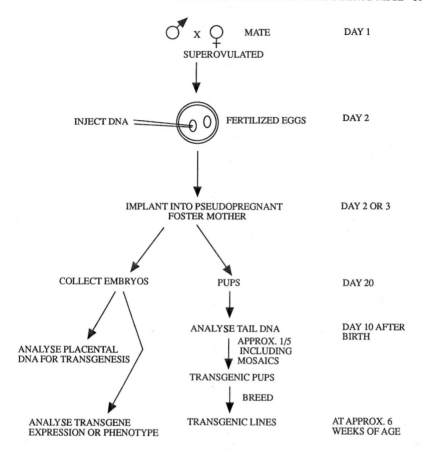

Fig. 1. Scheme for the production of transgenic mice. For experimental details, see Hogan _et al._ (1986).

constructs are introduced by microinjection into the germ line of mice. Pronuclei of mouse zygotes are injected with a few hundred copies of the gene of interest and transferred into the oviducts of pseudopregnant females, where they are left to develop to term (Brinster _et al._, 1985; Hogan _et al._, 1986). The efficiency of the technique in producing transgenic mice varies between different laboratories and for different gene constructs. In skilled hands, as many as >90% of the injected embryos survive microinjection, approximately 20% of these develop to term and approximately 25% of these are found to carry the injected DNA in tail biopsies and are therefore transgenic.

Integration of the transgene is believed to occur at a random site in the mouse genome. Either a single copy or, more often, a tandem, head-to-tail array of transgenes is found to be inserted at a single chromosomal locus. More rarely, insertion can occur at two or more positions in the mouse genome. Upon breeding, transgenic DNA is inherited as a simple Mendelian trait. If this is not observed, it is assumed that integration of the transgene occurred after the first round of replication in the single fertilized egg cell and that the resulting transgenic mouse is mosaic for the transgene, i.e. only a subset of cells carries the transgene. In practice, it has been observed that as many as 30% of transgenic founder mouse lines are mosaic (Wilkie *et al.*, 1986). At this level, mosaicism should be considered as a major cause of discrepancies in quantitative gene expression experiments or in detecting different degrees of pathological alterations in mice carrying lethal or deleterious transgenes. On the other hand, mosaicism can prove to be a means by which one can study development of pathology in early lethal phenotypes. For example, progeny from a transgenic founder which survives a lethal transgene due to mosaicism can provide enough biological material for the study of the lethal phenotype.

Since it has been shown that prokaryotic sequences inhibit the expression of some transgenes (Hammer *et al.*, 1985; Krumlauf *et al.*, 1985; Townes *et al.*, 1985), it has become common practice not to include any prokaryotic DNA sequences in the DNA fragments to be injected. In addition, linear fragments with non-blunt ends seem to give better integration efficiencies (Brinster *et al.*, 1985). As far as expression efficiency is concerned, it has been experimentally deduced that the presence of introns facilitates transcription of microinjected genes (Brinster *et al.*, 1988).

Finally, transcriptional efficiency is almost always (see below) influenced by the activation status of the chromatin neighbouring the transgene, a phenomenon which, in many cases, does not affect the specificity of transgene expression in the 'correct' tissues, but often causes expression in the 'wrong' tissues.

Transgenic mice have been generated routinely with fragments up to 45 kb and injection of fragments of this size presents no particular problems. There appears to be no difference in the rate of obtaining transgenic animals when small or large fragments are used and the upper limit of 45 kb is only because this is the largest fragment that can be cloned in a cosmid vector. Although fragments of several hundred kilobases can be cloned into yeast artificial chromosome (YAC) vectors, attempts to generate transgenic mice carrying YACs by microinjection has so far been unsuccessful. This is due to the very large size, when shearing of the DNA becomes a problem, and the difficulty of preparing sufficient amounts of purified YAC inserts at the required concentration. Two novel approaches can be used to obtain transgenes of a size between cosmids and YACs. The first relies on coinjection of overlapping cosmid fragments, which will result in homologous recombination between the fragments to restore one larger fragment (H. de Boer, personal communication; S. Eccles and F. Grosveld, unpublished). This method is very useful when expression *per se* of a gene product has to be obtained but has severe disadvantages in gene regulation studies because non-homologously combined fragments are cointegrated into the host genome and these can influence the expression of the gene (N. Dillon and P. Fraser, unpublished). We have recently developed a second alternative method for generating large fragments of the human β-globin locus for injection by joining together two cosmid inserts (J. Strouboulis *et al.*, submitted for publication). It involves cloning short oligonucleotides into the cosmid inserts at the position that they will be joined together. These oligonucleotides are then used to generate single-stranded complementary tails which can be annealed and ligated together very efficiently (Fig. 2). The large fragment is purified by gel electro-phoresis and, after microinjection, appears to give a high rate of transgenic animals carrying the intact large fragment ($> 10\%$ of animals born). At present it is not known what the upper size limit is for microinjection, although there almost certainly will be one due to shearing of the concentrated fragments during passage through the microinjection needle.

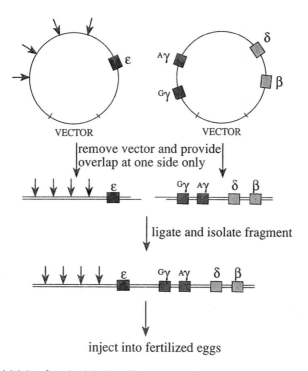

Fig. 2. Cosmid joining for microinjection. The two cosmid clones are derived from the human β-globin locus (Taramelli *et al.*, 1986). Genes are indicated by boxes. The drawing is not to scale.

3. Gene expression in transgenic mice:
locus control regions (LCRs)

Factors affecting the specificity of transcription have been the target of molecular genetics for many years. Most of our current knowledge on the role of *cis*-acting DNA elements in regulated gene expression comes from gene transfer experiments in cultured cells. However, the ability to manipulate gene expression in the context of a living organism provided access to questions concerning developmental regulation and tissue specificity of gene expression.

Numerous experiments with transgenic mice have shown that sequence information sufficient for tissue-specific and developmentally regulated expression is usually localized in the promoter and enhancer elements within a few kilobases from the gene. However, if a gene containing all proximal control regions is introduced into transgenic mice, it is rarely expressed at the same level as its endogenous counterpart, nor is there any strict correlation of integrated gene copies and level of expression. Moreover, some expression of the transgene is usually found in heterologous tissues and these can differ between different founder animals. This is thought to be caused by two different phenomena. Firstly, the transgene fragment that is injected does not contain all the regulatory regions required for full expression and, secondly, random integration of the transgene in the mouse genome results in so-called 'position effects'. The latter means that regulatory regions present at, or near, the site of integration of the transgene can act on that gene and influence its pattern of expression (see also Chapter 1). Similar effects are observed when genes are introduced into cultured cells by transfection or retroviral infection. Recent results obtained with transgenic mice and *Drosophila* have begun to address these questions in detail.

The genes of the human β-globin locus were considered to be very position-sensitive, showing none or only a few per cent of their normal levels of expression when introduced into transgenic mice (Magram *et al.*, 1985; Townes *et al.*, 1985; Kollias *et al.*, 1986, 1987). The globin genes were shown to have proximal elements which included the promoter, an intragenic sequence and a 3' enhancer which were sufficient to drive tissue-specific expression, but this was at very low levels (< 1%) and dependent on the integration site. High levels of regulated expression were achieved only when an additional distal regulatory region located 50–60 kb upstream of the β-globin gene was used. This LCR (Fig. 3) can confer high-level, gene copy number-dependent expression of the human β-globin gene in transgenic mice (Grosveld *et al.*, 1987) and mouse erythroleukaemia cells (Blom van Assendelft *et al.*, 1989), independent of the site of integration in the host genome. The β-globin locus LCR is also capable of inducing high-level erythroid expression of α-globin (Hanscombe *et al.*, 1989; Ryan *et al.*, 1989a) and of heterologous genes such as the murine *Thy-1* gene and the

Fig. 3. The human β-globin locus. The numbers are in kilobases and arrows indicate developmentally stable DNAase I-hypersensitive sites (Tuan *et al.*, 1985; Forrester *et al.*, 1987; Grosveld *et al.*, 1987). Genes are indicated by a black boxes, a pseudogene by a white box.

herpes thymidine kinase promoter (Blom van Assendelft *et al.*, 1989; Talbot *et al.*, 1989).

Position independence and copy number dependence can theoretically be explained by at least two independent mechanisms; either positive activation by the LCR is always achieved and can result in very high levels of expression, which obscures small position effects that may still be present in the background, or (and) the region contains elements that insulate it from neighbouring regions, providing a locus border element (LBE). Scaffold-attachment regions (SARs; Gasser and Laemmli, 1986; Jarman and Higgs, 1988) or 'A' elements (Stief *et al.*, 1989; Bonifer *et al.*, 1990; see Chapter 1) could be LBEs and we intially speculated that these were part of the LCR in addition to sequences activating expression (Grosveld *et al.*, 1987). However, preliminary experiments indicate that this is not the case and that such a border may be located further upstream. This is based on the fact that the DNAase I sensitivity of chromatin in isolated nuclei is strongly decreased in the sequences 25–30 kb upstream of the LCR (Fig. 3; Kioussis *et al.*, 1983; Forrester *et al.*, 1990). In the downstream direction the chromatin remains sensitive under the control of the LCR for at least 150 kb (Forrester *et al.*, 1990), suggesting that no such sequences are present for a very considerable distance). It therefore appears that the position independence we observe is (at least in part) due to the fact that the LCR achieves activation of transcription in some dominant fashion, perhaps by creating very stable interactions between the LCR and the genes. Consequently, positive position effects, if present at all, would only be present as part of the background and only become apparent in situations where the linked gene is suppressed (Dillon and Grosveld, 1991). Interestingly, position effects are not observed when low levels of expression are obtained by the use of part of the LCR or mutations in the LCR. This indicates that the interaction between the (part) LCR and the promoter is dominant except when the promoter is suppressed (Forrester *et al.*, 1989; Ryan *et al.*, 1989b; Talbot *et al.*, 1989, Collis *et al.*, 1990; Fraser *et al.*, 1990; Dillon and Grosveld, 1991).

Further fine mapping of individual elements within the LCR was based on the notion (Tuan *et al.*, 1985; Forrester *et al.*, 1987; Grosveld *et al.*, 1987) that it contains four tissue-specific DNAase I-hypersensitive sites (DHSs) over a region of 20 kb (Fig. 1). This region was still fully functional when reduced to a recombinant size of 6.5 kb containing only the four DHSs (Talbot *et al.*, 1989). Deletion experiments (Ryan *et al.*, 1989a; Collis *et al.*, 1990; Fraser *et al.*, 1990) demonstrated a considerable degree of functional redundancy between DHSs; individual sites could still give copy number-dependent expression of the linked transgene, albeit at lower levels. As has been shown for many other regulatory elements in other systems, each of the hypersensitive regions contains a number of nuclear-protein-binding sites, some of which bind erythroid-specific proteins and others which bind ubiquitously expressed proteins. Although not yet understood at the molecular level, these presumably perform different functions but synergize with each other to give the final effect.

Other LCR seqeuences have been described, e.g. for the CD2 gene, which codes for a T cell-specific cell surface protein (Greaves *et al.*, 1989) and the chicken lysozyme gene, which is expressed in macrophages (Bonifer *et al.*, 1990). In the latter case, the transgene construct also contained border elements which appear to prevent the expression of the lysozyme gene in any other tissues (for discussion see Chapter 1).

A number of interesting applications may be conceived for further experimental use of the properties of LCR sequences. First, by including LCR sequences in gene constructs designed for somatic gene therapy protocols, correctly regulated expression of genes is now a possible achievement. For example, in the case of thalassaemias, inclusion of globin LCR sequences in retroviral vector systems carrying the human β-globin gene may solve the problem of the dependence of expression on the site of integration and this may form the basis for gene therapy protocols by gene addition to haematopoietic stem cells (see also Chapter 10). Preliminary experiments in mice (F. Meyer, personal communication) indicate that this is indeed the case. Secondly, as a result of position effects in all gene expression experiments in transgenic mice and cultured cells, it has been very difficult to quantify gene expression and thus examine cooperation between different regulatory elements. This problem can be overcome by position-independent expression due to the inclusion of LCR sequences and border elements on the transgene constructs.

4.   Developmental gene expression studies in transgenic mice

Three very good examples of the advantages that transgenic mice have been provided by the study of the developmental regulation of the *Hox* genes, the albumin and α-fetoprotein genes and when the role of the LCR in the developmental expression of the globin genes was studied. The albumin and α-fetoprotein genes are both liver-specific genes which share a great deal of homology. However, the α-fetoprotein gene is switched off in the adult animal, whereas the albumin gene remains fully active. By exchanging the upstream control region of these genes, Camper and Tilghman (1989) were able to show that this also exchanged the developmental expression pattern of the genes in transgenic mice, implying that the upstream region of the α-fetoprotein gene contained a region responsible for suppression of expression in the adult stage. By further deletion experiments this could be localized to a particular sequence which, when deleted from the α-fetoprotein locus, led to continued expression of the gene in the adult. This very strong evidence was subsequently used to clone and characterize the suppressor factor (Vacher and Tilghman, 1990; S. Tilghman, personal communication).

A more complex situation has been studied in other multigene families. The *Hox* genes are part of a large gene family which is divided in several clusters on different chromosomes. These genes are involved in the laying down of the body pattern of the developing embryo and each of the genes shows a very complex but characteristic temporal- and lineage-specific pattern of expression. In the neuroectoderm the anterior boundaries of expression of the genes correlate in general with the order of the genes within a locus (Kessel and Gruss, 1990; Hunt and Krumlauf, 1991). It would almost be impossible to study the regulation of these genes by manipulating gene expression in culture systems. However, by using combinations of regulatory elements that flank the *Hox-1.1* or *Hox-2.6* gene and a reporter gene in transgenic mice, it has been possible to show that different DNA elements contribute different parts of the overall expression pattern of these genes, opening up the possibility to study each of these elements in detail and assess their contribution to the overall pattern (Puschel *et al.*, 1990; Whiting *et al.*, 1992).

A different type of experiment was carried out using the human globin locus to study the role of the LCR. The locus consists of five active genes, arranged in the order $5'\text{-}\varepsilon\text{-}^G\gamma\text{-}^A\gamma\text{-}\delta\text{-}\beta\text{-}3'$ over a DNA region of approximately 50 kb (Fig. 3). The embryonic ε-globin, fetal γ-globin and adult β-globin genes are tightly regulated during development. The activity of the LCR sequences on individual globin genes has been tested directly in transgenic mice carrying LCR sequences linked to the embryonic ε-globin, fetal γ-globin or adult β-globin gene (Behringer *et al.*, 1990; Enver *et al.*, 1990; Lindenbaum *et al.*, 1990; Raich *et al.*, 1990; Shih *et al.*, 1990; Dillon and Grosveld, 1991). In the absence of LCR the human ε-globin gene is not expressed (Shih *et al.*, 1990), the γ-globin gene is expressed exclusively in embryonic stages (Kollias *et al.*, 1986; Chada *et al.*, 1985), while the human β-globin gene is expressed specifically in fetal/adult erythroid tissue (Magram *et al.*, 1985; Townes *et al.*, 1985; Kollias *et al.*, 1986). When the LCR is linked to each individual gene, the ε-globin gene is fully active in the embryonic yolk sac but is switched off in the fetal liver (Raich *et al.*, 1990; Shih *et al.*, 1990; Lindenbaum *et al.*, 1990). The γ-globin gene is also activated in the yolk sac, but is gradually switched off in the fetal liver and is completely silenced in the adult (Dillon and Grosveld, 1991). This shows that each of these genes contains sufficient information in the flanking regions to be correctly activated and suppressed during development. In contrast, the β-globin gene is activated incorrectly at the embryonic state but is correctly expressed during the fetal and adult stages. Suppression of the β-globin gene at the embryonic stage and activation at the correct time in the fetal liver can be achieved by placing the γ- and β-globin genes in competition for one LCR (Behringer *et al.*, 1990; Enver *et al.*, 1990; Hanscombe *et al.*, 1991). This competition is dependent on the relative position of each of the genes to the LCR (Hanscombe *et al.*, 1991), in agreement with the results obtained from genetics. This led to the formulation of a novel model of the regulation of this locus (Fig. 2), which is primarily based on the specificity of each of the genes and their DNA protein-binding sites, but to which an extra level of complexity is added due to the position of the genes relative to the LCR, which results in a different ability of each of the genes to compete.

5.   Defining novel developmentally regulated genes in transgenic mice

Studies with transgenic mice can also be used to search for genes or regulatory regions that are expressed or cause expression in particular tissues at particular times of development. This method was developed in *Drosophila* (O'Kane and Gehring, 1987) and is based on the detection of position effects and easy visualization of the product of a bacterial β-galactosidase (β-gal) gene. Two variations heve been used in transgenic mice. In one approach a tansgene construct containing a weak promoter and an easily detectable marker gene such as β-galactosidase is introduced into mouse eggs and integrates into the mouse genome. The weak promoter will not allow expression of the β-gal marker gene unless the promoter is near an enhancer sequence present at or near the site of integration (Fig. 4(A)), resulting in an expression pattern of the transgene that reflects the specificity of the enhancer. In the alternative approach the β-gal gene is coupled to an RNA splice acceptor site (Fig. 4(B)). This will lead to expression of the β-gal marker if the transgene integrates in an intron of a mouse gene and will therefore be expressed with the same developmental pattern as that gene (Gossler *et al.*, 1989). In both cases, the cloning of the flanking region of the transgenic β-gal gene that shows an interesting developmental pattern of expression will result in the isolation of a gene or its flanking region that normally shows that same pattern of expression. Obviously the opposite, i.e. linking a marker β-gal gene to a known gene of interest, will allow easy detection of the pattern of expression of that gene.

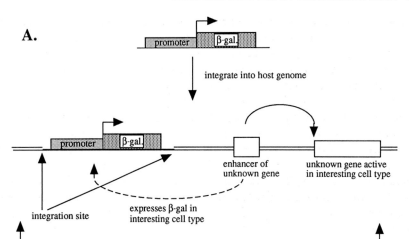

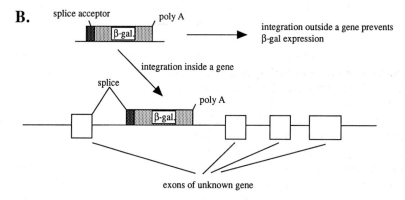

Fig. 4. Enhancer and/or traps for transgenic mice: (A) enhancer trap by distant transcription effect; (B) gene trap by splicing.

6. Manipulation of gene expression in transgenic mice

The ability to manipulate the developmental timing, tissue specificity and level of transgene expression is particularly valuable in the analysis of gene function in transgenic mice. Ectopic, deregulated or correctly regulated transgene expression can be established in transgenic mice by the use of specific promoters or enhancers driving homologous or heterologous genes to express in specific cell-types. Immunological tolerance, mammalian development, and the role of oncogenes in growth and differentiation are the best examples of processes which have been studied by this approach (Hanahan, 1989).

The analysis of the function of dominant, disease-causing or lethal genes in transgenic mice is often very difficult, since mouse lines cannot be established. In such cases a transgene construct with inducible properties is highly desirable. Designing an inducible transgene presents major problems, mainly because many mammalian 'inducers' are usually already found in mouse tissues. Invertebrate inducers are an obvious alternative but it remains to be tested whether they can be used in transgenic systems.

Based on viral transactivation mechanisms, some investigators were successful in devising a two-tiered methodology for use in inducible transgene regulation (Nerenberg et al., 1987; Khillan et al., 1988; Vogel et al., 1988; Byrne and Ruddle, 1989). According to this idea, two transgenic mouse lines should be constructed. One, the transresponder, carries the gene of interest under the influence of a transresponding viral promoter. The second, the transactivator, contains a viral transactivator gene under the influence of chosen tissue-specific regulatory elements. Cross-breeding of these two mouse lines gives double transgenic progeny with expession patterns for the gene of interest, dependent on the predesigned pattern of expression of the transactivator protein.

Two main considerations should be taken into account when using this strategy. First, the transresponder construct should have very low levels of expression in any transgenic tissue and, secondly, the transactivator construct should not produce any deleterious effects in the tissues that it will express. In two cases (Nerenberg et al., 1987; Khillan et al., 1988) when the long terminal repeats (LTRs) from the human immunodeficiency virus (HIV) and the human T cell lymphotropic virus type I (HTLV-I) were transactivated by the products of the TAT and tax genes, respectively, both the above-described criteria were not satisfied. The transresponder transgenes showed high basal levels of expression and the transactivators when expressed produced oncogenic phenotypes.

In a third case, Bryne and Ruddle (1989) used the herpes simplex virus type 1 (HSV-1) transactivation mechanism to obtain inducible transgene expression. Transresponder cis elements from an immediate early HSV-1 gene linked to the chloramphenicol acetyltransferase (CAT) gene produced very low basal levels of CAT expression. The HSV-1 transactivator VP16 gene, driven by regulatory elements of a neurofilament gene, could induce CAT expression in brain, spinal cord and heart of double transgenic mice. At least for these tissues the VP16 transgene did not induce any pathological abnormalities.

Another interesting approach for inducible activity of transgene products in mice can be developed on the basis of the findings of Picard et al. (1988) and Eilers et al. (1989), who showed that the activity of E1A and myc proteins can be subjected to hormonal regulation when E1A or myc proteins are fused to the ligand-binding domain of a steroid receptor. Such fusion proteins are found to be inactive in the absence of hormone. Upon administration of hormone in cells

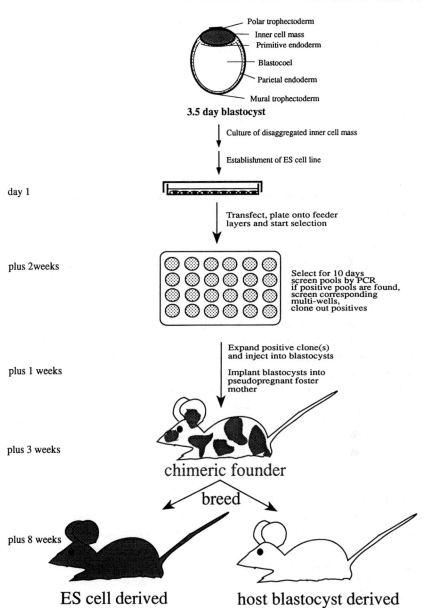

Fig. 5. ES cell derivation and application.

transfected with these hybrid genes, full activity of E1A and *myc* can be observed. It is conceivable, therefore, that a similar experimental design can be followed in transgenic mice when inducible activity of a specific transprotein is desirable. The use of a heterologous hormone–receptor system (such as ecdysone) may help to eliminate the complexities that may arise due to endogenous hormone production when mammalian hormone–receptor combinations are used.

7.   Inhibition or altering of gene function

The most powerful technique to inhibit gene function has become the use of ES cells (Fig. 5). These cells are derived from the inner cell areas of early blastocysts (Bradley *et al.*, 1984) and can be kept in culture in an undifferentiated state. After reintroduction into blastocysts, they can contribute to all the tissues of the mice derived from the reimplanted blastocysts, including germ line cells. Subsequent breeding of the (ES) mosaic founder mice can result in mice that are entirely derived from the cultured ES cells. Fortunately, ES cells can be easily transfected in culture to allow manipulation of the ES cell genome. This technology has been used to obtain mutagenesis by random integration, resulting in the development of mutant phenotypes with an overall efficiency of 5–6% (Jaenisch, 1988). Obviously, the most interesting application of ES cells is the ability to interrupt specific genes by the process of homologous recombination (see Fig. 6 and Chapter 2). This has provided a means to generate heterozygous null mutants for a particular gene and thereby gain insight into the function of that gene in the context of the entire organism when the animals are bred to homozygosity. A number of very interesting genes have already been 'knocked out' by this technique, confirming definitive evidence for the function of those genes during development. For example, the knock out of the *wnt1* gene, which codes for a growth factor, leads to malformation of the midbrain (Thomas *et al.*, 1991). However, in many of the cases the 'knock out' leads to gross alterations and lethality which can prevent the analysis of the function of the gene in different tissues or at later stages. One solution to this problem could come from the introduction of specific transgenes into the ES cells of mice heterozygous for the null mutation. By a careful choice of regulatory sequences present on the transgene construct, the missing gene product could be re-established in particular tissues or at particular developmental stages of the animal. Perhaps even more interesting is the possibility to introduce more subtle mutations by a double selection procedure in ES cells (Hasty *et al.*, 1991), or to interfere with particular regulatory sequences (D. Melton, personal communication). Since it would take less time and have technical advantages, homologous recombination to mutate particular genes has also been tested by direct injection of DNA into fertilized eggs (Brinster *et al.*, 1989). Unfortunately, this experiment required an enor-

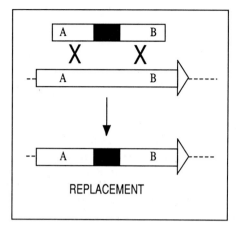

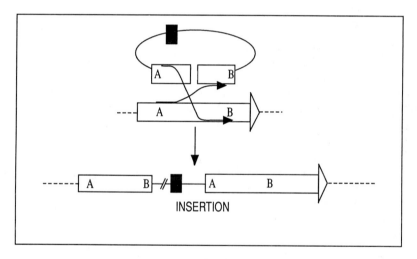

Fig. 6. Homologous recombination (see also Chapter 2) The replacement vector is recombined into the genome by a homologous double crossover or gene conversion (indicated by ×) into the mouse genome (indicated by open arrows). The insertion vector is recombined into the host genome by a crossover (indicated by arrows). The letters indicate regions of homology; the selectable marker is indicated by a black box.

mous number of microinjections, but recent experiments in ES cells show that this may have been due to the particular gene that was tested. Two groups have recently reported that they were unsuccessful in obtaining homologous recombination in ES cells, unless they used isogenic DNA, i.e. the identical gene to that present in the ES cells, on their target vector (A. Berns, personal communication; R. Jaenisch, personal communication). For polymorphic genes this reduces the amount of non-homology between the incoming DNA and the resident ES gene, thereby increasing the role of recombination (see also Chapter 2).

In particular cases, inactivation or interference with a gene product can also be achieved by direct microinjection. To inactivate the function of multimeric proteins a dominant negative approach has been proposed (Herskowitz, 1987). This involves the inhibition of the function of a wild-type gene product by the additional expression of an inhibitory variant of the same product. For example, when a mutated pro-α(I) collagen gene was introduced into the germ line of mice, a dominant perinatal lethal phenotype evolved, resembling the homologous human disease osteogenesis imperfecta II (Stacey *et al.*, 1988). Interestingly, expression levels of as little as 10% mutant RNA of the total collagen RNA were enough to cause this disease. Another good example of a dominant effect which does not require high levels of expression is provided by the use of oncogenes (see Chapter 9). Unfortunately, low-level expression of transgenes is often not sufficient to cause disease. Generation of a transgenic mouse model of human sickle cell anaemia by the use of a mutated human β-globin gene required high-level expression, which could only be achieved by the use of the globin LCR (Greaves *et al.*, 1989). High-level expression might also be a prerequisite to effectively inhibit gene expression by antisense RNA produced *in vivo* (see also Chapter 8).

Future developments in transgenic technology, coupled with broader understanding of the molecular mechanisms regulating gene expression, will provide the fine tools needed for the analysis of development and physiology.

References

Behringer, R. R., Ryan, T. M., Palmiter, R. D., Brinster, R. L. and Townes, T. M. (1990). Human γ- to β-globin gene switching in transgenic mice. *Genes Dev.* **4**, 380–389.

Blom van Assendelft, G., Hanscombe, O., Grosveld, F. and Greaves, D. R. (1989). The β-globin domain control region activates homologous and heterologous promoters in a tissue-specific manner. *Cell* **56**, 969–977.

Bonifer, C., Vidal, M., Grosveld, F. and Sippel, A. E. (1990). Tissue-specific and position-independent expression of the complete gene domain for chicken lysozyme in transgenic mice. *EMBO J.* **9**, 2843–2848.

Bradley, A., Evans, M., Kaufman, M. H. and Robertson, E. (1984). Formation of germline chimaeras from embryo derived teratocarcinoma cells. *Nature* **309**, 255–258.

Brinster, R. L., Braun, R. E., Lo, D., Avarbock, M. R., Oram, F. and Palmiter, R. D. (1989). Targeted correction of a major histocompatibility class II Eα gene, by DNA microinjected into mouse eggs. *Proc. Natl Acad. Sci. USA* **86**, 7087-7091.

Brinster, R. L., Chen, H. Y., Trumbauer, M. E., Yagle, M. K. and Palmiter, R. D. (1985). Factors affecting the efficiency of introducing foreign DNA into mice by microinjecting eggs. *Proc. Natl Acad. Sci. USA* **82**, 4438-4442.

Brinster, R. L., Allen, J. M., Behringer, R. R., Gelinas, R. E. and Palmiter, R. D. (1988). Introns increase transcriptional efficiency in transgenic mice. *Proc. Natl Acad. Sci. USA* **85**, 836-840.

Byrne, W. and Ruddle, F. H. (1989). Multiplex gene regulation: a two-tiered approach to transgene regulation in transgenic mice. *Proc. Natl Acad. Sci. USA* **86**, 5473-5477.

Camper, S. and Tilghman, S. (1989). Postnatal repression of the α fetoprotein gene is enhancer independent. *Genes Dev.* **3**, 537-546.

Chada, K., Magram, J., Raphael, K., Radice, G., Lacy, E. and Costantini, R. (1985). Specific expression of a foreign β-globin gene in erythroid cells of transgenic mice. *Nature* **314**, 377-380.

Collis, P., Antoniou, M. and Grosveld, F. (1990). Definition of the minimal requirements within the human β-globin gene and the dominant control region for high level expression. *EMBO J.* **9**, 233-240.

Dillon, N. and Grosveld, F. (1991). Human γ-globin genes silenced independently of other genes in the β-globin locus. *Nature* **350**, 252-254.

Dillon, N. and Grosveld, F. (1992). *Transcriptional Regulation Using Transgenic Animals.* Oxford University Press, Oxford (in press).

Eilers, M., Picard, D., Yamamoto, K. and Bishop, J. M. (1989). Chimaeras of myc oncoprotein and steroid receptors cause hormone-dependent transformation of cells. *Nature* **340**, 66-68.

Enver, T., Raich, N., Ebens, A. J., Papayannopoulou, T., Costantini, F. and Stamatoyannopoulos, G. (1990). Developmental regulation of human fetal-to-adult globin gene switching in transgenic mice. *Nature* **344**, 309-313.

Forrester, W. C., Takegawa, S., Papayannopoulou, T., Stamatoyannopoulos, G. and Groudine, M. (1987). Evidence for a locus activation region: the formation of developmentally stable hypersensitive sites in globin-expressing hybrids. *Nucl. Acids Res.* **15**, 10159-10177.

Forrester, W., Novak, U., Gelinas, R., Groudine, M. (1989). Molecular analysis of the human β-globin locus activation region. *Proc. Natl Acad. Sci. USA* **86**, 5439-5443.

Forrester, W., Epner, E., Driscoll, C., Enver, T., Brice, M., Papayannopoulou, T. and Groudine, M. (1990). A deletion of the human β-globin locus activation region causes a major alteration in chromatin structure and replication across the entire β-globin locus. *Genes Dev.* **4**, 1637-1649.

Fraser, P., Hurst, J., Collis, P. and Grosveld, F. (1990). DNaseI hypersensitive sites 1, 2 and 3 of the human β-globin dominant control region direct position-independent expression. *Nucl. Acids Res.* **18**, 3503-3508.

Gasser, S. M. and Laemmli, U. K. (1986). Cohabitation of scaffold binding regions with upstream/enhancer elements of three developmentally regulated genes of D. melanogaster. *Cell* **46**, 521-530.

Greaves, D. R., Fraser, P., Vidal, M. A., Hedges, M. J., Ropers, D., Luzzatto, L. and Grosveld, F. (1990). A transgenic mouse model of sickle cell disorders. *Nature* **343**, 183–185.

Gordon, J. W. and Ruddle, F. H. (1983). Gene transfer into mouse embryos: production of transgenic mice by prenuclear injection. *Methods Enzymol.* **101**, 411–433.

Gordon, J. W., Scangos, G. A., Plotkin, D. J., Barbosa, J. A. and Ruddle, F. H. (1980). Genetic transformation of mouse embryos by microinjection of purified DNA. *Proc. Natl Acad. Sci. USA* **77**, 7380–7384.

Gossler, A., Joyner, A. L., Rossant, J. and Skarnes, W. C. (1989). Mouse embryonic stem cells and reporter constructs to detect developmentally regulated genes. *Science* **244**, 463–465.

Greaves, D. R., Wilson, F. D., Lang, G. and Kioussis, D. (1989). Human CD2 3′-flanking sequences confer high-level, T-cell specific, position independent gene expression in transgenic mice. *Cell* **56**, 979–986.

Grosveld, F., Blom van Assendelft, G. B., Greaves, D. R. and Kollias, G. (1987). Position-independent high level expression of the human β-globin gene in transgenic mice. *Cell* **51**, 975–985.

Hammer, R. E., Brinster, R. L. and Palmiter, R. D. (1985). Usse of gene transfer to increase animal growth. *Cold Spring Harbor Symp. Quant. Biol.* **50**, 379–387.

Hanahan, D. (1989). Transgenic mice as probes into complex systems. *Science* **246**, 1266–1275.

Hanscombe, O., Vidal, M., Kaeda, J., Luzzatto, L., Greaves, D. R. and Grosveld, F. (1989). High-level, erythroid-specific expression of the human α-globin gene in transgenic mice and the production of human hemoglobin in murine erythrocytes. *Genes Dev.* **3**, 1572–1581.

Hanscombe,O., Whyatt, D., Fraser, P., Yannoutous, N., Greaves, D., Dillon, N. and Grosveld, F. (1991). Importance of globin gene order for correct developmental expression. *Genes Dev.* **5**, 1387–1394.

Hasty, P., Ramirez-Solis, R., Krumlauf, R. and Bradley, A. (1991). Introduction of a subtle mutation into the Hox26 locus in embryonic stem cells. *Nature* **350**, 243–245.

Herskowitz, I. (1987). Functional inactivation of genes by dominant negative mutations. *Nature* **329**, 219–222.

Hogan, B., Costantini, E. and Lacy, E. (1986). *Manipulating the Mouse Embryo: A Laboratory Manual.* Cold Spring Harbor Laboratory, Cold Spring Harbor.

Hunt, P. and Krumlauf, R. (1991). Deciphering the Hox code: clues to patterning bronchial regions of the head. *Cell* **66**, 1075–1078.

Jaenisch, R. (1976). Germ line integration and Mendelian transmission of the exogenous Moloney leukemia virus. *Proc. Natl Acad. Sci. USA* **73**, 1260–1264.

Jaenisch, R. (1988). Transgenic animals. *Science* **240**, 1468–1474.

Jahner, D., Haase, K., Mulligan, R. and Jaenisch, R. (1985). Insertion of the bacterial gpt gene into the germ line of mice by retroviral infection. *Proc. Natl Acad. Sci. USA* **82**, 6927–6931.

Jarman, A P. and Higgs, D. R. (1988). Nuclear scaffold attachment sites in the human globin gene complexes. *EMBO J.* **7**, 3337–3344.

Kessel, M. and Gruss, P. (1990). Murine developmental control genes. *Science* **249**, 374–379.

Khillan, J. S., Sweet, R. W., Westphal, H., Yu, S. H., Deen, K. C. and Rosenberg, M. (1988). Gene transactivation mediated by the TAT gene of human immunodeficiency virus in transgenic mice. *Nucl. Acids Res.* **16**, 1423–1430.

Kioussis, D., Vanin, E., deLange, T., Flavell, R. A. and Grosveld, F. (1983). β-Globin gene inactivation by DNA translocation in γ-thalassaemia. *Nature* **306**, 662–666.

Kollias, G., Wrighton, N., Hurst, J. and Grosveld, F. (1986). Regulated expression of human $^Aγ$-,β-, and hybrid γβ-globin genes in transgenic mice: manipulation of the developmental expression patterns. *Cell* **46**, 89–94.

Kollias, G., Hurst., J., deBoer, E. and Grosveld, F. (1987). A tissue and developmental specific enhancer is located downstream from the human β-globin gene. *Nucl. Acids Res.* **15**, 5739–5747.

Krumlauf, R., Hammer, R. E., Brinster, R. L., Chapman, V. M. and Tilghman, S. (1985). Regulated expression of α-fetoprotein genes in transgenic mice. *Cold Spring Harbor. Symp. Quant. Biol.* **50**, 371–378.

Lindenbaum, M. and Grosveld, F. (1990). An *in vitro* globin gene switching model based on differentiated embryonic stem cells. *Genes Dev.,* **4**, 2075–2085.

Magram, J., Chada, K. and Costantini, F. (1985). Developmental regulation of a cloned adult β-globin gene in transgenic mice. *Nature* **315**, 338–340.

Nerenberg, M., Hinrichs, S. H., Reynolds, R. K., Khoury, G. and Jay, G. (1987). The TAT gene of human T-lymphotropic virus type 1 induces mesenchymal tumors in transgenic mice. *Science* **237**, 1324–1329.

O'Kane, C. J. and Gehring, W. J. (1987). Detection in situ of genomic regulatory elements in Drosophila. *Proc. Natl Acad. Sci. USA* **84**, 9123–9127.

Palmiter, R. D. and Brinster, R. L. (1986). Germline transformation of mice. *An. Rev. Genet.* **20**, 465–499.

Picard, D., Salser, S. J. and Yamamoto, K. R. (1988). A movable and regulable inactivation function within the steroid binding domain of the glucocorticoid receptor. *Cell* **54**, 1073–1080.

Puschel, A., Balling, R. and Gruss, P. (1990). Separate elements cause lineage restriction and specific boundaries of Hox1 expression. *Development* **112**, 279–288.

Raich N., Enver T., Nakamoto B., Josephson B., Papayannopoulos T. and Stamatoyan-nopoulos G. (1990). Autonomous developmental control of human embryonic globin switching in transgenic mice. *Science* **250**, 1147–1149.

Ryan, T. M., Behringer, R. R., Martin, N. C., Townes, T. M., Palmiter, R. D. and Brinster, R. L. (1989a). A single erythroid specific DNaseI superhypersensitive site activates high levels of human β-globin gene expression in transgenic mice. *Genes Dev.* **3**, 314–323.

Ryan, T. M., Behringer, R. R., Townes, T. M., Palmiter, R. D. and Brinster, R. L. (1989b). High-level erythroid expression of human α-globin genes in transgenic mice. *Proc. Natl Acad. Sci. USA* **86**, 37–41.

Shih, D., Wall, R. and Shapiro, S. (1990). Developmentally regulated and erythroid specific expression of the human embryonic ε-globin gene in transgenic mice. *Nucl. Acids Res.* **18**, 5465–5472.

Stacey, A., Bateman, J., Choi, T., Mascara, T., Cole, W. and Jaenisch, R. (1988). Perinatal lethal osteogenesis imperfecta in transgenic mice bearing an engineered mutant pro-α(I) collagen gene. *Nature* **332**, 131–136.

Stief, A., Winter, D. M., Stratling, W. H. and Sippel, A. E. (1989). A nuclear DNA attachment element mediates elevated and position-independent gene activity. *Nature* **341**, 343–345.

Talbot, D., Collis, P., Antoniou, M., Vidal, M., Grosveld, F. and Greaves, D. R. (1989). A dominant control region from the human β-globin locus conferring integration site independent gene expression. *Nature* **338**, 352–355.

Taramelli, R., Kioussis, D., Vanin, E., Bartram, K., Groffen, J., Hurst, J. and Grosveld, F. G. (1986). γδβ-Thalassaemias 1 and 2 are the result of a 100 kbp deletion in the human β-globin cluster. *Nucl. Acids Res.* **14**, 7199–7212.

Thomas, K. R., Musci, T. S., Neumann, P. E. and Capecchi, M. R. (1991). Swaying is a mutant allele of the proto-oncogene wnt-1. *Cell* **67**, 969–976.

Townes, T. M., Lingrel, J. B., Chen, H. Y., Brinster, R. L. and Palmiter, R. D. (1985). Erythroid-specific expression of human β-globin genes in transgenic mice. *EMBO J.* **4**, 1715–1723.

Tuan, D., Solomon, W., Qiliang, L. S. and Irving, M. L. (1985). The 'β-like-globin' gene domain in human erythroid cells. *Proc. Natl Acad. Sci. USA* **32**, 6384–6388.

Vacher, J. and Tilghman, S. (1990). Dominant negative regulation of the mouse α fetoprotein gene in adult liver. *Science* **280**, 1732–1735.

van der Putten, H., Botteri, F. M., Miller, A. D., Rosenfeld, M. G. *et al.* (1985). Efficient insertion of genes into the mouse germ line via retroviral vectors. *Proc. Natl Acad. Sci. USA* **82**, 6148–6152.

Vogel, J., Hinrichs, S. H., Reynolds, R. K., Luciw, P. A. and Jay, G. (1988). The HIV tat gene induces dermal lesions resembling Kaposi's sarcoma in transgenic mice. *Nature* **335**, 606–611.

Whiting, J., Marshall, H., Cook, M., Krumlauf, R., Rigby, P., Stott, D. and Allemann, R. (1992). Multiple spatially-specific enhancers are required to reconstruct the pattern of Hox2.6 gene expression. *Genes Dev.* **5**, 2048–2059.

Wilkie, T. M., Brinster, R. L. and Palmiter, R. D. (1986). Germline and somatic mosaicism in transgenic mice. *Dev. Biol.* **118**, 9–18

5
Genome imprinting

WOLF REIK

Department of Molecular Embryology, Institute of Animal Physiology and Genetics Research, Babraham, Cambridge CB2 4AT, UK

1. Introduction

In this chapter, properties of transgenes are examined that have – at least in the early days of transgenesis – mostly been ignored. The transgenes to be examined fall into the general class of transgenes that do not behave properly and, since for some time it was not clear whether there was any underlying biological significance to their misbehaviour, they have caused some degree of irritation and confusion. It was recognized, early on, that the same transgene construct integrated into different chromosomal environments could result in different expression phenotypes (reviewed by Jaenisch, 1988; Kothary *et al.*, 1989). It was not widely appreciated, however, that the same transgene construct in the same chromosomal location, that is, the same genetic locus, could show variability in expression. An early glimpse of this phenomenon was obtained by Palmiter *et al.* (1984), who showed that within one transgenic line carrying a metallothionein thymidine kinase construct there was wide variation of expression levels between different individuals of that pedigree. The factors responsible

TRANSGENIC ANIMALS
ISBN 0-12-304530-4

for this variation remained obscure at the time. Thus, in the most general sense the biological phenomenon that we are concerned with in this chapter is the ability of single genetic loci (transgenes in our case) to produce multiple phenotypes (e.g. different levels of expression in different individuals). This is illustrated in Fig. 1, where two fetuses of the same transgenic mouse strain have different expression levels of the transgene, in this case in response to unlinked modifying genes (see below).

Of course these observations remind us of a genetical phenomenon known for a long time, namely that a lot of mutations can have variable penetrance or expressivity when segregating in a pedigree. Not surprisingly, therefore, it has been suggested that variable expressivity of transgenes can be useful as a genetic model for variable expressivity of genetic traits, or for 'dominance modification' (Reik, 1989; Sapienza, 1989).

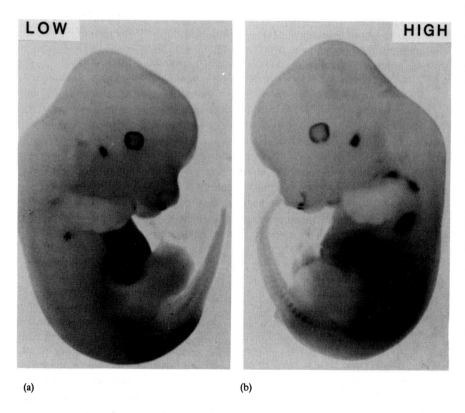

(a)   (b)

Fig. 1. High- and low-expressing fetuses of the TKZ751 strain. In this strain, a *lacZ*-containing trans-gene is expressed at different levels depending on genotype-specific modifier alleles. (From Allen *et al.* (1990).)

Interest in variability of transgene expression has been rekindled in recent years with the demonstration that patterns of DNA methylation and expression of transgenes could apparently be influenced by their inheritance from mother or father (reviewed by Surani et al., 1988). Parental inheritance was thus one of the factors involved in influencing phenotype variability. This observation has created particular interest since it demonstrates the molecular behaviour expected of parentally imprinted genes, whose existence was inferred from nuclear transplantation and genetic studies in the mouse. In the absence, until recently, of any knowledge of endogenous gene loci that undergo parental imprinting, imprinted transgenes have served as a model for both the molecular and the genetic components of imprinting.

In this chapter we examine some of the properties of imprinted transgenes as well as the more general aspects of variable expressivity. We ask how useful this approach is to the identification of some of the molecular and genetic factors involved in imprinting. We also address the question of whether transgenes are useful probes to identify endogenous gene loci that undergo imprinting.

2. Evidence for imprinting of the mammalian genome

The term 'imprinting' was first used by Helen Crouse (1960) to describe the peculiar behaviour of parental chromosomes in *Sciara*. Paternal autosomes and X chromosomes are inactivated and later selectively eliminated; this chromosome inactivation process determines sex, as all zygotes in which it occurs will develop into males. Parental chromosomes therefore seem to carry imprints indicative of their parent of origin. Importantly, however, the egg seems to be able to either recognize or to ignore these imprints, as the same female is able to produce both male and female offspring. In mammals the first evidence for chromosome imprinting came with the realization that, in the female mouse embryo, the paternal X chromosome is preferentially inactivated in the extra-embryonic tissues (Takagi and Sasaki, 1975; West et al., 1977). This is probably also true of the human embryo (Harrison, 1989). In marsupials, non-random X inactivation is also observed in somatic tissues (VandeBerg et al., 1978).

Two lines of experimentation show that in the mouse, autosomes are also imprinted. First, the lethality of parthenogenetic and androgenetic embryos, that is, of embryos that develop with maternal or paternal chromosomes only, is of nuclear origin (McGrath and Solter, 1984; Mann and Lovell-Badge; 1984, Surani et al., 1984). Secondly, uniparental disomy or chromosome duplication from one parent and loss from the other, of specific chromosomal segments, causes particular phenotypes in the offspring; the phenotypes of maternal and paternal disomy of the same region are sometimes complementary (Cattanach and Kirk, 1985; Searle and Beechey, 1985). Most of the chromosomes of the mouse have now been tested for parental imbalance in function in this way and a map of imprinted regions of the mouse genome has been compiled (Searle et al., 1989; Cattanach and Beechey, 1990). The imprinted regions presumably contain specific imprinted genes whose dosage causes the phenotypes observed in disomic mice. This, however, does not allow the conclusion that regions that are phenotypically neutral do not contain imprinted genes (Cattanach and Beechey, 1990). Many of the phenotypes that arise in parthenogenetic and andro-genetic cells and in disomic embroys seem to be associated with disturbances in embryonic growth, and so embryonic growth factors might be seen as prime candidates for imprinted genes.

The biological significance of having this distinction between maternal and paternal chromosome function in mammals is a matter of speculation (Solter, 1988; Sapienza, 1989; Hall, 1990; Surani et al., 1990a); an intelligent suggestion has recently been made based on the different interests that maternal and paternal genes should have in resource acquisition of the embryo from maternal tissues (Haig and Westoby, 1989; Moore and Haig, 1991).

In addition to the experimental evidence, coming mainly from the mouse, there are many other observations from genetic studies which suggest that imprinting is involved (reviewed by Monk, 1988; Solter, 1988; Reik, 1989; Sapienza, 1989; Hall, 1990). Perhaps the most striking ones concern pathological conditions in the human; the complete hydatidiform mole is a conceptus in the human that goes wrong, with very poor if any fetal development but often invasive proliferation of trophoblast tissue. Chromosome analysis of these concept uses shows a diploid androgenetic constitution, that is all chromosomes are paternally inherited (Lawler et al., 1982). Prader–Willi syndrome in the human can be caused by either a paternally inherited deletion of chromosome 15, or by a maternal disomy of chromosome 15 (Nicholls et al., 1989). These are very persuasive cases for imprinting in the human genome, but many other examples exist which include the sex-of-parent-specific phenotypes of mono-genetic disorders, as in Huntington's chorea where the juvenile onset form is frequently the result of paternal transmission of the mutation. They also extend to the non-random retention of parental chromosomes in recessive tumour syn-dromes, e.g. the preferential loss of maternal chromosomes in Wilms tumour and rhabdomyosarcoma. Further manifestations of imprinting may be the occur-rence of non-reciprocal phenotypes in crosses between different strains or dif-ferent species (discussed in more detail later).

All of these experiments and observations suggest that there are genes in the mammalian genome whose final level of expression is influenced by whether they have been contributed by the mother or father. It is likely that this involves 'epigenetic modifications' of these genes, with essentially two properties. First, that such modifications are introduced during gametogenesis or very early embryogenesis and thus are sex-of-parent-specific. Secondly, that they are clonally stable and influence gene activity, presumably at the level of transcription, at later stages of development. Often, when chromosome imprinting is referred to in a more general sense, it also includes stably modified phenotypic states that are not necessarily influenced by parental origin. Parental imprinting is thus a particular class of chromosome imprinting.

It may be appropriate at this point to clarify the terms 'modification' and 'modifying genes', as the use of them next to each other is sometimes confusing. The term 'modifying genes' is an expression with a long-standing tradition in genetics, describing changes of phenotype at one locus in response to alleles at a different locus, the modifier (Haldane, 1941). Distinct from this, the term 'modification' means epigenetic alterations of DNA, such as DNA methylation, that are usually heritable as well as reversible. It will be seen later that one way that 'modifying genes' can act is by affecting 'modification' of DNA.

3. Imprinting of transgenes

In the last section I tried to point out that the epigenetic state of genes, pieces of chromosomes or whole chromosomes can be determined by parental transmission. The types of heritable modifications to be considered are mainly DNA methylation, and variations in chromatin structure or compaction. So far, only the role of DNA methylation has been addressed experimentally, but chromatin compaction has recently attracted attention because of its involvement in position effect variegation in *Drosophila melanogaster*, and the overt similarities in genetic control between the two systems (see below).

In pursuit of the molecular determinants of imprinting, one must, in a normal individual, distinguish maternal and paternal chromosomes to then ask whether there are epigenetic modifications that mark the two chromosomes. Transgenes provide such polymorphisms by creating hemizygous loci in the mouse and, furthermore, allow for fairly non-subjective experimentation since their integration into the genome is assumed to be random. In addition, it is known that patterns of transgene expression and methylation can be influenced by the site of their insertion into the mouse genome (reviewed by Jaenisch, 1988) and, hence, it was thought that transgenes could provide positional probes to identify and, indeed, access imprinted regions in the genome.

There are several examples now where, indeed, DNA methylation and expression of transgenes is influenced by parental transmission (Table 1, Fig. 2). A wide variety of sequences can apparently undergo imprinting and it is estimated that around 10–20% of all transgenic strains will show signs of imprinting. The sequence of the transgene itself may have some influence as one out of seven strains carrying a CAT–IgH–SV40 construct showed imprinting (Reik *et al.*, 1987), but three out of four carrying a troponin I construct were imprinted

Table 1. Properties of imprinted transgenes

	Location	Imprinting	Influence of modifier loci	Methylation differences in male germ line	Mosaicism	References
CAT17		m > p	+	+		Reik et al. (1987, 1990)
Troponin I 379		m > p	+	−		⎱ Sapienza et al. (1987, 1989)
Troponin I		m > p				⎰
Troponin I		p > m				
RSV-myc		m > p				Swain et al. (1987)
HbsAg	Chr 13	m > p				Hadchouel et al. (1987)
TKZ751		m(B/c) > p(B/c)	+	+	−	Allen et al. (1990)
OX1-5	Chr 7	m > p	+			Reik et al. (1990)
MPA434	Chr 11, A5	m > p	−			Sasaki et al. (1991)
Tg4	Chr 1		+		+	McGowan et al. (1989)
CMZ12		m(B/c) > p(B/c)	+		+	Surani et al. (1990b)

m > p: maternal transmission of the transgene results in higher methylation than does paternal transmission. m(B/c) > p(B/c): maternal transmission of BALB/c modifier results in higher methylation of the transgene than does paternal transmission. Note that in the case of CMZ12 no differences in methylation are detected (see text for details).

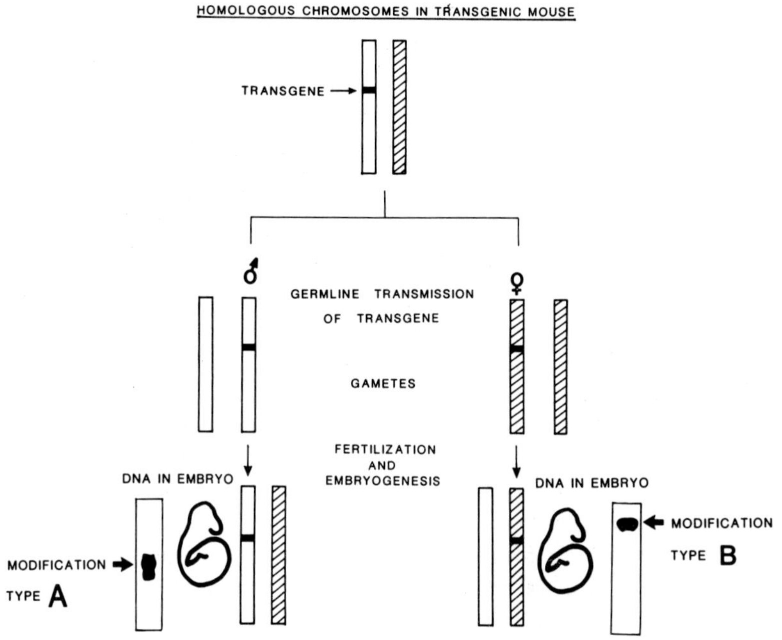

HOMOLOGOUS CHROMOSOMES IN TRANSGENIC MOUSE

Fig. 2. Maternal and paternal transmission of a transgene can result in different methylation phenotypes in the offspring. The transgene is maternally or paternally inherited; DNA from the offspring was subjected to analysis by restriction with methylation-sensitive enzymes and Southern blotting. Paternal transmission results in a lower level of methylation than does maternal transmission. Modification in this case refers to DNA methylation.

(Sapienza *et al.*, 1987). The RSV–*myc* construct studied by Swain *et al.* (1987) may also be particularly prone to becoming imprinted. However, there has to be an influence of the site of insertion into the genome since not all insertions of any one construct result in imprinting (but see, Chaillet, 1992). There are a number of possible explanations for the relatively high frequency of transgene imprinting, including the possibility that integration into the genome may be random. If so, insertion will occur at an appreciable frequency into or near heterochromatic regions and so subject the transgene to variegating position effects analogous to those observed in *Drosophila*.

Do transgenes that are imprinted map to the imprinted regions of the genome? Progress in determining chromosomal position has been slow, but there are at least two examples of imprinted transgenes that appear to lie outside the imprinted regions (Hadchouel *et al.*, 1987; Reik *et al.*, 1990). There is only one example of a transgene that lies firmly within an imprinted region (Sasaki *et al.*

1991). Hence, it appears possible that imprinting of transgenes can occur irrespective of whether or not they lie in imprinted regions as defined by uniparental disomy. Indeed, imprinting could possibly occur in a large number of locations in the genome but only cause visible phenotypes when dosage of the genes involved is of major importance to the development and survival of the animal. Supporting this contention, subtle deficiencies in complementation have recently been found with a number of chromosomes not previously thought to be imprinted (Cattanach and Beechey, 1990).

Another interesting pattern emerges from comparing all the available results on transgene imprinting. Mostly, with only one exception, it is with paternal transmission of the transgene that undermethylation (and in some cases expression) is associated, but maternal inheritance leads to repression and high levels of methylation (Table 1). Interestingly, with the two chromosome regions where reciprocal phenotypes are produced with maternal and paternal disomy it is the larger mice (Chromosome 11) and the hyperkinetic mice (Chromosome 2) that are characteristic of paternal duplication, suggesting that paternal transmission is associated with higher levels of expression of genes in these regions (Cattanach and Kirk, 1985). Hence there may be a preponderance of imprinted traits whose expression is accentuated by paternal derivation. But there must also exist imprinted traits that behave in the opposite way: the T^{hp} deletion, for example, results in a much more severe phenotype when maternally inherited than when paternally derived (Johnson, 1974, 1975). This suggests that there are genes on proximal Chromosome 17 whose expression is reduced when paternally contributed. This has recently been borne out by molecular analysis (see below). The possible preponderance of paternally overexpressed traits is explained by evoking different classes and numbers of 'imprinting genes' (Sapienza, 1990) (discussed below). Also, depending on the products that imprinted genes encode (e.g. placental growth factors), they can help or inhibit the transfer of nutrient resources from the mother to the fetus (Moore and Haig, 1991). Paternally derived genes are therefore expected to be expressed more highly if they encode products that *aid* the conceptus in resource acquisition from the mother, but are expected to be expressed at a lower level if they produce proteins that *inhibit* resource acquisition.

Detailed analyses of the precise regions and CpG sites of transgenes that experience methylation changes are lacking, mainly because such an analysis is not straightforward in multicopy transgene inserts. But recently an analysis of a single-copy imprinted transgene has been made (Sasaki *et al.*, 1991). The locus consisted of pUC vector, metallothionein promoter, and transthyretin coding sequences. Methylation changes were clearly confined to testable CpG sites in the vector and in the promoter sequences, but methylation was constant in the coding region (Sasaki *et al.*, 1991). It is interesting to note that the region that was affected has a higher CpG frequency than the coding region, thus perhaps forming a 'CpG island-like' structure that can, under appropriate circumstances, no longer be protected from becoming methylated (following maternal transmission in this case). In the case of another imprinted transgene, however, variable methylation occurred outside the transgene in a cellular HpaII recognition site (Reik *et al.*, 1990).

We mentioned earlier that transgene imprinting must be, at least in part, regarded as a position effect. What is the nature of this position effect? Studies have only begun to address this point but one interesting result has emerged. In an analysis of an imprinted integrant, Sasaki *et al.* (1991) have cloned the insertion site of the transgene and asked whether or not methylation differences exist between the parental chromosomes in the absence of the transgene, that is, in the wild-type situation. No methylation imprinting was detected in four different methylatable sites tested. This indicates that methylation imprints on the transgene are not simply a reflection of methylation imprinting in the region of integration. The situation might be more complex. Ignoring the possibility that methylation imprinting *does* occur in the native chromosomes, but at other sites than the CpG sites tested so far, we are left with the possibility that other types of epigenetic modifications exist on the parental chromosomes which lead to a differential methylation of the transgene. Such modifications could involve, for example, chromatin compaction as in position effect variegation in *Drosophila*, and it might be interesting to compare the DNAse I sensitivity of the parental chromosomes in the absence of the transgene. Alternatively, there may be no epigenetic differences in the parental chromosomes in their native state, but transgene insertion, together with the frequently encountered deletions and rearrangements of cellular DNA, may actually disturb a pre-existing chromatin domain and thus result in variable heterochromatization of the transgene. Finally, some particular transgene–mouse DNA arrangements may even be looked upon as nucleation sites for heterochromatin and hence create heterochromatic regions *de novo*. It might be rewarding to reinsert cloned imprinted transgenes into the mouse genome to see whether they become imprinted invariably and independently of position of insertion.

Whatever the precise mechanism of transgene imprinting, the notion appears to be emerging that what goes on in the transgene or in its immediate vicinity may not be an accurate picture of what goes on in the native chromosome region where the transgene is inserted. This contention is also supported by observations on transgenes integrated on the X chromosome, where repression of transgene expression when carried on the inactive X chromosome does not always occur (Krumlauf *et al.*, 1986; Goldman *et al.*, 1987).

Is methylation a primary imprinting signal or the consequence of a different primary mechanism? This question cannot be answered at present, but may best be approached by looking at the developmental changes of methylation on the transgene, particularly as they occur in germ cells. One of the critical points to be addressed is whether, at the time of fertilization, methylation differs in the sperm and the oocyte genome. In all imprinted transgenes where paternal transmission leads to undermethylation, methylation in sperm is as low as or lower than the somatic paternal transmission pattern. The male germ line pattern is usually independent of whether the transgene has been maternally or paternally inherited (by the male bearing the transgene). Also, when this was looked at, the germ line-specific pattern was already established at the spermatogonial stage of gametogenesis (Sasaki et al., 1991). There is only one analysis so far of methylation in the female germ line and during early embryogenesis (Rous Sarcoma Virus (RSV) myc; Chaillet et al. 1991), in which it was shown that in ovulated oocytes the transgene was highly methylated, and became so during oogenesis, presumably during growth or maturation of the oocyte. In addition, day 14 primordial germ cells of both sexes were neutral in that they were undermethylated irrespective of parental inheritance of the transgene (Chaillet, et al. 1991). Thus, the impression gained from this analysis is that methylation differences are being established during the later stages of gametogenesis and, in particular, that de novo methylation occurs in oogenesis. Following fertilization, these differences are propagated in the same or in similar form in the somatic tissues, but are lost in germ cells of either sex.

It is interesting to compare this pattern to the methylation of endogenous sequences in oocytes and sperm, where it appears that the oocyte genome is relatively undermethylated in comparison to the sperm (Monk et al., 1987; Sanford et al., 1987). However, we have recently found that some endogenous sequences become de novo methylated during later stages of oogenesis, and that germinal vesicle stage and unfertilized oocytes have a high level of methylase protein (Howlett and Reik, 1991). Differential methylation during gametogenesis must therefore be regarded as a potential component of any imprinting process.

There may, however, be transgenes that behave differently. The CMZ-12 transgene is expressed during preimplantation development with a variegated phenotype and the expression level is influenced by parental inheritance of a modifier gene (Surani et al., 1990b) (see next section). Preliminary analysis indicates that where genotypes with different expression levels are compared, the transgene does not differ in methylation. Moreover, expression at postimplantation stages is not influenced by parental origin or genotype (Surani et al., 1990b), suggesting that whatever epigenetic mechanism is responsible for the preimplantation phenotype, it is relatively short lived. Interestingly, while the imprinting of the RSV myc transgenic strain seems to be relatively independent of genetic constitution and persists on an essentially homozygous background (R. Chaillet, personal communication), the imprinting of CMZ-12 is only observed with heterozygous combinations of modifier alleles and is therefore a postfertilization process. This important difference will be more fully elaborated in the following section.

4.   Genetic control of imprinting: modifier genes and imprinting genes

The genetic control of transgene imprinting has received a lot of attention over the last 2 years, obviously because it is thought that, by identifying genetic factors, and ultimately genes, that influence imprinting, substantial insight into the molecular mechanism of imprinting may be gained. The general observation is that genes exist that influence the methylation and expression phenotypes of transgene loci. Thus, in the case of the troponin I379 transgene, paternal transmission of the transgene will result in two different phenotypes in the offspring, depending on maternal genotype (Sapienza et al., 1989). A high-methylation phenotype is observed when the mother contributes a C57BL/6 modifier allele, as compared with a lower-methylation phenotype when the mother contributes a DBA/2 modifier allele. In this case, the different degrees of methylation of the paternally derived alleles must be brought about by the action of modifier alleles after fertilization. Similar observations have been made with other transgene loci (Reik et al. 1990).

How imprinting can be brought about by modifier alleles is best illustrated by the behaviour of the transgene locus TKZ751 (Allen et al., 1990). Crossing the transgene into the BALB/c strain produced a high-methylation type and crossing it into the DBA/2 strain resulted in a low-methylation phenotype (methylation polymorphism). This was inversely correlated with expression levels, hence the expressivity of the transgene locus is controlled by modifiers, different alleles of which segregate in inbred strains of mice (Figs 1 and 3; the same is true of the Tg4 locus, see Table 1). In an essentially homozygous background, whether BALB/c or DBA/2, the transgene phenotype was not influenced by parental transmission (Fig. 3(a)). However, when a TKZ–DBA (low-methylation) male is crossed with a BALB/c female, an increase in methylation is seen in the offspring while in the offspring of the reciprocal cross methylation stays low (Fig. 3(b)). Therefore, differences in phenotype are observed as a result of reciprocal crosses between the two genotypes (see also Fig. 4). Clearly, the phenotype of the transgene is not dependent upon its own parental origin, but is dependent upon the parental origin of its modifying locus (Fig. 3). A maternally derived BALB/c allele at this locus (or at these loci, for it is at present unclear how many there are) will increase methylation of the transgene in its bearer. As pointed out before, this will only result in parental origin effects in a heterozygous, or segregating, population.

The maternally derived BALB/c allele that increases methylation can either be a nuclear gene whose expression is parent-specific, or the activity can result from nucleocytoplasmic interactions involving BALB/c egg cytoplasm *and* a BALB/c nuclear gene. Further experiments show that, at least in the case of the transgenic strain TKZ751, it is likely that the cytoplasm is involved and that

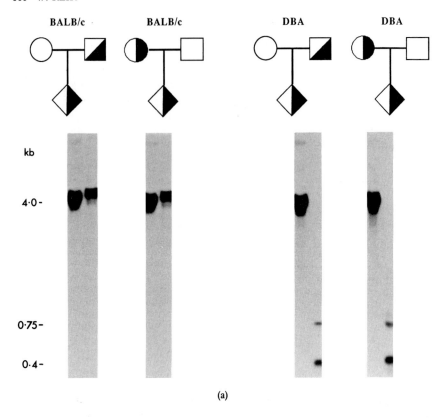

(a)

Fig. 3. Genetic control of imprinting at the TKZ751 locus. (a) Within a BALB/c background, both maternal and paternal transmission of the transgene result in high methylation (all DNAs were digested with SacI, first lane, and with SacI and HpaII, second lane). In a DBA background, both modes of transmission result in a low-methylation phenotype.

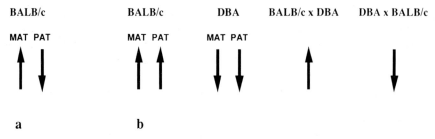

Fig. 4. Two different types of transgenic imprinting. (a) On a homozygous background (inbred over several generations), maternal and paternal transmission result in different methylation phenotypes of the transgene in offspring (↑, high methylation; ↓, low methylation). (b) On a homozygous background, there is no effect of parental origin on methylation phenotype of the transgene. However,

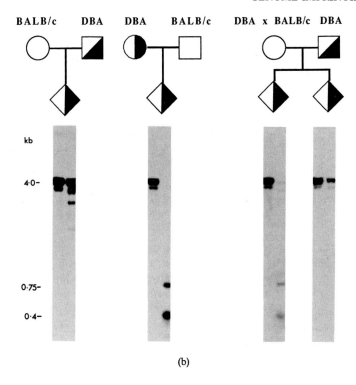

Fig. 3. (b) Reciprocal crosses between the two genotypes lead to different methylation types, with high methylation of the transgene when BALB/c is the maternal genotype. When the mother is a hybrid between BALB/c and DBA, the maternally inherited BALB/c allele segregates in the offspring, resulting in progeny with high-methylation types and progeny with low-methylation types. (Adapted from Allen *et al.* (1990).)

it can successfully interact with either a maternally or paternally derived chromosomal BALB/c allele (Surani *et al.*, 1990b). There are overt similarities with the DDK mutant in the mouse, in which non-reciprocal lethality is the result of a cross between DDK females and 'alien' (other inbred strain) males (Wakasugi, 1974). Studies have shown that this phenotype is the result of an early interaction of DDK-type cytoplasm with a nuclear gene from other inbred

---

the methylation phenotypes are different in response to different backgrounds ('methylation poly-morphism'). When reciprocal crosses between the different background strains are carried out, different methylation phenotypes are observed at the transgene locus. This indicates that maternal and paternal transmission of a modifier locus controls the methylation phenotype at the transgene locus. (From Reik *et al.* (1990).)

strains, but this interaction only occurs when the nuclear allele is *paternally* inherited (Wakasugi, 1974; Babinet *et al.*, 1990). So here we seem to have both a cytoplasmic influence *and* the effect of parental imprinting of a chromosomal gene. We are led to believe that at some point following fertilization, egg cytoplasm and both sets of chromosomes interact to set up the genome for future function. Any aberrant interaction at this stage, perhaps induced by bringing a 'wrong' combination of modifier alleles together, will be unhealthy. We believe that this mechanism may underlie the many problems seen in interspecific hybridizations once fertilization has taken place, so-called postzygotic isolation mechanisms. Seen as an early step in speciation, postzygotic isolation is often non-reciprocal, producing sterile or non-viable offspring in one, but not in the reciprocal, cross (Charlesworth *et al.*, 1987; Orr and Coyne, 1989). Indeed, it has been shown in *Drosophila* that such phenotypes are the result of aberrant nucleocytoplasmic interactions in the hybrids (Orr, 1989). Also, there are many genetic traits that show differences in phenotype in reciprocal crosses involving different strains within the same species that could be brought about by this mechanism (reviewed by Shire, 1989). A well-known example of the possible 'variability' of imprinting is the T^{hp} deletion in the mouse, where phenotypes are produced with both maternal and paternal transmission, but maternal transmission results in a much more severe phenotype, usually death at fetal stages. However, the time of death is highly variable and dependent on modifier genes, and the strain can be selected so that occasional fetuses with maternal transmission of the mutation survive (Winking, 1986).

As pointed out before, there are also transgenic strains in which the genetic background has very little or no effect on the imprinting of the transgene (Table 1). Hence, imprinting persists on essentially homozygous or inbred backgrounds (Fig. 4). Clearly, this imprinting cannot be caused by the mechanism described above, but these transgenes behave according to their own parental origin in an invariant way. This is interesting because we have to deduce from observations on parthenogenetic and androgenetic embryos that a class of imprinted genes exist whose function is essentially independent of genetic background and exclusively governed by parental origin. This is because parthenogenetic embryos made with many different inbred strains show similar phenotypes (Surani *et al.*, 1990a), and the phenotype of parthenogenetic cells in chimeras is largely independent of genetic background (R. Fundele, personal communication). In addition, while the phenotype of some monoparental disomies seems to be influenced by background genes, others appear to be constant (Cattanach and Beechey, 1990). The genetic control of this invariant type of imprinting, both for transgenes and for any putative endogenous imprinted genes, remains to be unravelled. Some interesting suggestions have been made, however, and we will turn to these in the next section.

5.   The nature of modifiers and imprinting genes

While the impression may be gained that a lot of what is said in this chapter is speculative, this is clearly the most speculative of all sections and very little is substantiated by experimental evidence. We emphasized before that although methylation is a useful indicator of phenotype, so far there is no evidence for or against methylation being a primary imprinting mechanism. However, it is possible that modifier genes represent different alleles of methyltransferase genes, or different alleles that code for proteins that bind to and modify the activity of methyltransferase. It is difficult to imagine how specificity could be regulated in this system, although there are suggestions that activity of methylase versus replication timing of genes during the cell cycle could introduce specificity. But other epigenetic modifications that are clonally stable must also be considered – they include dosage-dependent chromatin assembly processes that lead to variable heterochromatization, thought to play a role in, for example, position effect variegation in *Drosophila* (reviewed by Eissenberg, 1989; Tartof and Bremer, 1990). There are overt similarities between transgene imprinting and position effect variegation, where usually a gene whose phenotype can be monitored is placed (by a translocation) near heterochromatin and is thus variably expressed. The extent of variegation can be influenced by parental transmission (Spofford, 1976; Kuhn and Packert, 1988). There are also examples of variegating chromosome arrangements in the mouse (Cattanach and Perez, 1970). Often expression varies from cell to cell in a mosaic fashion and overall expression phenotype is controlled by the extent of variegation. Interestingly, a large number of gene loci in the *Drosophila* genome have been identified, variant alleles of which modify the phenotypic response (Locke et al., 1988). Some of these modifier genes have recently been isolated and shown to encode proteins that possibly participate in the formation of heterochromatic domains and chromatin compaction. A mass action model based on cooperative binding effects between various products of modifier genes has been put forward (reviewed in Tartof and Bremer, 1990).

With these concepts in mind, Laird (1990) has suggested a possible explanation for the parental origin effect in Huntington's disease, and Sapienza (1990) has suggested a general model for how 'imprinting' genes may act to imprint transgenes and endogenous genes. Both models are based on the notion of having modifier genes on the sex chromosomes, in particular on the X chromosome, and thus having a dosage difference at those modifier genes when male and female gametes are compared. Hence, autosomal loci become differently imprinted during male and female gametogenesis because of this dosage difference, and their imprints, that is, their degree of chromatin compaction, is heritable for many cell generations. The Sapienza model also suggests that there exist many more enhancers of variegation than suppressors of variegation, which could explain why imprinted transgenes are often more methylated and less highly expressed following maternal inheritance. DNA methylation differences observed at relatively late stages of embryogenesis would simply be a consequence of differences in chromatin compaction at earlier stages.

This model is appealing for its simplicity, but there is one major criticism that can be applied. If all the imprinting is carried out by dosage-sensitive modifiers on the X chromosome, the imprinting status of autosomes in a XY gamete should be equivalent to that of the autosomes in an XO gamete. That means that offspring of XO females should resemble closely androgenetic embryos, because they should have two paternally imprinted genomes despite one of them originating in an oocyte. This is not the case. Although XO females in the mouse are of reduced fertility, they can produce normal-looking offspring, albeit at reduced numbers (Burgoyne and Biggers, 1976). Hence, the imprinting status of a genome originating from an XO gamete is probably different from a genome originating from an XY gamete.

Interestingly, a sequence motif of a *Drosphila* heterochromatin protein has recently been found in a family of genes in mouse and humans (Singh *et al.*, 1991). Some members of this family map to the X chromosome (Hamvas *et al.*, 1992).

6. Modifications that involve the germ line: epigenetic inheritance

We pointed out earlier that, under normal circumstances, germ line passage of any imprinted trait is expected to 'neutralize' the imprints, and to re-establish new imprints that are specific for the sex of the carrier. However, this is not always the case. If the germ line fails to erase all epigenetic modifications, grandparental effects and, indeed, multigenerational influences may be witnessed (Jablonka and Lamb, 1989; Surani *et al.*, 1990a). These have some interesting implications for how natural selection may act on modifier and modified alleles. Whenever more phenotypes at a transgene locus are observed than can be explained by the number of different modifier alleles in the population, it must be suspected that the target gene is epigenetically marked in the germ line. For example, low-somatic-methylation phenotype males of the troponin I 379 strain when crossed with B6 and DBA females, will give rise to intermediate-methylation and low-methylation progeny, respectively, whereas high-somatic-methylation phenotype males will give rise to high-, and intermediate-, methylation progeny in the same crosses (Sapienza *et al.*, 1989). Hence, an epigenetic difference must exist at the transgene locus in germ cells of males with different somatic phenotypes. This epigenetic modification serves as a template for the modifier genes to act upon in the next generation. The simplest explanation is that the transgene is not only modified in somatic cells, but also in germ cells,

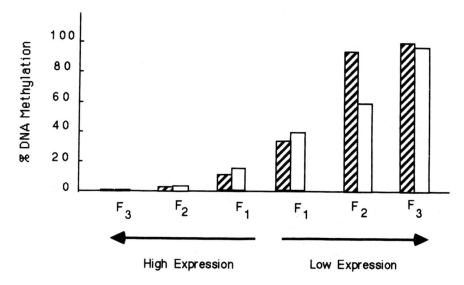

Fig. 5. Cumulative changes in DNA methylation over several generations involve the germ line. Methylation can increase or decrease over several generations in somatic tisues of the TKZ751 strain, depending on the inbred strain used for the breeding. Methylation in sperm is increased or decreased in parallel to the somatic changes, suggesting that the methylation phenotype depends both on genotype-specific modifiers and on the epigenetic state of the gene as it is inherited. (From Surani *et al.* (1990a).)

for example, by an early methylation event before allocation of the germ cell lineage, and subsequent failure of germ cells to erase this imprint. While no apparent methylation differences were observed in germ cells of low- and high-methylation phenotype males in the troponin I 379 pedigree, such differences have been seen in males of the TKZ751 and of the CAT17 pedigree (Figs 5 and 6) (Allen *et al.*, 1990; Reik *et al.*, 1990; Surani *et al.*, 1990a). Hence, in these cases the epigenetic modification that persists in the germ line to produce different phenotypes in the next generation appears to be methylation itself. This can lead to cumulative changes in phenotype over several generations, and may be manifest as grandparental effects. Interestingly, grandparental effects have been observed in the expression of the parental origin effect in Huntington's chorea (Ridley *et al.*, 1988). Also, the observation of 'anticipation', that is, of earlier and earlier onset phenotypes of some genetic diseases over several generations, may be explained by the cumulative loss or gain of epigenetic modification (Höweler *et al.*, 1989).

Modifier genes can possibly act on a large number of loci and so collective changes in phenotype are possible with a small number of mutations that

Sperm

low   high

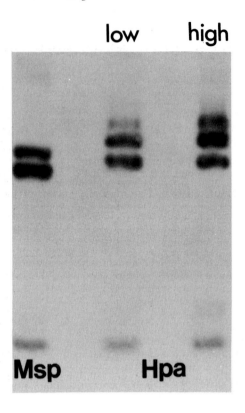

Fig. 6. Methylation of the CAT17 locus in sperm DNA is dependent on the somatic methylation phenotype. Methylation was assayed by HpaII digestion and was compared between two carrier males who have low and high somatic methylation, respectively. Note the subtle difference in methylation between the two types of sperm DNA. (From Reik *et al.* (1990).)

produce new alleles at modifier loci. This can lead to rapid adaptive changes, as envisaged by Gould: 'Instead, I envisage a potential saltational origin of the essential features of key adaptions' (Gould, 1980). If all epigenetic information were lost in the germ line, natural selection for phenotypes produced by specifically modified alleles would select for mutations in modifier loci. However, experiments show that some epigenetic modifications are stable over several generations without being subject to change by different modifier alleles (Hadchouel *et al.*, 1987; Allen *et al.*, 1990). Indeed, particular genes may be 'locked' into certain epigenetic states in an irreversible fashion, a process that has been termed 'epimutation' (Holliday, 1987). Clearly, under these circumstances natural selection would operate on the modified allele itself, instead of on the modifier gene.

7. Mechanism of expression of multiple phenotypes

Different phenotypes of methylation or expression of a transgene can in principle be the result of two different mechanisms that can be used exclusively or in combination. First, the transgenome can have different states of activity in every cell of a tissue under examination (Fig. 7). Secondly, the tissue can be composed of varying numbers of cells that themselves have different states of activity (cellular mosaicism) (Fig. 7). This type of phenotype control is well known from the study of position effect variegation in flies (reviewed by Spofford, 1976; Eissenberg, 1989; Tartof and Bremer, 1990). The classical example of variegating arrangements of the white locus shows that the eye of adult flies is a mosaic of cells that either express or do not express the white gene. Active or inactive phenotypes of individual cells are possibly determined early in development, at around the blastoderm stage. These states of activity are then clonally inherited upon cell division and so clonally related 'patches' of expression phenotypes are

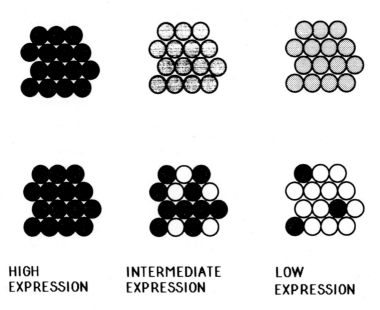

HIGH   INTERMEDIATE   LOW
EXPRESSION   EXPRESSION   EXPRESSION

Fig. 7. Models of variable expressivity. Different levels of collective gene expression in a tissue can be achieved by gradual changes of expression in all cells, or by variable composition of the tissue containing cells that either express or do not express the gene (mosaicism).

observed in adult tissues. Mosaic phenotypes are also observed in some experimental situations in yeast (Pillus and Rine, 1989; Gottschling *et al.*, 1990). Similarly, some transgenes in the mouse show mosaic patterns of expression. Indeed, genetic control of expression phenotype has been shown to involve the variable composition of a tissue from 'phenotypic units', that is, cells that are either in an 'on' state or an 'off' state of expression (McGowan *et al.*, 1989). Methylation mosaicism has also been demonstrated in endogenous sequences, and can be under genetic control by modifier alleles (McGowan *et al.*, 1989; Reik *et al.*, 1990). Recently, the mosaic expression of an endogenous gene has been demonstrated in chick embryos (Stern *et al.*, 1990). It has been suggested that a mosaic pattern of expression or methylation may be established at an early stage of embryogenesis in the mouse (McGowan *et al.*, 1989). If expression patterns are clonally stable from that stage, the final composition of any one tissue of expressing and non-expressing cells will be determined by cell sampling, mixing and clonal growth during development. It will thus be interesting to compare patterns of mosaic patches to patches arising in chimeras made at an early developmental stage by aggregating or injecting cells of discernible genotypes (McLaren, 1976). This comparison may shed some light on how and when mosaicism is generated.

Mosaic inactivation of genes can have particularly dramatic phenotypic consequences when tumour suppressor genes are involved. This can increase greatly the chances of a cell to become a cancer cell. A parental transmission-specific mosaic inactivation of these genes can indeed explain the non-random loss and retention of parental chromosomes in certain cancers (Reik and Surani, 1989; Scrable *et al.*, 1989; Ferguson-Smith *et al.*, 1990).

Not all transgenes with variable expressivity are expressed in a mosaic way. For example, the TKZ751 transgenic insert, carrying a *lacZ in situ* marker, is expressed at higher or lower levels in all cells in which expression can occur (Allen *et al.*, 1990). Thus, a graded response at the level of the cell to modifiers is produced here. This may be achieved by switching on or off variable numbers of the tandemly repeated multicopy transgene.

It remains to be seen how variable phenotypes are expressed in normal (non-mutated) cellular genes. Some of the mechanisms described for transgenes may only apply to variegation-type arrangements, either involving abnormal transgene arrangements, or abnormal arrangements of cellular genes with heterochromatin.

8.   Endogenous imprinted genes

I pointed out before that the original promise of the transgene experiments seemed to be the ability to access imprinted regions and to find and clone imprinted genes. For all the reasons given above, this seems to be less straight-forward than it first appeared. For one reason, the number of imprinted genes can theoretically be very small. A minimum estimate can be obtained from the number of imprinted regions of the mouse genome, since it is possible that within any one region there exists only one imprinted gene. Also, in the mouse there exists only one clear-cut parental effect mutation (T^{hp}), although other loci can show parental effects when background genes are heterozygous and segregate (Reik, 1989). However, since at least three endogenous genes have recently been found to be imprinted, we can look at how they have been iden-tified and whether or not there is any promise for a general strategy of finding imprinted genes.

The first one was identified accidentally. A mutation was made by homologous recombination in the mouse insulin-like growth factor II (Igf) gene. When pater-nally transmitted, this mutant (null) Igf2 allele had a phenotypic effect, in that offspring that carried this allele were smaller than their normal littermates. Sur-prisingly, transcription of the maternally inherited wild-type Igf2 allele was not 50% of normal diploid as expected, but was greatly reduced (de Chiara *et al.*, 1990). Hence, it appears that maternal transmission of the Igf2 gene leads to at least partial repression of transcription (DeChiara *et al.*, 1991; Ferguson-Smith *et al.*, 1991).

Near the Igf2 gene, inseparable by recombination and perhaps very close to it, is another imprinted gene, H19 (Bartolomei *et al.*, 1991). Interestingly, this gene whose protein product is at present unknown is imprinted in the opposite way. Hence, it is expressed from the maternal chromosome but repressed on the paternal one.

The third gene was found by asking what genes map to the T^{hp} deletion that, as pointed out before, shows transmission-specific phenotypes in the offspring. Amongst other genes, the *receptor* for Igf2 maps to this region and is only expressed when carried on the maternally derived chromosome 17. However, other genes in the deletion may also show parental dosage, and so it is not clear whether or not the Igf:2r gene is responsible for the T^{hp} phenotype (Barlow *et al.*, 1991). It is intriguing, however, that ligand and receptor appear to be under opposite imprinting influence. We note that if the main function of the receptor is sequestration of the ligand, rather than signal transduction, ligand and receptor have somewhat opposing functions. Following the resource acquisi-tion model of imprinting (Moore and Haig, 1991), opposite imprinting of ligand and receptor is not entirely unexpected.

So far, none of the approaches to identifying imprinted genes have been systematic, so the quest for a general strategy remains. Two potential approaches come to mind immediately: differential cDNA libraries, and differential protein expression as monitored by two-dimensional protein electrophoresis. Both methods can be applied to comparisons of parthenogenetic, androgenetic, normal, and all types of disomic embryos. Indeed, some early studies suggested the possibility of differences in some proteins between parthenogenetic and normal embryos (Petzoldt and Hoppe, 1980). cDNA subtraction cloning is being tried in a number of laboratories, but it must be borne in mind that differences in gene expression may be quantitative rather than absolute, and that dosage differences may be difficult to pick up with this method. In that respect, looking for differences in protein expression seems to be more promising, although it is by no means trivial to clone the genes that are responsible for differences in expression. However, we have recently found proteins that are differently expressed in reciprocal crosses between inbred strains of mice, and have identified some of the genes involved (J. Klose, I. Gurtmann, T. Vogel and W. Reik, unpublished).

9. Conclusions

In a remarkable study published in 1978, Searle and Beechey pointed out 'the possibility that haploid expression of particular maternal or paternal genes is important for normal mouse development'. This is all the more remarkable since at the time the view prevailed that uniparental diploid mice could develop normally (Hoppe and Illmensee, 1977). However, in the 10 or so years since Searle and Beechey's and also Lyon and Glenister's (1977) suggestion, the notion of genes that are expressed according to parental origin has become firmly established in mammalian biology. Indeed, we are now witnessing the identification of the first imprinted genes in the mouse, and characterization of 'imprinting' genes may also not be far distant.

The development of these concepts has clearly been aided by studies in transgenic mice. Transgenic mice have first provided molecular evidence for imprinting, and have shown that DNA methylation may be one of the components involved in imprinting and epigenetic inheritance, although other epigenetic mechanisms, particularly those involved in chromatin compaction, are also being implicated. Transgenic studies have shown how multiple methylation and expression phenotypes can be produced at single loci, and that modifier genes, different alleles of which are present in inbred strains of mice, must be involved in controlling epigenetic marking of chromosomes and genes. Transgenes will continue to be useful in the identification and characterization of some of these modifier loci. Importantly, it has been noticed that at least two different types of transgenic imprinting exist: one is retained on an essentially homozygous background, while the other requires heterozygosity at some modifying loci in the genome and is observed as differences in phenotype in reciprocal crosses. It is likely that these two types of imprinting also occur in endogenous genes.

Transgenic studies will also be extremely valuable in trying to understand variegating chromosome arrangements in mammals. For example, a transgenic mouse strain has recently been identified where a dominant skeletal malformation phenotype has been caused by transgene insertion. Interestingly, the mutant phenotype is only expressed following paternal transmission of the transgene (DeLoia and Solter, 1990). Apart from the possibility that an endogenous imprinted gene has been mutated, it is also possible that a variegating arrangement of the transgene causes an endogenous gene to be regulated in an inappropriate way. This would provide an interesting model for a number of human genetic disorders where it is thought that the genetic lesion may be at some distance from the gene whose inappropriate expression causes the phenotype (Laird, 1990).

Finally, it is not yet clear whether transgenes are useful probes to identify endogenous imprinted genes or whether other, more general, approaches will be more powerful.

Acknowledgements

I thank Azim Surani, Anne Ferguson-Smith, Sarah Howlett, Nick Allen and Cristina Rada for many discussions and helpful comments on the manuscript. I also thank Richard Chaillet and Reinald Fundele for communication of results prior to publication. The author is a fellow of the Lister Institute of Preventive Medicine. Work in the authors' laboratory is also supported by the AFRC and Combat Huntington's Chorea.

References

Allen, N. D., Norris, M. L. and Surani, M. A. H. (1990). Epigenetic control of transgene expression and imprinting by genotype-specific modifiers. *Cell* **61**, 853–861.

Babinet, C., Richoux, V., Guenet J.-L. and Renard, J.-P. (1990). The DDK inbred strain as a model for the study of interactions between parental genomes and egg cytoplasm in mouse preimplantation development. *Development* Suppl., 81–88.

Barlow, D. P., Stoger, R., Herrmann, B. G., Saito, K. and Schweifer, N. (1991). The mouse Igf-II receptor maps to the *maternal-effect* locus on chromosome 17 and is expressed only from the maternally inherited chromosome. *Nature* **349**, 84–87.

Bartolomei, M. S., Zemel, S. and Tilghman, S. M. (1991). Parental imprinting of the mouse H19 gene. *Nature* **351**, 153–155.

Burgoyne, P. S. and Biggers, J. D. (1976). The consequences of X-dosage defiency in the germ-line: impaired development *in vitro* of pre-implantation embroys from XO mice. *Dev. Biol.* **51**, 109–117.

Cattanach, B. M. and Beechey, C. V. (1990). Autosomal and X-chromosome imprinting. *Development* Suppl., 63–72.

Cattanach, B. M. and Kirk, M. (1985). Differential activity of maternally and paternally derived chromosome regions in mice. *Nature* **315**, 496–498.

Cattanach, B. M. and Perez, J. N. (1970). Parental influence on X-autosome translocation-induced variegation in the mouse. *Genet. Res.* **15**, 43–53.

Chaillet, J. R. (1992). DNA methylation and mammalian genomic imprinting. In: Surani, M. A. and Reik, W. (eds). *Genetic imprinting in mice and man. Seminars in Developmental Biology.* W. B. Saunders, Philadelphia (in press).

Chaillet, J. R., Vogt, T. F., Beier, D. R. and Leder, P. (1991). Parental specific methylation of an imprinted transgene is established during gametogenesis and progressively changes during embryogenesis. *Cell* **66**, 77–84.

Charlesworth, B., Coyne, J. A. and Barton, N. H. (1987). The relative rates of evolution of sex chromosomes and autosomes. *Am. Nat.* **130**, 113–146.

Crouse, H. V. (1960). The controlling element in sex chromosome behaviour in *Sciara*. *Genetics* **45**, 1429–1443.

DeChiara, T. M., Efstratiadis A. and Robertson, E. J. (1990). A growth deficiency phenotype in heterozygous mice carrying an insulin-like growth factor II gene disrupted by targetting. *Nature* **345**, 78–80.

DeChiara, T. M., Robertson, E. J. and Efstratiadis, A. (1991). Parental imprinting of the mouse insulin-like growth factor II gene. *Cell* **64**, 849–859.

Deloia, J. A. and Solter, D. (1990). A transgene insertional mutation at an imprinted locus in the mouse genome. *Development* Suppl., 73–80.

Eissenberg, J. C. (1989). Position effect variegation in *Drosophila*: towards a genetics of chromatin assembly. *BioEssays* **11**, 15–17.

Ferguson-Smith, A. C., Cattanach, B. M., Barton, S. C., Beechey, C. V. and Surani, M. A. (1991). Embryological and molecular investigations of parental imprinting on mouse chromosome 7. *Nature* **351**, 667–670.

Ferguson-Smith, A., Reik, W. and Surani, M. A. (1990). Genomic imprinting and cancer. *Cancer Surveys* **9**, 487–503.

Goldman, M. A., Stokes, K. R., Idzerda, R. L., McKnight, C. S., Hammer, R. E., Brinster, R. L. and Gartler, S. M. (1987). A chicken transferrin gene in transgenic mice escapes X-chromosome inactivation. *Science* **236**, 593–595.

Gottschling, D. E., Aparicio, O. M., Billington, B. L. and Zakian, V. A. (1990). Position effect at *S. cerivisiae* telomeres: reversible repression of Pol II transcription. *Cell* **63**, 751–762.

Gould, S. J. (1980). Is a new and general theory of evolution emerging? *Paleobiology* **6**, 119–130.

Hadchouel, M., Farza, H., Simon, D., Tiollais, P. and Pourcel, C. (1987). Maternal inhibition of hepatitis B surface antigen gene expression in transgenic mice correlates with *de novo* methylation. *Nature* **329**, 454–456.

Haig, D. and Westoby, M. (1989). Parent-specific gene expression and the triploid endosperm. *Am. Nat.* **134**, 147–155.

Haldane, J. B. S. (1941). The relative importance of principal and modifying genes in determining some human diseases. *J. Genet.* **41**, 149–157.

Hall, J. G. (1990). Genomic imprinting: Review and relevance to human diseases. *Am. J. Hum. Genet.* **46**, 857–873.

Hamvas, R. M. J., Reik W., Gaunt, S. J., Brown, S. D. M. and Singh P. B. (1992). Mapping of a mouse homolog of a heterochromatin protein gene to the X chromosome. *Mamm. Genome* **2**, 72–75.

Harrison, K. B. (1989). X-Chromosome inactivation in the human cytotrophoblast. *Cytogenet. Cell Genet.* **52**, 37–41.

Holliday, R. (1987). The inheritance of epigenetic defects. *Science* **238**, 163–171.

Hoppe, P. C. and Illmensee, K. (1977). Microsurgically produced homozygous-diploid uniparental mice. *Proc. Natl Acad. Sci. USA* **74**, 5657–5661.

Höweler, C. J., Busch, H. F. M., Geraedts, J. P. M., Niermeijer, M. F. and Staal, A. (1989). Anticipation in myotonic dystrophy: fact or fiction? *Brain* **12**, 779–797.

Howlett, S. K. and Reik, W. (1991). Methylation levels of maternal and paternal genomes during preimplantation development. *Development* **113**, 119–127.

Jablonka, E. and Lamb, M. (1989). The inheritance of acquired epigenetic variations. *J. Theor. Biol.* **139**, 69–83.

Jaenisch, R. (1988). Transgenic animals. *Science* **240**, 1468–1474.

Johnson, D. R. (1974). Hairpin-tail: a case of post-reductional gene action in the mouse egg? *Genetics* **76**, 795–805.

Johnson, D. R. (1975). Further observations on the hairpintail (T^{hp}) mutation in the mouse. *Genet. Res.* **24**, 207–213.

Kothary, R. K., Allen, N. D. and Surani, M. A. H. (1989). Transgenes as molecular probes of mammalian developmental genetics. In: Maclean, N. (ed.) *Oxford Surveys on Eukaryotic Genes*, pp. 145–178. Oxford University Press, Oxford.

Krumlauf, R., Chapman, V. M., Hammer, R. E, Brinster, R. and Tilghman, S. M. (1986). Differential expression of α-fetoprotein genes on the inactive X chromosome in extraembryonic and somatic tissues of a transgenic mouse line. *Nature* **319**, 224–226.

Kuhn, D. T. and Packert, G. (1988). Paternal imprinting of inversion Uab^1 causes homeotic transformations in *Drosophila*. *Genetics* **118**, 103–107.

Laird, C. D. (1990). Proposed genetic basis of Huntington's disease. *Trends Genet.* **6**, 242–247.

Lawler, S. D., Povey, S., Fisher, R. A. and Pickthal, V. J. (1982). Genetic studies on hydatidiform moles. II. The origin of complete moles. *Ann. Hum. Genet.* **46**, 209–222.

Locke, J., Kotarski, M. A. and Tartof, K. D. (1988). Dosage-dependent modifiers of positional effect variegation in *Drosophila* and a mass action model that explains their effect. *Genetics* **120**, 181–198.

Lyon, M. F. and Glenister, P. (1977). Factors affecting the observed number of young resulting from adjacent-2 disjunction in mice carrying a translocation. *Genet. Res.* **29**, 83–92.

McGowan, R., Campbell, R., Peterson, A. and Sapienza, C. (1989). Cellular mosaicism in the methylation and expression of hemizygous loci in the mouse. *Genes Dev.* **3**, 1669–1676.

McGrath, J. and Solter, D. (1984). Completion of mouse embryogenesis requires both the maternal and paternal genomes. *Cell* **37**, 179–183.

McLaren, A. (1976). *Mammalian Chimearas*. Cambridge University Press, Cambridge.

Mann, J. R. and Lovell-Badge, R. H. (1984). Inviability of parthenogenones is determined by pronuclei, not egg cytoplasm. *Nature* **310**, 66–67.

Monk, M. (1988). Genomic imprinting. *Genes Dev.* **2**, 921–925.

Monk, M., Boubelik, M. and Lenhart, S. (1987). Temporal and regional changes in DNA methylation in the embryonic, extraembryonic and germ cell lineages during mouse development. *Development* **99**, 371–382.

Moore, T. and Haig, D. (1991). Genomic imprinting in mammalian development: a parental tug-of-war. *Trends Genet.* **7**, 45–49,

Nicholls, R. D., Knoll, J. H. M., Butler, M. G., Karam, S. and Lalande, M. (1989). Genetic imprinting suggested by maternal heterodisomy in non-deletion Prader Willi syndrome. *Nature* **342**, 281–285.

Orr, H. A. (1989). Genetics of sterility in hybrids between two subspecies of *Drosophila*. *Evolution* **43**, 180–189.

Orr, H. A. and Coyne, J. A. (1989). The genetics of postzygotic isolation in the *Drosophila virilis* group. *Genetics* **121**, 527–537.

Palmiter, R. D., Wilkie, T. M., Chen, H. Y. and Brinster, R. L. (1984). Transmission distortion and mosaicism in an unusual transgenic mouse pedigree. *Cell* **36**, 869–877.

Petzoldt, U. and Hoppe, P. C. (1980). Spontaneous parthenogenesis in *Mus musculus*: comparison of protein synthesis in parthenogenetic and normal preimplantation embryos. *Mol. Gen. Genet.* **180**, 547–552.

Pillus, L. and Rine, J. (1989). Epigenetic inheritance of transcriptional states in *S. cerevisiae*. *Cell* **59**, 637–647.

Reik, W. (1989). Genomic imprinting and genetic disorders in man. *Trends Genet.* **5**, 331–336.

Reik, W. and Surani, M. A. (1989). Genomic imprinting and embryonal tumours. *Nature* **338**, 112–113.

Reik, W., Collick, A., Norris, M. L., Barton, S. C. and Surani, M. A. H. (1987). Genomic imprinting determines methylation of parental alleles in transgenic mice. *Nature* **328**, 248–251.

Reik, W., Howlett, S. K. and Surani, M. A. (1990). Imprinting by DNA methylation: from transgenes to endogenous gene sequences. *Development* Suppl., 99–106.

Renard, J. P. and Babinet, C. (1986). Identification of paternal development effect on the cytoplasm of one-cell-stage mouse embryos. *Proc. Natl Acad. Sci. USA* **83**, 6883–6886.

Ridley, R. M., Frith, C. D., Crow T. J. and Conneally, P. M. (1988). Anticipation in Huntington's disease is inherited through the male line but may originate in the female. *J. Med. Genet.* **25**, 589–595.

Sanford, J. P., Clark, H. J., Chapman, V. M. and Rossant, J. (1987). Differences in DNA methylation during oogenesis and spermatogenesis and their persistence during early embryogenesis in the mouse. *Genes Dev.* **1**, 1039–1046.

Sapienza, C. (1989). Genome imprinting and dominance modification. *Ann. NY Acad. Sci.* **564**, 24–38.

Sapienza, C. (1990). Sex-linked dosage-sensitive modifiers as imprinting genes. *Development* Suppl., 107–114.

Sapienza, C., Tran, T. H., Paquette, J., McGowan, R. and Peterson, A. (1987). Degree of methylation of transgene is dependent on gamete of origin. *Nature* **328**, 251–254.

Sapienza, C., Paquette, J., Tran, T. H. and Peterson, A. (1989). Epigenetic and genetic factors affect methylation imprinting. *Development* **107**, 165–168.

Sasaki, H., Hamada, T., Ueda, T., Seki, R., Higashinakagawa, T. and Sakaki, Y. (1991). Inherited type of allelic methylation variations in a mouse chromosome region where an integrated transgene shows methylation imprinting. *Development* **111**, 573–581.

Scrable, H., Cavanee, W., Ghavimi, F., Lovell, M., Morgan, K. and Sapienza, C. (1989). A model for embryonal rhabdomyosarcoma tumourigenesis that involves genome imprinting. *Proc. Natl Acad. Sci. USA* **86**, 7480–7484.

Searle, A. G. and Beechey, C. V. (1978). Complementation studies with mouse translocations. *Cytogenet. Cell Genet.* **20**, 282–303.

Searle, A. G. and Beechey, C. V. (1985). Non-complementation phenomena and their bearing on nondisjunctional effects. In Dellarco, V. L., Voytek, P. E. and Hollaender, A., (eds) *Aneuploidy*, pp. 363–376. Plenum Press, New York.

Searle, A. G., Peters, J., Lyon, M. F., Hall, J. G., Evans, E. P., Edwards, J. H. and Buckle, V. F. (1989). Chromosome maps of man and mouse. IV. *Ann. Hum. Genet.* **53**, 89–140.

Shire, J. G. M. (1989). Unequal parental contributions: genomic imprinting in mammals. *New Biologist* **1**, 115–120.

Singh, P. B., Miller, J. R., Pearce, J., Kothary, R., Burton, R. D., Paro, R., James, T. C. and Gaunt, S. J. (1991). A sequence motif found in a *Drosophila* heterochromatin protein is conserved in animals and plants. *Nucleic Acids Res.* **19**, 789–794.

Solter, D. (1988). Differential imprinting and expression of maternal and paternal genomes. *An. Rev. Genet.* **22**, 127–146.

Spofford, J. B. (1976). Position-effect variegation in *Drosophila*. In: Ashburner, M. and Novitski E. (eds) *The Genetics and Biology of Drosophila*, pp. 955–1018. Academic Press, London.

Stern, C. D. and Canning, D. R. (1990). Origin of cells giving rise to mesoderm and endoderm in chick embryo. *Nature* **343**, 273–275.

Surani, M. A. H., Barton, S. C. and Norris, M. L. (1984). Development of reconstituted mouse eggs suggests imprinting of the genome during gametogenesis. *Nature* **308**, 548–550.

Surani, M. A., Reik, W. and Allen, N. D. (1988). Transgenes as molecular probes for genomic imprinting. *Trends Genet.* **4**, 59–62.

Surani, M. A., Allen, N. D., Barton, S. C., Fundele, R., Howlett, S. K., Norris, M. L. and Reik, W. (1990a). Developmental consequences of imprinting of parental chromosomes by DNA methylation. *Phil. Trans. R. Soc. Lond.* B **326**, 313–327.

Surani, M. A., Kothary, R., Allen, N. D., Singh, P. B., Fundele, R., Ferguson-Smith, A. C. and Barton, S. C. (1990b). Genome imprinting and development in the mouse. *Development* Suppl., 89–98.

Swain, J. L., Stewart, T. A. and Leder, P. (1987). Parental legacy determines methylation and expression of an autosomal transgene: a molecular mechanism for parental imprinting. *Cell* **50**, 719–727.

Takagi, N. and Sasaki, M. (1975). Preferential inactivation of the paternally derived X-chromosome in the extraembryonic membranes of the mouse. *Nature* **256**, 640–641.

Tartof, K. D. and Bremer, M. (1990). Mechanisms of the construction and developmental control of heterochromatin formation and imprinted chromosome domains. *Development* Suppl., 35–46.

Vandeberg, J. L., Robinson, E. S., Samallow, P. B. and Johnston, P. G. (1987). X-linked gene expression and X chromosome inactivation: marsupials, mouse and man compared. In: Markert, C. L. (ed.) *Genetics, Development and Evolution. Isozymes: Current Topics in Biological and Medical Research*, vol. 15, pp. 225–253. New York: A. R. Liss.

Wakasugi, N. (1974). A genetically determined incompatibility system between spermatozoa and eggs leading to embryonic death in mice. *J. Reprod. Fert.* **41**, 85–96.

West, J. D., Frels, W. I., Chapman, V. M. and Papaioannou, V. E. (1977). Preferential expression of the maternally derived X-chromosome in the mouse yolk sac. *Cell* **12**, 873–882.

Winking, H. (1986). T/t Mutationen und Robertson' sche Translokationschromosomen der Hausmaus Mus musculus linnaeus: Charakterisierung neuer t-mutationen sowie deren Anwendung auf Gebieten der experimentellen Genetik und Zytogenetik. Habilitationsschrift at the Medical University of Lübeck.

6
Engineering cellular deficits in transgenic mice by genetic ablation

MARTIN L. BREITMAN and ALAN BERNSTEIN

Division of Molecular and Developmental Biology, Mount Sinai Hospital Research Institute, 600 University Avenue, Toronto, Ontario, Canada M5G 1X5 and Department of Medical Genetics, University of Toronto, Toronto, Ontario, Canada

1. Introduction

In higher organisms, organogenesis and formation of the overall body plan are orchestrated through the orderly development and interaction of cells of different lineages. Cells may become committed to a particular fate through inheritance of a developmentally restricted genetic programme, or through interaction with extrinsic environmental cues, the nature of which are genetically determined

TRANSGENIC ANIMALS
ISBN 0–12–304530–4

within the developing embryo. Elucidation of the molecular mechanisms controlling these process is central to the understanding of development.

In some organisms, notably *Drosophila* and *Caenorhabditis elegans*, the molecular mechanisms governing cellular determination can be studied readily by a combination of embryological, genetic, and molecular approaches. In mammals, however, these same experimental approaches are not often technically feasible or are of limited practical application, due either to the inaccessibility of the developing embryo to direct experimental manipulation or to the relative difficulty in conducting classical Mendelian genetics on larger experimental animals. Consequently, it has been necessary to devise or rely heavily upon alternative experimental strategies for the study of mammalian development.

The ability to manipulate the mammalian germ line by the introduction of foreign DNA provides a powerful experimental approach for creating mutant transgenic animals and for examining the molecular processes that govern developmental decisions and ultimately the patterns of gene expression in the developing embryo. Such technology has been used effectively not only to analyse genetic regulatory elements utilized by specific cell types in directing lineage-specific gene expression, but also to manipulate cell and organ physiology and to evaluate the contribution of an ectopically or inappropriately expressed gene to the implementation or execution of a particular developmental programme. In the latter case, the power of this technology has been vividly demonstrated by two recent studies which showed that aberrant expression of particular homeobox-containing genes in transgenic mice leads to homeotic mutations affecting gut or skeletal structures (Wolgemuth *et al.*, 1989; Kessel *et al.*, 1990). To date, however, transgenic mouse technology has been used primarily to generate mutant mice whose phenotypes result from a gain of function. The purpose of this chapter is to review a relatively new application of transgenic technology which has as its goal the generation of animals exhibiting a loss of function due to the elimination or destruction of specific cell types. This approach, which we have termed genetic ablation (Breitman *et al.*, 1987), mediates selective cell killing by the targeted expression of a potentially toxic gene to specific cell types with appropriate transcriptional regulatory elements. As such, it represents an extension of classical approaches to evaluating the developmental role of specific cells using surgical or laser ablation. However, in contrast to these earlier approaches, genetic ablation is readily applicable for use in developmental studies of the mammalian embryo: because the ablation of specific cell types is programmed by genetic means, genetic ablation circumvents the problems normally posed by inaccessibility of the mammalian embryo and the plasticity of its developmental processes. As such, it provides a novel and relatively straightforward method of generating mutant animals that lack specific cell types or an entire lineage. As outlined in this chapter, genetic ablation can be used to create animal models of particular human diseases and to formulate and address very specific questions about the functions, origins, and inter-dependencies of specific cell types either in the developing embryo or in the adult. Below we review some of the approaches that have recently been used for generating and analysing mutant ablation mice and discuss some of the technical and conceptual issues that are related to the use of these technologies.

2. Methods of genetic ablation

The ability to ablate specific cell types in transgenic animals by genetic methods derives from the fact that virtually all differentiated cell types in the body can be distinguished by the expression of one or more specialized proteins that are unique to that cell type. As a result, it is possible through the use of gene transfer to define transcriptional regulatory elements located in or around the structural regions of genes that are sufficient for determining that developmentally regulated genes are switched on or off in the proper cell type at the appropriate times during embryological development or during cellular differentiation in the adult. Once identified, these cis-acting regulatory sequences can be separated from the coding regions of a gene and used to direct expression of heterologous sequences in a developmentally similar manner in transgenic mice. Thus, when coupled to genes encoding cytotoxic or potentially cytotoxic proteins, such cis-acting regulatory elements can be used to programme the ablation of specific cell types in transgenic animals.

At present, two distinct and complementary strategies have been devised for achieving genetic ablation in transgenic mice. One strategy is based on the expression of genes encoding highly toxic proteins, such as the A chains of diphtheria toxin or ricin (Breitman et al., 1987; Palmiter et al., 1987; Landel et al., 1988). In this approach, cell killing is constitutive, in that cell ablation is initiated as an obligate consequence of toxin gene expression, provided that threshold levels of intracellular toxin are attained. The second strategy that has been developed is based on expression of the thymidine kinase (tk) gene of the herpes simplex virus (Borrelli et al., 1988; Heyman et al., 1989). The enzyme activity encoded by the viral tk gene is not toxic to animals or animal cells except in the presence of certain nucleoside analogues that cannot be metabolized by normal cells that do not contain the herpes enzyme. Thus, expression of the herpes tk gene constitutes a conditional or inducible system for achieving cell killing in vivo. Each of these strategies is discussed below in relation to their operational advantages and disadvantages and in the contex of the selective ablation of specific cell types in transgenic animals.

2.1. Ablation with the A chain of diphtheria toxin

Diphtheria toxin is a highly potent toxin encoded by the *tox* gene, which is carried by lysogenic corynephages of pathogenic strains of *Corynebacterium diphtheriae*. The product of the *tox* gene is secreted by *C. diphtheria* as a precursor polypeptide that is subsequently cleaved enzymatically into two chains of 193 and 342 amino acids, designated A and B, that remain joined together by an interchain disulphide bond (reviewed in Pappenheimer, 1977). The B chain interacts with a specific membrane receptor at the cell surface and is essential for the delivery of the A chain into the cytoplasm. The A chain is active in the absence of the B chain and catalyses the NAD-dependent ADP ribosylation of a modified histidine residue on elongation factor 2, causing it to detach from ribosomes, which results in the inhibition of protein synthesis and consequent cell death (Collier, 1975; Moehring *et al.*, 1984; Kohno *et al.*, 1986). To enable the diphtheria toxin A chain (DT-A) to be expressed in eukaryotic cells, the sequences encoding the signal peptide of native DT were removed and the coding sequences of DT-A were provided with an initiation codon from the human metallothionein IIA gene and a polyadenylation signal from the early region of simian virus 40 (SV40) (Maxwell *et al.*, 1986). Expression of DT-A in eukaryotic cells in the absence of the B chain therefore leads to specific cell killing, as any extracellular DT-A protein that may be released by dying cells is unable to gain entry into neighbouring cells. In this way, targeted expression of DT-A by cell type-specific transcriptional regulatory elements can be used to programme the suicide of specific cell populations in transgenic mice.

2.2. Ablation with the A chain of ricin

A very similar approach to ablating cells with DT-A has also been described using the extremely toxic plant lectin ricin. Ricin is produced by the castor bean (*Ricinus communis*) and, in its native form, has an A and B subunit structure analogous to that of diptheria toxin. The B chain of ricin binds specifically to galactose-containing carbohydrates present at the cell surface and facilitates penetration of the lipid bilayer by the toxic A chain (R-A). The R-A is a highly specific ribonuclease that inactivates ribosomes by cleaving the 28S ribosomal RNA molecule at A_{4365}, thereby arresting protein synthesis (Endo and Tsurugi, 1987). As in the case of DT-A, extracellular R-A is ineffective at entering cells (Pastan *et al.*, 1986) and, thus, its targeted expression also allows for selective and efficient cell killing without affecting non-expressing bystander cells.

2.3. Inducible ablation with the tk gene of the herpes simplex virus

A completely different strategy for generating transgenic mice lacking specific cell types involves a system of conditional cell ablation in which expression of the targeted gene product is not by itself harmful to mammalian cells but is capable of rendering cells selectively sensitive to the killing action of certain drugs that cannot be metabolized by normal cells. Thus, expression of the herpes *tk* gene is not itself deleterious, but, because the viral enzyme, in contrast to mammalian thymidine kinase, is capable of phosphorylating certain nucleoside analogues, it has been possible to develop a number of drugs such as acyclovir, gancyclovir and 1-(2-deoxy-2-fluoro-β-D-arabinofuranosyl)-5-iodouracil (FIAU), that kill only dividing cells expressing the herpes *tk* gene. The high degree of selectivity of the viral thymidine kinase for these nucleoside analogues ensures that only herpes *tk*-expressing cells will incorporate them into DNA and be killed due to the resulting inhibition of DNA synthesis. Hence, expression of the herpes *tk* gene under the control of a cell type-specific promoter, followed by drug administration, constitutes an inducible system for achieving cell ablation in transgenic mice (Heyman *et al.*, 1989) and, as discussed below, provides a useful complement to genetic ablation with naturally occurring toxins.

3.   Constitutive versus inducible ablation: practical considerations

The two methods of genetic ablation that have been developed each have distinct operational advantages and disadvantages that in many cases bear directly on their suitability for generating and studying particular cellular deficits in transgenic mice. For example, DT-A and R-A, which act through inhibition of protein synthesis, theoretically provide the means to ablate virtually any cell type in the body provided that promoter/enhancer elements of appropriate specificity have been defined. Moreover, because these agents kill cells directly, ablation of specific cell types with appropriate transcriptional regulatory elements can be readily achieved in the developing embryo, providing the opportunity to study cell lineage relationships and cell–cell interactions during organogenesis and other morphogenetic processes in the developing organism. One potential drawback of ablation with DT-A or R-A, however, is that it does not allow definition of the cell population expressing the toxin gene prior to ablation and, as such, difficulties could arise in distinguishing those aspects of aberrant development that are secondary to toxin gene expression from those that are directly due to toxin-induced cell killing. A second potential drawback associated with using DT-A or R-A relates to their extreme toxicity. For example, in the case of DT-A, as little as one molecule per cell has been estimated to be sufficient to effect killing (Yamaizumi *et al.*, 1987). Thus, the extreme sensitivity of eukaryotic cells to these toxins necessitates that there is extreme lineage or cell-type specificity of the targeting gene regulatory elements in transgenic mice, regardless of the chromosomal site of integration. It is possible, however, that this latter problem may be partly overcome by the use of suitably attenuated toxins (see below).

In contrast to ablation with DT-A and R-A, conditional ablation with the herpes *tk* gene is likely to be effective only for actively dividing cell populations, as drug-induced cell killing is mediated through inhibition of DNA synthesis. Thus, many potential target cell populations, such as non-cycling stem cells and terminally differentiated cell types, are likely to be refractory to the toxic effects of administered drug, despite significant expression of the herpes *tk* gene. In addition, methodologies have not yet been described for conditional ablation with *tk* in the developing embryo, which will undoubtedly require that it be possible to deliver to the embryo selectively toxic levels of drug transplacentally. Despite these limitations, conditional ablation by targeted expression of the herpes *tk* gene affords the tremendous practical advantage of allowing stable transgenic pedigrees to be established prior to the initiation of ablation studies. Also, it provides the means to control the timing and duration of toxic insult and to assess the potential for recovery of an ablated cell population within the transgenic animal. Thus, both approaches have distinct attributes that, in certain cases, render one more relevant than the other for use in ablating specific cell populations in transgenic mice.

4. Systems of genetic ablation

4.1. Lens

4.1.1. With native toxin

Perhaps, not surprisingly, the eye was one of the first developmental systems in which cellular interactions during morphogenetic processes were examined by genetic ablation. The vertebrate eye has long been used as a model system by embryologists for investigating the factors that influence tissue development and differentiation. As early as 1781, regeneration of the amphibian eye was studied upon physical removal of the lens (reviewed in Harding and Crabbe, 1984). In mammals, formation of the eye is known to involve a precise sequence of complex, genetically determined interactions among several distinct and highly specialized ocular tissues, including the optic nerve, neural retina, cornea, ciliary epithelium, iris and ocular lens. Understanding the developmental inter-relationships among these different structures poses a major challenge, since each ocular tissue has the potential to influence normal morphogenesis and, as such, aberrations in any one developing tissue may profoundly alter the growth and differentiation of surrounding structures. In the mouse, where microdissection techniques are not readily applicable, ocular morphogenesis has traditionally been studied through developmental analysis of naturally occurring mutations. Unfortunately, natural mutation, while having the advantage of being non-disruptive, is not often understood at the molecular level. Thus, a number of mutations have been described that affect early development of the murine eye, such as eyeless-1 (Chase and Chase, 1941), eyeless-2 (Chase and Chase, 1944) and aphakia (Varnum and Stevens, 1968), for which the molecular basis of the genetic defect is unknown. Consequently, analysis of these mutants has not allowed the interrelationships of the components of the developing eye to be unequivocally defined.

One inroad to studying ocular morphogenesis by genetic methods is provided by the ability to ablate cells or specific cell types of the mammalian ocular lens by targeted expression of toxin genes in transgenic mice. The lens is particularly well suited to such studies since its development and the relationships of its component cell types have been extensively studied (reviewed in McAvoy, 1978). The mature lens is a transparent spheroidal structure consisting of essentially two morphologically distinct lens cell types: terminally differentiated fibre cells, which are longitudinally arrayed and make up the main body of the lens, and morphologically undifferentiated epithelial cells, which are mitotically active and form a monolayer at the anterior surface of the lens. During development, the lens originates from a small number of cells of the head ectoderm which, under local influence of the optic vesicle, commence elongating to form the lens placode. The lens placode subsequently invaginates to form the lens vesicle and, within a few hours, cells at the posterior of the lens vesicle begin to elongate into primary fibre cells; these migrate anteriorly, filling in the lumen of the lens vesicle, forming the lens nucleus. Thereafter, and throughout life, growth of the lens occurs by formation of secondary fibre cells, which differentiate continuously from the anterior lens epithelium and are deposited in successive layers about the lens nucleus, forming the lens cortex. Development of the lens and elongation and terminal differentiation of the lens fibre cells are characterized by differential expression of the α-, β- and γ-crystallins, the major water-soluble proteins of the lens which, through short-range ordering (Delaye and Tardieu, 1983), are thought to mediate transparency of the lens and to confer upon the organ its delicate light-refractive properties. Differential regulation of the crystallin genes results in differences in the spatial distribution of individual crystallin proteins across the lens as well as in differences in the classes of crystallin synthesized at different stages of lens cell differentiation. Thus, α-crystallins are expressed earliest in development and are found in all cells of the adult lens, whereas γ-crystallins are expressed later during organogenesis and only in terminally differentiating fibre cells. Taken together, these developmental features of the lens provide an attractive and experimentally amenable system with which to address a number of questions related to the role that intrinsic and inductive factors play in the development of neighbouring tissues and in the programmed expression of developmentally regulated genes. Such questions include: Is a given lens fibre cell instructed to express its repertoire of crystallins as a function of position in the organ or is its profile of crystallin gene expression intrinsically programmed as a consequence of commitment decisions that have occurred within earlier progenitor cells? Is development of the lens required for induction or maturation of neighbouring ocular structures or does development of these structures proceed independently of the lens? The ability to generate transgenic mice ablated of the lens, or of specific lens cell subpopulations, by targeted expression of toxin genes, provides a genetically defined experimental system with which to investigate these questions.

Using a 804 bp 5 ' flanking segment of the mouse γF (previously γ2) crystallin gene, Breitman and coworkers (1987) targeted the expression of the DT-A gene to nuclear fibre cells in the lenses of transgenic mice. The resulting animals were visibly microphthalmic as early as 3 weeks of age and contained lenses that were considerably reduced in size. The remainder of the eye was unremarkable and there was no evidence of cytotoxic effects outside of the lens. By crossing mice hemizygous for the γF–DT-A gene with animals homozygous for the *lacZ* reporter gene driven by the identical γF promoter sequences, it was possible to visualize *in situ* the precise cells in the lens ablated by the γF–DT-A gene following staining of *lacZ*-expressing cells in the lenses of the microphthalmic F_1 progeny (Breitman *et al.*, 1989). The results showed that while the majority of γF crystallin-expressing cells in the lens nucleus were ablated by the γFDT-A gene, lens organogenesis was often able to proceed by formation of a new lens nucleus composed primarily of a later population of (normally cortical) fibre cells that efficiently expresses the γB gene and not the γF gene. Thus, phenotypically and developmentally distinct populations of lens fibre cells were able to contribute to the lens nucleus during organogenesis, suggesting that morphogenetic commitment of cell fate is not inextricably linked to the developmental programming of crystallin gene expression. The above studies also revealed that occasional γF-expressing cells fail to be ablated by the γF–DT-A transgene and that there is some variability in the number of these cells among individual animals or even between lenses of the same transgenic mouse. Similar observations have also been reported in other systems of genetic ablation (see below). The variable penetrance of ablation events observed within the lens was subsequently found by mating studies to be profoundly influenced by dosage of the transgene (Breitman *et al.*, 1989). Thus, when microphthalmic animals hemizygous for the γF–DT-A transgene were crossed to generate homozygous offspring, the majority of the homozygotes exhibited a much more severe ocular phenotype, characterized by a lack of bulge in the eye socket and a failure of the eyes to open at 3 weeks after birth. The more severe microphthalmia manifest by the latter animals was not anticipated: because one molecule of DT-A has been demonstrated to be sufficient to kill a cell (Yamaizumi *et al.*, 1987), homozygous and hemizygous animals were expected to show identical phenotypes. However, the correlation between transgene dosage and increased severity of ocular phenotype may be explained if occasional cells in hemizygous animals escape ablation through failure of the transgene to become transcriptionally activated. In this case, the relationship between gene dosage and extent of ablation would result from cells homozygous for the transgene having a significantly greater likelihood of activating at least one of their two DT-A alleles. Results consistent with this hypothesis have also been obtained in experiments utilizing an attenuated form of the DT-A gene (see below).

Sectioning of the vestigial eyes of adult γF–DT-A homozygotes revealed that they contained little or no lens tissue in conjunction with extensive convolution of the sensory retina, malformation of the ciliary epithelium, and visible absence of the iris (aniridia) and aqueous and vitreous humours. These profound ocular abnormalities, which were not displayed by microphthalmic animals hemizygous for the transgene, were also characteristic of transgenic mice in which complete or extensive ablation of the lens had been achieved following targeted expression of R-A or a DT-A/human growth hormone (hGH) hybrid gene by the promoter of the mouse αA crystallin gene, which is expressed in all cell types of the lens (McAvoy, 1978; Van Leen *et al.*, 1987; Landel *et al.*, 1988; Kaur *et al.*, 1989). The multiple anomalies observed in the eyes of these animals thus vividly illustrate the importance of the lens to normal ocular morphogenesis and raise fundamental questions about the precise role the lens plays in early eye development and the extent to which aberrant ocular morphogenesis is secondary to or concurrent with lens ablation. To address these questions, we have undertaken a developmental study of transgenic mice completely ablated of the lens as the result of introduction into the mouse germ line of the DT-A gene driven by αA crystallin promoter sequences (Harrington *et al.*, 1991). Histological studies revealed that, during embryological development, the optic cup and lens vesicle form normally and that ablation of the lens occurs as a gradual degenerative process beginning between day 12 and 13.5 of gestation. Degeneration of the lens vesicle coincides with retarded growth and development of the neuroretina, sclera and cornea as well as failure of the iris to form. In contrast, retinal folding, and degeneration of the ciliary epithelium, are secondary to lens deterioration and occur only upon complete destruction of the lens at birth. Taken together, these results suggest that development of a functional lens from embryonic day 12 onwards is not critical for formation of most surrounding ocular structures, but is important for formation of the iris and for appropriate growth, development and maintenance of morphology of the retina, cornea and ciliary epithelium. The results also showed that a detailed description of ablation events that occur during development is necessary for a full understanding of the morphogenetic abnormalities observed in the adult, particularly since the delay between onset of toxin gene expression and cell death can have important implications for both the time course of ablation events in the target tissue and the interpretation of the significance of a lingering tissue or cell type for the development or differentiation of surrounding structures.

4.1.2. With attenuated toxin

As previously mentioned, one potential problem associated with using DT-A, or R-A, to ablate specific cell populations in transgenic mice stems from its extreme cytotoxicity. Although this ensures that even weakly expressing cells in a 'targeted' cell population will be killed, it also means that any 'leakiness' of the transcriptional regulatory elements that results in trace levels of ectopic toxin gene expression could lead to embryo lethality or to deleterious effects that could compromise efforts to analyse or interpret the phenotype of the resulting transgenic animal. Leakiness of the targeting promoter element could pose a major problem, particularly if exacerbated by chromosomal position effects, which are known to result in variable levels of expression among randomly integrated transgenes (Palmiter and Brinster, 1986). We therefore investigated whether there might be an experimental advantage in using an attenuated toxin gene for genetic ablation studies, since the significantly higher level of expression required to kill a cell might mitigate the problem of promoter leakiness, yet at the same time permit the intended target cell population to be selectively and efficiently destroyed. To this end, we evaluated the relative efficacy of a mutant, attenuated form of the DT-A gene for ablating fibre cells in the lens following targeted expression in transgenic mice by γF crystallin promoter and enhancer sequences (Breitman et al., 1990). The mutant DT-A gene, tox176, is about 30-fold less cytotoxic than the wild type due to a point mutation that results in a Gly-to-Asp amino acid substitution at residue 128 (Maxwell et al., 1987). Following introduction of γF–tox176 into the mouse germ line, we established transgenic lines that could be classified into two groups on the basis of ocular phenotype. One group of animals showed predominantly severe microphthalmia, whereas the other group displayed predominantly cataracts with occasional slight microphthalmia, characterized by the presence of necrotic or degenerating fibre cells in the central nuclear region of the lens. These two disparate phenotypes constituted an apparent paradox since they were at the same time both less severe and more severe than the microphthalmia consistently observed amongst lines of transgenic mice hemizygous for the wild-type γF–DT-A gene. The two different ocular phenotypes manifest by the γF–tox176 mice did not correlate with differences in transgene copy number, indicating that chromosomal position effects play a significant role in modulating expressivity of the transgene. Thus, the ability to generate severely microphthalmic animals with γF–*tox176* and not with γF–DT-A suggests that the *tox176* gene, unlike the more potent DT-A gene, is able to integrate into chromosomal sites that are favourable for expression without compromising viability of the embryo. Microphthalmic lines carrying the DT-A gene would therefore be analogous to cataractous lines carrying *tox176* in that both phenotypes result from integration events at chromosomal sites that are not conducive to transgene expression. Taken together, these results demonstrate the considerable advantage of using the *tox176* gene for genetic ablation studies; they also demonstrate that it should be possible to use attenuated toxin genes to create transgenic models of a variety of degenerative disorders, in which the intracellular levels of toxin produced are not acutely lethal but, rather, lead to eventual cell death over a considerable period. Such technology might be particularly useful when applied to cellular components of the nervous system and could potentially be used to provide information subsets complementary to those obtained from conventional cellular ablation studies.

4.2. Pancreas

Pancreatic development in the mouse involves the orderly development of several different cell types, including duct, islet and acinar cells. The progenitor cells necessary for complete development of the pancreas have been shown to be present within two small patches of epithelial cells along the gut of the day 8 embryo (Wessells, 1964; Wessells and Cohen, 1967; Parsa *et al.*, 1969; Richardson and Spooner, 1977). However, the exact lineage relationships among duct, islet and acinar cells are unknown, nor is information available concerning possible interdependencies among these cell types that might govern their differentiation and extent of proliferation. One possible approach to investigating these questions is provided by the ability to ablate specific cell types in the developing pancreas. Elastase I is a digestive enzyme produced exclusively in the exocrine pancreas. A 205 bp 5 ' flanking segment of the rat elastase I gene, consisting of the promoter and an adjacent enhancer element, has been shown to direct specific expression of heterologous genes in acinar cells of transgenic mice, beginning at about 14 days of development (Orinitz *et al.*, 1985a, b). The identification of these *cis*-acting regulatory sequences thus affords the opportunity to ablate acinar cells selectively and examine the consequent effects on pancreatic development. To this end, Palmiter and coworkers (1987) generated transgenic mice carrying a construct in which DT-A expression was driven by elastase I promoter and enhancer sequences. Neonatal founder animals carrying this construct were either normal and failed to show detectable expression of the transgene, or contained only a small pancreatic rudiment that resembled the embryonic pancreas at 14 days of development. These small pancreatic rudiments were composed almost exclusively of duct-like cells and developing islet cells, although occasional rudiments contained very small numbers of acinar cells that had escaped ablation by the transgene. The underrepresentation of both duct and islet cells in these rudiments compared to age-matched control pancreases suggests that all three cell types may arise from a common pluripotent stem cell that expresses low levels of elastase prior to differentiation and, therefore, is killed by the transgene during pancreatic development. An entirely different explanation is that the orderly development and proliferation of duct and islet cells depend in some way on physical or biochemical cues normally emanating from developing acinar cells. Unfortunately, these questions could not be analysed further due to the neonatal lethality of the elastase I/DT-A construct.

4.3. Pituitary

The studies described above have all involved experimental systems in which genetic ablation was used to obtain new insights into cell lineage relationships or cell–cell interactions within a developing organ. However, the ability to engineer specific cellular deficits in transgenic mice also provides an experimental approach for creating animal models of a variety of human disorders. Experiments simultaneously capitalizing on both aspects of this technology have been described in which specific hormone-producing cell populations within the anterior pituitary were ablated by targeting the expression of either the DT-A gene or the herpes *tk* gene.

The anterior pituitary gland is comprised of somatotropes, lactotropes, gonadotropes, corticotropes and thryotropes, each of which is characterized by the production of one or more specific hormones. Whether some or all of these endocrine cells form from a common progenitor or arise independently as distinct lineages is unclear. It has been suggested that growth hormone (GH)-producing somatotropes and prolactin (PRL)-producing lactotropes may be derived from a common stem cell, since cells and tumour cell lines expressing both hormones have been described (Bancroft, 1981; Hoeffler *et al.*, 1985; Frawley and Clark, 1986). In mice, GH production precedes PRL synthesis by about 2 weeks (Slabaugh *et al.*, 1982). Thus, one way to investigate the lineage relationship between somatotropes and lactotropes would be to examine the relative proportion of these two cell types that form in the anterior pituitary of transgenic animals carrying an ablation construct driven by GH gene regulatory elements. Using this strategy, Behringer *et al.* (1988) generated transgenic mice in which expression of DT-A was targeted by a 310 bp 5′ flanking segment of the rat GH gene. Animals expressing DT-A displayed dwarfism and were markedly deficient in somatotropes and lactotropes, suggesting that both cell types develop from a common progenitor. The pituitary, while reduced in size, appeared otherwise normal morphologically, consistent with the fact that GH synthesis and, hence, presumably DT-A expression, commence at a time when morphogenesis of the organ is essentially complete. Whereas almost all somatotropes were ablated in these animals, there was a small, but considerably larger, number of residual lactotropes. The few surviving somatotropes (10 against 2×10^5 in normal mice) presumably corresponded to cells that escaped ablation, as was discussed earlier in crystallin/DT-A transgenic mice. The low, but significantly higher, number of residual lactotropes, however, raised the possibility that a minor subpopulation of lactotropes was able to form either by an alternative pathway that is independent of GH production, or from occasional GH-expressing cells that rapidly specialize towards PRL synthesis prior to succumbing to the lethal effects of DT-A expression.

A similar study involving a conditional approach was carried out by Borrelli *et al.* (1989), who generated lines of transgenic mice in which expression of the herpes *tk* gene was driven by either the promoter of the rat GH or the rat PRL gene. Animals expressing GH–*tk* and treated with FIAU developed dwarfism and showed a near total depletion of both somatotropes and lactotropes. In contrast, mice expressing PRL–*tk* and administered FIAU were unaffected in pituitary development and showed no reduction in the number of somatotropes or lactotropes. These results suggested that somatotropes and lactotropes arise from a common proliferating stem cell that expresses GH but not PRL, and that commitment to PRL expression and lactotrope differentiation represent post-mitotic events, since mice expressing PRL–*tk* were unaffected by treatment with FIAU. When FIAU treatment of GH–*tk* mice was delayed until 2, 5 or 10 days after birth, adult animals showed a progressive increase in the number of somatotropes, but not lactotropes, suggesting that terminal differentiation of somatotropes occurs shortly after birth. Moreover, when GH–*tk* mice exposed to FIAU for 8 weeks were allowed to recover from the drug, the animals showed a dramatic recovery in both body weight and the number of GH$^+$ cells, indicating that, even after 8 weeks, there remained a significant population of cells that had escaped ablation and that were capable of repopulating the anterior pituitary with GH-producing progeny.

4.4. Lymphoid system

The haematopoietic system is comprised of a large and diverse array of cell types whose origins can all be traced through an ill-defined hierarchy of committed progenitor cells to a common pluripotent stem cell. Elucidation of the functions and interrelations of these cell types is of fundamental importance not only from a developmental perspective but also in terms of understanding haematological disease. The ability to generate transgenic pedigrees ablated or transiently depleted of specific cell types within either the myeloid (macrophages, granulocytes, platelets, erythrocytes) or lymphoid (B and T cells) lineages would be of considerable practical advantage. Such mice would provide not only new animal models of myeloid dysfunctions or immune deficiencies found in humans, but also genetically defined experimental systems in which to analyse cell lineage relationships, the self-renewing capacity of specific stem cells, and the role of specific cell types in the large and complex array of cell–cell interactions that occur within the haematopoietic system.

Using the approach of conditional ablation, Heyman *et al.* (1989) generated transgenic mice carrying the herpes *tk* gene driven by a hybrid regulatory element consisting of the mouse μ heavy-chain enhancer and the κ light-chain promoter. The immunoglobulin regulatory sequences directed expression of *tk* in these animals within the lymphoid system, resulting in significant levels of viral thymidine kinase activity in spleen, lymph nodes, bone marrow and thymus. Following 1 week of treatment with the nucleoside analogue gancyclovir, transgenic, but not control, animals showed marked atrophy of the spleen, mesenteric lymph nodes, and thymus, in association with extensive cellular destruction. The atrophied thymus was virtually devoid of the cortex and showed marked hypocellularity of the medulla, consistent with the idea that cellular ablation was most extensive in regions of the thymus most densely populated with rapidly proliferating cells. Flow cytometry of peripheral blood revealed a 95% reduction in B cells and a 65% reduction in T cells. The residual T cell population showed a greater relative depletion of double-positive L3T4$^+$, Lyt2$^+$ cells than either single-positive or double-negative cells, suggesting that the more mature and presumably less rapidly dividing thymocytes were less susceptible to the toxic effects of the drug. When treatment with gancyclovir was stopped for 7 days, transgenic animals regained the normal complement of T cells in the thymus, suggesting that even after 7 days of continuous exposure to the drug, residual cells that had escaped ablation were capable of fully reconstituting the thymus. In a similar study, we generated transgenic mice in which expression of DT-A was driven by promoter and enhancer elements derived solely from the mouse κ light-chain gene (M. Breitman, C. Paige and A. Bernstein, unpublished). When analysed at 3 weeks of age, the transgenic founder animals showed no apparent change in T cells, but a 95% reduction in the number of colony-forming B cells (CFU-B) capable of synthesizing either heavy-chain or κ light-chain RNAs. Unexpectedly, when analysed at 6 months, these same animals contained normal levels of CFU-B, suggesting that the virtually unlimited capacity of the haematopoietic system for self-renewal inevitably leads to the emergence and outgrowth of immunoglobulin-producing cells that do not functionally express the transgene. The continuous proliferation and differentiation of cells within the haematopoietic system therefore has important implications for both the design and interpretation of genetic ablation studies that are directed to target cell populations that have unlimited capacity for self-renewal throughout adult life.

4.5. Nervous system

One of the broader applications of genetic ablation is likely to lie in the use of this technology to study the function and interrelations of specific cells within the nervous system. The nervous system is composed of a large array of cell types of remarkably diverse function. Efforts to elucidate the embryonic origins of these cells and their functional interactions in the adult have been hampered by both the inherent complexity of the mammalian nervous system and the experimental difficulties associated with manipulating the central nervous system (CNS) either *in vivo* or in cell culture. Classical approaches to investigation of these problems have included analysis of naturally occurring mutants, such as jimpy and shiverer, which display hypomyelination due to mutations in the genes for myelin-specific proteolipid protein (PLP) and myelin basic protein (MBP), respectively (Roach *et al.*, 1985; Dantigny *et al.*, 1986), and exposure of the nervous system to a wide variety of toxic molecules in an attempt to ablate specific cells. An alternative experimental strategy is provided by the ability to ablate selectively specific cell types within the nervous system by targeted expression of toxic genes in transgenic mice. We have generated transgenic animals in which expression of the mutant, attenuated form of the DT-A gene *tox176*, is programmed by a 5′ flanking segment of the gene for MBP. MBP is one of the major structural components of the myelin sheath and is expressed in oligodendroglia and Schwann cells after their initial association with axons, but before the elaboration of compact myelin. In the shiverer mouse, where MBP is absent (Roach *et al.*, 1983), Schwann cells elaborate relatively normal myelin sheaths only slightly reduced in relative thickness, whereas oligodendroglia produce only a few turns of uncompacted myelin lamellae (Peterson and Bray, 1984). The latter defects lead to action tremors about 2 weeks after birth, progressive convulsions and, ultimately, death of the animals within 2–4 months.

Transgenic mice bearing the MBP–*tox176* construct displayed runting and at about 12 days of age developed profound behavioural abnormalities that ranged from mild tremors and slight incoordination of the hind limbs to severe convulsions that occasionally culminated in death 2–3 weeks after birth. Histological and ultrastructural analyses of these animals revealed extensive ablation of oligodendroglial cell populations in all tracts of the CNS and only negligible or relatively minor involvement of myelinating Schwann cells in the parasympathetic nervous system, suggesting either that Schwann cells and oligodendroglia have markedly different tolerance levels to reduced protein synthetic capacity or, more likely, that there is differential transcriptional regulation of MBP in the two cell types, such that oligodendroglia achieve lethal levels of attenuated *tox176* protein while most Schwann cells do not. No changes were observed in the number of astrocytes, indicating that, in absolute terms, the extent of proliferation of these cells was unaffected by oligodendrocyte ablation. Moreover, whereas in certain experimental preparations, Schwann cells are capable of ensheathing and myelinating CNS axons, there was no indication of Schwann cell invasion and proliferation within the CNS. The MBP–*tox176* mice thus demonstrate the utility of the *tox176* gene for preferentially ablating specific cell types within composite populations of transgene-expressing cells; they also provide a new model of demyelinating disease in which ablation of myelinating cells results by a defined mechanism – reduced protein synthetic capacity. As such, these animals provide an important complement to our understanding of demyelination *per se* and of the organism's response to such insult at both the behavioural and cellular levels.

5. Summary and conclusions

The studies described above document the utility of genetic ablation for creating mutant mice that lack specific cell types. Transgenic animals derived by this approach can be used to study cell lineage relationships, the self-renewing capacity of stem cells, and the role of cell–cell interactions during organogenesis and other physiological processes in the developing embryo or adult. Also, genetic ablation provides a relatively straightforward approach for creating animal models of various human disorders, such as those typified here by the mutant ablation mice displaying microphthalmia, dwarfism, immunodeficiency and demyelination. The ability to engineer these and other loss-of-function phenotypes in transgenic mice by defined genetic mechanisms could facilitate interpretation of the cellular basis of similar developmental or physiological defects in humans. However, it is important to point out that any biological conclusions derived from studies employing genetic ablation must take into account a number of conceptual and technical issues related to the use of this technology. For example, in many of the mutant ablation mice that have been described, a small residual population of cells escapes ablation that is clearly capable of functionally expressing the transcriptional elements used to target expression of the toxic gene. The significance of these residual cell populations with respect to inductive processes or cellular interdependencies may thus be difficult to evaluate and, in certain cases, could render it inappropriate to draw conclusions about the precise role of a particular cell type in developmental processes.

Similarly, the delay between onset of toxin gene expression and cell obliteration could have important implications with respect to the significance of a lingering tissue or cell type and the transmission of instructive environmental cues to neighbouring tissues. Each of these issues needs to be carefully assessed when interpreting or intuiting the function of a particular ablated cell population. Moreover, simultaneous ablation of more than one cell type could indicate that there is ectopic expression of the transgene, expression of the transgene by a common progenitor cell, or an interdependence or inductive relationship amongst the various ablated cell populations. Distinguishing amongst these possibilities requires that there is detailed knowledge of both the tissue and developmental profile of expression of the targeting transcriptional regulatory elements. These potential caveats notwithstanding, genetic ablation provides a powerful new approach for investigating physiological and developmental processes in transgenic animals. Undoubtedly, further refinement of conditional ablation strategies, in conjunction with more precise definition of the *cis*-acting regulatory elements necessary for directing position-independent, cell-specific gene expression, will be required before the full potential of this approach can be realized.

Acknowledgements

We thank Frances Hogue and Lynda Woodcock for their excellent secretarial assistance. Work in the authors' laboratories is supported by the Medical Research Council and the National Cancer Institute of Canada. A. B. and M. L. B. are McLaughlin and MRC Scientists, respectively.

References

Bancroft, F. C. (1981). GH cells: functional clonal lines of rat pituitary tumor cells. In: Sato G. (ed.) *Functionally Differentiated Cell Lines*, pp. 47–55 Alan R. Liss, New York.

Behringer, R., Mathews, L. S., Palmiter, R. D. and Brinster, R. L. (1988). Dwarf mice produced by genetic ablation of growth hormone-expressing cells. *Genes Dev.* **2**, 453–461.

Borrelli, E., Heyman, R., Hsi, M. and Evans, R. M. (1988). Targeting of an inducible toxic phenotype in animal cells. *Proc. Natl Acad. Sci. USA* **85**, 7572–7576.

Borrelli, E., Heyman, R., Arias, C., Sawchenko, P. E. and Evans, R. M. (1989). Transgenic mice with inducible dwarfism. *Nature* **339**, 538–541.

Breitman, M. L., Clapoff, S., Rossant, J., Tsui, L.-C., Glode, L. M., Maxwell, I. H. and Bernstein, A. (1987). Genetic ablation: targeted expression of a toxin gene causes microphthalmia in transgenic mice. *Science* **238**, 1563–1565.

Breitman, M. L., Bryce, D. M., Giddens, E., Clapoff, S., Goring, D., Tsui, L.-D., Klintworth, G. K. K. and Bernstein, A. (1989). Analysis of lens cell fate and eye morphogenesis in transgenic mice ablated for cells of the lens lineage. *Development* **106**, 457–463.

Breitman, M. L., Rombola, H., Maxwell, I. H., Klintworth, G. K. K. and Bernstein, A. (1990). Genetic ablation in transgenic mice with an attenuated diphtheria toxin A gene. *Mol. Cell. Biol.* **10**(2), 474–479.

Chase, H. B. and Chase, E. B. (1941). Studies on an anophthalmic strain of mice. I. Embryology of the eye region. *J. Morph.* **68**, 279–301.

Chase, H. B. and Chase, E. B. (1944). Studies on an anophthalmic strain of mice. IV. A second major gene for anophthalmia. *Genetics* **29**, 264–269.

Collier, R. J. (1975). Diphtheria toxin: mode of action and structure. *Bacteriol. Rev.* **39**, 54–85.

Dantingy, A., Mattei, M.-G., Morello, D., Alliel, P. M., Pham-Dinh, D., Amar, L., Arnaud, D., Simon, P., Mattei, J.-F., Guenet, J.-L., Jolles, P. and Avner, P. (1986). The structural gene coding for the myelin-associated proteolipid protein is mutated in jimpy mice. *Nature* **321**, 867–869.

Delaye, M. and Tardieu, A. (1983). Short-range order of crystallin proteins accounts for eye lens transparency. *Nature* **302**, 415–417.

Endo, Y. and Tsurugi, K. (1987). RNA n-glycosidase activity of the ricin A chain. *J. Biol. Chem.* **262**, 8128–8130.

Frawley, L. S. and Clark, C. L. (1986). Ovine prolactin (PRL) and dopamine preferentially inhibit PRL release from the same subpopulation of rat mammotropes. *Edocrinology* **119**, 1462–1466.

Harding, J. J. and Crabbe, M. J. C. (1984). The lens: development, proteins, metabolism and cataract. In: Davson, H. (ed.) *The Eye*, pp. 207–492. Academic Press, London.

Harrington, L., Klintworth, G. K., Secor, T. E. and Breitman, M. L. (1991). Developmental analysis of ocular morphogenesis in αA-crystallin/diphtheria toxin transgenic mice undergoing ablation of the lens. *Dev. Biol.* **148**, 508–516.

Heyman, R. A., Borrelli, E., Lesley, J., Anderson, D., Richman, D. D., Baird, S. M., Hyman, R. and Evans, R. M. (1989). Thymidine kinase obliteration: creation of transgenic mice with controlled immune deficiency. *Proc. Natl Acad. Sci. USA* **86**, 2698–2702.

Hoeffler, J. P., Boockfor, F. R. and Frawley, L. S. (1985). Ontogeny of prolactin cells in neonatal rats: initial prolactin secretors also release growth hormone. *Endocrinology* **117**, 187–195.

Kaur, S., Key, B., Stock, J., McNeish, J. D., Akeson, R. and Potter, S. S. (1989). Targeted ablation of alpha-crystallin-synthesizing cells produces lens-deficient eyes in transgenic mice. *Development* **105**, 613–619.

Kessel, M., Balling, R. and Gruss, P. (1990). Variation of cervical vertebrae after expression of a Hox 1.1. transgene in mice. *Cell* **61**, 301–308.

Kohno, K., Uchida, T., Ohkubo, H., Nakanishi, S., Nakanishi, T., Fukui, T., Ohtsuka, E., Ikehara, M. and Okada, Y. (1986). Amino acid sequence of mammalian elongation factor 2 deduced from the cDNA sequence: homology with GTP-binding proteins. *Proc. Natl Acad. Sci. USA* **83**, 4978–4982.

Landel, C. P., Zhao, J., Bok, D. and Evans, G. A. (1988). Lens-specific expression of recombinant ricin induces developmental defects in the eyes of transgenic mice. *Genes Dev.* **2**, 1168–1178.

McAvoy, J. W. (1978). Cell division, cell elongation, and distribution of alpha-, beta-, and gamma-crystallins in the rat lens. *J. Embryol. Exp. Morph.* **44**, 149–165.

Maxwell, F., Maxwell, I. H. and Glode, L. M. (1987). Cloning, sequence determination, and expression in transfected cells of the coding sequence for the *tox176* attenuated diphtheria toxin A chain. *Mol. Cell. Biol.* **7**, 1576–1579.

Maxwell, I. H., Maxwell, F. and Glode, L. M. (1986). Regulated expression of a diphtheria toxin A-chain gene transfected into human cells: a possible strategy for inducing cancer cell suiside. *Cancer Res.* **46**, 4660–4664.

Moehring, T. J., Danley, D. E. and Moehring, J. M. (1984). In vitro biosynthesis of diphthamide, studied with mutant chinese hamster ovary cells resistant to diphtheria toxin. *Mol. Cell. Biol.* **4**, 642–650.

Orinitz, D. M. Palmiter, R. D., Hammer, R. E., Brinster, R. L. Swift, G. H. and MacDonald, R. J. (1985a). Specific expression of an elastase–human growth hormone fusion gene in pancreatic acinar cells of transgenic mice. *Nature* **313**, 600–603.

Orinitz, D. M., Palmiter, R. D., Messing, A., Hammer, R. E., Pinkert, C. A. and Brinster, R. L. (1985b). Elastase I promoter directs expression of human growth hormone and SV40 T antigen genes to pancreatic acinar cells in transgenic mice. *Cold Spring Harbor Quant. Biol.* **50**, 389–409.

Palmiter, R. D. and Brinster, R. L. (1986). Germ line transformation of mice. *An. Rev. Genet.* **20**, 465–499.

Palmiter, R. D., Behringer, R. R., Quaife, C. J., Maxwell, F., Maxwell, I. H. and Brinster, R. L. (1987). Cell lineage ablation in transgenic mice by cell-specific expression of a toxin gene. *Cell* **50**, 435–443.

Pappenheimer Jr, A. M. (1977). Diphtheria toxin. *An. Rev. Biochem.* **46**, 69–94.

Parsa, I., Marsh, W. H. and Fitzgerald, P. J. (1969). Pancrease acinar cell differentiation. I. Morphological and enzymatic comparisons of embryonic rat pancreas and pancreatic anlage grown in organ culture. *Am. J. Pathol.* **57**, 457–487.

Pastan, I., Willingham, M. C. and FitzGerald, D. J. P. (1986). Immunotoxins. *Cell* **47**, 641–648.

Peterson, A. C. and Bray, G. M. (1984). Hypomyelination in the peripheral nervous system of shiverer mice and in shiverer–normal chimera. *J. Comp. Neurol.* **227**, 348–356.

Richardson, K. E. Y. and Spooner, B. S. (1977). Mammalian pancreas development: regeneration and differentiation in vitro. *Dev. Biol.* **58**, 402–420.

Roach, A., Boylan, K., Horvath, S., Pruisner, S. B. and Hood, L. E. (1983). Characterization of cloned cDNA representing rat myelin basic protein: absence of expression in brain of shiverer mutant mice. *Cell* **34**, 799–806.

Roach, A., Takahashi, N., Pravtcheva, D., Ruddle, F. and Hood, L. (1985). Chromosomal mapping of mouse myelin basic protein gene and structure and transcription of the partially deleted gene in shiverer mutant mice. *Cell* **42**, 149–155.

Slabaugh, M. B., Leiberman, M. E., Rutledge, J. J. and Gorski, J. (1982). Ontogeny of growth hormone and prolactin gene expression in mice. *Endoctrinology* **110**, 1489–1497.

Van Leen, R. W., Bruer, M. L., Lubsen, N. H. and Schoenmakers, J. G. G. (1987). Developmental expression of crystallin genes: in situ hybridization reveals a differential localization of specific mRNAs. *Dev. Biol.* **123**, 338–345.

Varnum, D. S. and Stevens, J. C. (1968). Aphakia, a new mutation in the mouse. *J. Heredity* **59**, 147–150.

Wessells, N. K. (1964). DNA synthesis, mitosis, and differentiation in pancreatic acinar cells in vitro. *J. Cell Biol.* **20**, 415–433.

Wessells, N. K. and Cohen, J. (1967). Early pancreas organogenesis: morphogenesis, tissue interactions, and mass effects. *Dev. Biol.* **15**, 237–270.

Wolgemuth, D. J., Behringer, R. R., Mostoller, M. P., Brinster, R. L. and Palmiter, R. D. (1989). Transgenic mice overexpressing the mouse homeobox-containing gene Hox 1.4 exhibit abnormal gut development. *Nature* **337**, 464–467.

Yamaizumi, M., Mekada, E., Uchida, T. and Okada, Y. (1987). One molecule of diphtheria toxin fragment A introduced into a cell can kill the cell. *Cell* **15**, 245–250.

7
Transgenic mice in immunology

ANDREW L. MELLOR

Division of Molecular Immunology, National Institute for Medical Research,
The Ridgeway, Mill Hill, London NW7 1AA, UK

1. Introduction

Immunological phenomena are notoriously difficult to study at the cell or molecular level, primarily because of their dependence on the integrity of the whole organism for complete expression of the phenomena being measured but

TRANSGENIC ANIMALS
ISBN 0–12–304530–4

also because of the innate complexity of immune responses. Progress in immunology has been rapid in the last two decades as increasingly sophisticated cellular and molecular biological techniques have been applied to the study of cells of the lymphoid system. The ability to establish and maintain cultured lymphoid cells as resources for biochemical and molecular biological studies has been of critical importance in determining the source of the enormous structural diversity in the immune system and in understanding lymphoid cell differentiation in relation to development of the immune system. However, little progress has been made in understanding how the immune system is regulated during development or during immune responses. Progress in this field of immunology is limited, firstly, by the sheer complexity of cellular interactions which take place and, secondly, by the lack of success in establishing cultured cell lineages (other than certain lymphoid cells) involved in these interactions. Indeed, it is extremely difficult to find out exactly where and when critical cell interactions take place which set in motion the sequence of events leading to the manifestation of immunological phenomena *in vivo*, let alone define the cells and factors involved in these interactions which may number two or more different types of cell and a number of possible soluble regulatory factors such as interleukins. Transgenic mice offer a potentially useful way forward because they provide a means of observing the effects of introducing subtle genetic and biochemical differences into the complete developmental programme of mice. Undoubtedly, the full versatility of this technique is yet to be realized as new experimental strategies are devised and applied to problems of an immunological nature. It is my aim in this chapter to explain how the use of transgenic mice has benefited immunologically based research and why this type of approach is considered to offer great potential for future work in the field. I will outline general strategies involving the use of transgenic mice which have found application in the immunology field to date and, subsequently, I will discuss a selection of actual experiments which relate to problems in immunology. Finally, by speculating and extrapolating I will attempt to predict how the use of transgenic mice may benefit the immunologically inclined investigator in the near future.

2. Why use transgenic mice for immunological research?

2.1. Technical considerations

In this section I will discuss how modifications to biological behaviour as a result of the introduction (or removal) of specific genes by transgenesis can aid immunologically based research. For easy reference I have compiled a list of applications for transgenic mice carrying different types of transgenes (Table 1).

Table 1. Uses of transgenic mice in immunology

Transgene	Applications
$\alpha\beta$ TCR	Allelic exclusion T cell development T cell repertoire selection Production of monoclonal lymphocytes
$\gamma\delta$ TCR	$\gamma\delta$ T cell development and function $\gamma\delta$ TCR ligand specificity and restriction
IgM(D)	Allelic exclusion B cell development B cell repertoire selection
MHC I	Structural requirements for function Allogeneic responses to MHC I antigens Modification of immune responses Induction of tolerance to non-thymic self
MHC II	T cell repertoire selection Alteration of immune response phenotype Induction of tolerance to non-thymic self
F^1 (MHC × TCR)	T cell selection or tolerance in the presence of modified MHC antigens or altered pattern of MHC expression
F^1 (Ag × Ig)	B cell selection or tolerance in the presence of specific antigen
CD2, CD4, CD8	Functional characterization of human CD surface markers
Oncogenes	Targeted oncogenesis Generation of specific cell lineages
Lymphokines	Regulation of immune responses Role in differentiation

Ag, ligand for specific Ig idiotype; Ig, immunoglobulin; F^1, cross between two transgenic mouse lineages; TCR, T cell receptor.

The methods used for creating transgenic mice have been discussed elsewhere in this volume. As with all new techniques, it is important to be aware of potential artefacts arising from the application of the technique. Transgenic mice are no exceptions in this respect and I will draw attention to some apparent artefacts where these have come to light.

One advantage of transgenesis to the immunologist is that the transgene is introduced at a very early stage in development so that the effects of transgene expression are manifested as part of the normal pattern of mouse development without surgical intervention. Moreover, once created, transgenic mice can be mated to pass on the transgene to their offspring so that many mice with the transgenic phenotype can be generated. Pioneers of the transgenesis technique used hybrid strains of mice to produce oocytes for microinjection, resulting in the creation of transgenic founder mice which inherited a mixture of genes derived from at least two haplotypes carried by standard laboratory inbred strains. This creates potential difficulties for immunologists who frequently need to measure very weak immunological effects (e.g. immune responses to minor histocompatibility antigens and tumour antigens) which are influenced profoundly by the genetic background. Nor can cell transfer experiments be carried out satisfactorily between mice segregating two haplotypes. Several laboratories have now been successful in manipulating oocytes derived from particular inbred strains so that mixing of genes from different haplotypes is avoided. Use of inbred mice to produce oocytes is sometimes more demanding technically, but it is often well worth the effort of using an inbred strain at the outset of a long-term immunological project since the potential benefits of having a well-defined genetic background nearly always outweigh the extra effort expended in obtaining transgenic founder mice. In some strains (e.g. CBA in our institute) the extra effort needed to generate transgenic mice is negligible, whereas other strains are extremely difficult to use. For example, DBA/2 mice at our institute are difficult to use because the supply of females for oocyte isolation is erratic since the colony breeds poorly, the oocytes are relatively fragile and DBA/2 females produce few pups and do not look after them well. However, colonies of the same inbred strain in different institutions can behave differently in this respect. Another technical consideration concerns the nature of the DNA used for transgenesis. In my experience, it is a good idea to use large fragments of cloned genomic DNA so that transgenes retain flanking sequences and normal intron/exon organization. This policy is the best insurance against 'aberrant' expression patterns in the resulting transgenic mice arising from the loss of essential regulatory elements from the surrounding DNA. Recombinant transgenes containing DNA elements from more than one cloned gene are also used to make transgenic mice. Care should be exercised in making sure, by all available means, that the pattern of expression of these recombinant transgenes is as close as possible to the expected pattern and that it is not altered in the recombinant genes. The copy number and site of integration of the transgene cannot be controlled and these factors may influence the pattern or level of expression of the transgene in particular founder mice. Thus, it is very important to establish more than one lineage carrying a particular transgene to avoid misleading results arising from aberrant expression in one lineage.

2.2.   The ultimate congenic mouse

Congenic mouse strains produced by repeated backcrossing and selection for the appropriate phenotypic marker have been of enormous value in immunogenetic research. Nevertheless, congenic strains retain variable, but sometimes extensive, stretches of adjacent linked DNA carried along with the gene under selection. Consequently, there is always some degree of uncertainty in assigning traits to particular genetic loci. One of the most useful features of transgenic mice is that they differ genetically from the parental mice used to produce the oocytes (or from the parental embryonic stem (ES) cell line) by a single gene. Thus, phenotypic differences between transgenic and matched non-transgenic mice can be interpreted as direct or indirect effects of the expression of a single extra gene. In this sense, transgenic mice represent perfect congenic strains since there can be no doubt about the genetic basis of particular phenotypic effects.

2.3. Monoclonal lymphoid populations

Because of the extraordinary receptor diversity generated during T and B lymphocyte development, the frequency of a given receptor structure (idiotype) amongst lymphocyte populations in mice is extremely low. One way round this is to introduce a productively rearranged receptor gene (obtained by molecular cloning of the gene from a mature T or B cell expressing a given receptor molecule) into mice by transgenesis. Due to the allelic exclusion effect (see below), T or B cell populations can be generated in which most of the lymphocytes express receptors of only one type, those encoded by the transgene. Transgenic mice carrying particular T cell receptors (TCRs) or immunoglobulin (Ig) receptors have been created in a number of laboratories using this approach and they have already proved to be very useful in immunological studies to several ends. However, the allelic exclusion effect is often not complete, so that the proportion of lymphocytes bearing the receptor molecule encoded by the transgene is less than 100%, although it is likely that this effect can be eliminated by backcrossing the transgene onto SCID (severe combined immunodeficiency syndrome) mice which are unable to rearrange endogenous receptor genes due to a defect in a DNA recombinase gene. In addition, unexpected distortions in the absolute number and distribution of lymphocyte subsets defined by surface markers (e.g. CD4 and CD8 for T cells) have also been reported in these types of transgenic mice and this could have unexpected functional consequences. Indeed, the fate of thymocytes and T cells in TCR transgenic mice has been used to provide evidence for events which influence T cell repertoire selection in several studies on tolerance induction whilst the expected functional consequences of these selection events are not manifested (E. Simpson, personal communication), possibly due to 'overloading' of the normal selection systems in TCR transgenic mice. Consequently, caution should be exercised when extrapolating from the results of studies on T cell fate in such mice unless functional tests are also carried out on the same mice. Despite, this, there is no doubt that TCR and

Ig transgenic mice have already been extremely important in immunological research, particularly in understanding how tolerance in the T and B cell compartments is induced and maintained. This strategy can also be used to generate large numbers of identical T or B cells for biochemical or cell biological studies *in vitro*.

2.4. Targeting specific genes or cell lineages

The ability to regulate the expression of particular molecules of interest by constructing recombinant genes incorporating heterologous transcriptional promoter and other regulatory elements has proved to be of enormous use in immunological research. Thus, the use of a tissue-specific promoter allows the investigator to target gene expression to particular cell lineages. A favourite promoter amongst many workers interested in the study of self-tolerance and autoimmunity is the insulin gene promoter which is only active in β islet cells of the pancreas (e.g. see Hanahan, 1985; Adams *et al.*, 1987). By using tissue-specific promoters to drive expression of MHC or viral structural genes it is possible to study the immunological consequences of this restricted pattern of gene expression on tolerance and/or on the development of autoimmune diseases such as diabetes as a result of destruction of the insulin-secreting β islet cells.

Using the ES cell route of transgenesis it is possible to inactivate a particular gene of interest by insertional mutagenesis and so generate transgenic mice which inherit a specific genetic defect. This technique has great potential in immunological research for determining cell lineage relationships and testing for immunological consequences of either removing specific types of cells from mice or of removing specific cell markers such as differentiation antigens and antigen receptors. Alternatively, specific cell lineages can be eliminated in transgenic mice by introducing transgenes which confer susceptibility to a cytotoxic drug into cells which express the transgene during development or in experimentally determined circumstances. Target cell lineages for cytotoxic drugs are selected using tissue- or cell type-specific transcriptional promoters to drive the structural gene conferring sensitivity to the drug. For example, administration of the nucleoside analogue gancyclovir to transgenic mice carrying a herpes simplex virus type 1 thymidine kinase (HSV-1 *tk*) gene driven by an Ig gene promoter causes massive destruction of B and T cell lineages, whereas there is no effect on non-lymphoid tissues which do not express the HSV-1 *tk*) gene (Heyman *et al.* 1989). The effect is reversible in the absence of gancyclovir, since a few progenitor cells are not affected by the drug treatment and the lymphoid populations reconstitute after 7 days. Similarly, T cell lineages only are targeted for ablation if the *Thy-1* T cell-specific promoter is used to drive transcription of the HSV-1 *tk* gene (E. Dzierzak, personal communication). This system provides the investigator with a very versatile system for studying the dynamics of complex systems such as the immune system since T cells can be eliminated at particular developmental stages by administering the drug gancyclovir during gestation or at any stage after the birth of transgenic animals. The application of this technique is restricted by the availability of genes which are expressed in a tightly regulated pattern so that a limited range of cell types or lineages are targeted. In addition, it is not clear whether drugs can be delivered in appropriate doses to target cells to cause effective elimination of all target cells. A variation of this approach is to use a tissue- or cell type-specific transcriptional promoter to drive expression of an oncogene in, for example, pancreatic β islet cells using the insulin gene promoter (Hanahan, 1985) or T cells using the *Thy-1* gene promoter (D. Kioussis, personal communication). In this way it is possible to engineer a high rate of tumour incidence into particular cell lineages in transgenic mice. This should help to generate new cell lines from tissues which have proved to be inaccessible to more conventional methods of establishing cell lines for cell biological studies. Progress in understanding a number of immunological phenomena could result from success using this approach to generate new cell lines. Another variation of this approach is to use temperature-sensitive oncogene products so that the time of onset of oncogenesis can be controlled by manipulation of temperature (P. Jat, personal communication).

2.5. Models for disease

Many diseases, unless they result directly from genetic defects, involve the immune system at some stage in their progression or amelioration. Immunologically mediated resistance to disease frequently displays a measurable genetic component but the extent of this is often difficult to assess because of the complexity of disease processes. This is particularly true in the study of human disease when individuals with very different genetic backgrounds have to be studied. The mouse is already a useful model system for a number of diseases with a marked genetic component but the ability to introduce new or mutated genes onto well-defined genetic backgrounds makes the use of transgenic mice a very attractive area for future studies into more complex diseases such as cancer (tumour growth and rejection) and progressive autoimmune disorders, particularly when there is evidence for a connection between the MHC and disease. For example, transgenic mice have proved useful in refining our understanding of autoimmune effects which lead to diabetes in the NOD (non-obese diabetic) mouse model system (Lund *et al.*, 1990).

3. Using transgenic mice in immunological research

3.1. Development of the immune system

T and B lymphocytes differentiate and mature during later fetal life. This involves complex cellular interactions in the fetal bone marrow, thymus, lymph nodes and spleen. Receptor diversity is generated by random rearrangements of genes coding for antigenic receptor molecules during lymphocyte differentiation and allelic exclusion ensures that a single type of receptor is expressed on each lymphocyte because productive rearrangement of a gene on one chromosome prevents rearrangements on the second chromosome. One of the most fascinating, and puzzling, aspects of the immune system is how the T and B cell effector systems cope with the dual requirements for (1) maximization of receptor diversity to generate specificity against a wide range of foreign agents and (2) lack of response to structures displayed on cells which are part of the organism (i.e. self structures). These seemingly opposed criteria are reconciled in the clonal deletion hypothesis in which it is proposed that naive, immature T and B cell precursors displaying TCR or Ig molecules are eliminated from the developing repertoire if they recognize self structures. There is good evidence, some of which involves the use of transgenic mice, for this model in both the T cell and B cell compartments. However, very little is known about the cellular interactions underlying the selection processes which are an integral part of the clonal deletion model.

3.1.1. Allelic exclusion

The molecular basis of allelic exclusion is not understood. It is assumed that expression of surface IgM or of the TCR/CD3 complex on the developing B or T cell precursors leads to the generation of a signal which blocks further rearrangements of DNA in the same cell. Introduction of cloned DNA containing productively rearranged Ig or TCR genes isolated from B or T cell clones results in blocking of rearrangements of endogenous Ig (e.g. see Goodnow *et al.*, 1988) or TCR genes (e.g. see Kisielow *et al.*, 1988) in transgenic mice. This effect is not complete, probably due to variations in the timing and/or pattern of transgene expression with respect to endogenous receptor genes. However, this system is extremely useful for studies designed to test for immunological consequences of the expression of particular Ig or TCR idiotypes of defined antigen specificity (see below). Potentially, this system can also be exploited to find out more about the underlying mechanisms controlling allelic exclusion. For example, if transgenic mice carrying a receptor gene in germ line configuration, but which undergoes productive rearrangement in somatic cells, is mated to another lineage of transgenic mice carrying a productively rearranged receptor transgene it should be possible to define the *cis*-acting control elements and, ultimately, the *trans*-acting factors which are involved in the process of allelic exclusion.

3.1.2. The T cell repertoire

One of the predictions of the clonal deletion hypothesis is that the repertoire of T cell antigen recognition specificities will be influenced by the complement of self-determinants present during T cell development. This prediction has been confirmed by, amongst other approaches, experiments with transgenic mice carrying MHC transgenes encoding an I-Eα gene in an I-E$^-$ host background (Marrack *et al.*, 1988). As a result of expression of the I-E molecule in the transgenic mice, T cells expressing certain V_β TCR chains are eliminated during development in the thymus in a process termed negative selection. Using a different approach, but addressing the same issue, it has been shown that productively rearranged TCR α and β chain genes, isolated from a T cell clone specific for the male-specific H-Y antigen in the context of H-2D^b MHC class I molecules, produce a high proportion of thymocytes expressing receptors encoded by the TCR $\alpha\beta$ chain genes when introduced into transgenic mice (Kisielow *et al.*, 1988). By following the fate of these thymocytes in male and female transgenic mice of H-2^b haplotype it was shown that the presence of the H-Y antigen in the thymus of male mice results in the elimination of thymocytes bearing transgenic TCRs whereas T cells with the same receptors were preferentially expressed on mature CD8$^+$ T cells in the peripheral, mature T cell population of female mice. This provides additional evidence for negative selection in the male thymus and evidence that positive selection, involving H-2D^b molecules, is required for maturation of thymocytes in female mice. The use of transgenic mice for this research allows unequivocal characterization of the I-E molecule as a determinant of repertoire selection during T cell development and, in the second example, effectively amplifies the proportion of T cells bearing one particular receptor idiotype to easily detectable levels. These are distinct advantages in a field of research which has had to rely on assumptions about the genetic differences between congenic mouse strains and on very careful and precise measurements of very small changes in the proportions of T cells bearing particular idiotypes in T cell populations. However, the pattern of TCR $\alpha\beta$ chain gene expression in transgenic mice is often aberrant since (1) not all T cells in the mice express the transgene TCR idiotype and (2) T cell subsets, as measured by the CD4 and CD8 surface markers, are often present in very different relative numbers compared to normal mice. These aberrations almost certainly arise from technical effects such as transgene copy number and/or the site of integration in the host DNA which leads to inappropriate temporal and/or spatial expression of the transgenic TCR $\alpha\beta$ chain genes in transgenic mice. Nevertheless, this does not alter the interpretation of experiments to test the clonal deletion hypothesis, since these derive significance from the fact that the presence of a particular self molecule causes profound alterations to the repertoire of T cell receptors which are displayed on mature T cells. It would be much more difficult to rely on evidence from transgenic mice if more subtle information on the development or regulation of T lymphocytes was sought. Improvements in knowledge about gene expression and in targeting transgenes to particular chromosomal sites should help remove these technical difficulties, enabling transgenic mice to be used for applications of this type in the future. Definition of the cellular and molecular interactions underlying repertoire selection depends on establishing cell lines *in vitro* which mimic these events. Progress in this direction is likely to follow from the creation of transgenic mice carrying oncogenes driven by T cell-specific transcriptional promoters such as *Thy-1* and CD2. Mice carrying these transgenes develop thymic tumours at high frequency from which both epithelial and lymphoid cell lines have been established in culture (D. Kioussis, personal communication). Some of the epithelial cell lines adhere to lymphoid cells and it is possible that further studies will reveal information about the cell interactions which cause positive and negative selection to take place *in vivo*.

3.1.3.   The B cell repertoire

A similar approach has been used to study the development of the B cell repertoire. As with $\alpha\beta$ TCR transgenic mice, Ig molecules with defined antigen-binding characteristics are expressed on a high proportion of B cells in transgenic mice carrying productively rearranged Ig genes. The fate of these B cells can be tested in different genetic strains of mice, some of which express the ligand bound by the transgene encoded Ig molecule, and the results compared to the situation in mice which do not express the same ligand. When the ligand concerned was surface H-2K^k molecules (Nemazee and Burki, 1989), all B cells expressing transgenic Ig receptor with H-2K^k binding specificity were eliminated at an immature stage in development, demonstrating that clonal deletion of immature lymphocytes occurs in the B cell compartment as well as in the T cell compartment when self antigens are present in the bone marrow or thymus compartments, respectively. This experiment was extended to find out what effect the biochemical nature of the antigenic molecule has on clonal deletion. Thus, when mated to transgenic mice expressing a soluble form of the H-2K^k molecule, clonal deletion did not take place in the bone marrow compartment (D. Nemazee, personal communication). Consequently, potentially self-reactive B cells were allowed to mature in these mice. Despite this, the B cells did not respond to self-soluble H-2K^k molecules encountered in the blood and so a state of tolerance was maintained by an alternative mechanism to clonal deletion. Other studies of a similar nature have helped shed light on this peripheral tolerance mechanism (see below).

3.1.4. TCR $\gamma\delta$ chain-bearing T cells

Despite superficial similarity between $\gamma\delta$ and $\alpha\beta$ TCR molecules, the role of T cells bearing $\gamma\delta$ TCR molecules during immune responses is not known. Recently, transgenic mice bearing $\gamma\delta$ TCR molecules on a large proportion of T cells have been created using productively rearranged $\gamma\delta$ TCR genes. Clonal deletion of these T cells takes place only in the presence of self molecules encoded in the Tla region closely linked to the H-2D region of the MHC (Dent A. L. *et al.*, 1990). The exact identity of the self molecule involved in this selection process is not known and awaits further study. The relationship between $\gamma\delta$ and $\alpha\beta$ chain-bearing T cell lineages has also been studied using these mice. The results imply that the two types of T cells develop as distinct lineages from an early common precursor.

3.2. Modification of immune responses

General effects on immune responsiveness due to the introduction of particular genes are readily demonstrated in transgenic mice. Many studies involve the introduction of novel MHC class I or class II genes into the germ line of specific inbred strains of mice in order to study the genetic control of immune responses directed against allogenic MHC molecules and against infectious agents such as viruses.

3.2.1. Responses to MHC class I molecules

Immune responses to murine H-2 and human HLA class I gene products have been studied in detail using transgenic mice. A whole range of immunological tests such as tissue grafting and T cell or B cell proliferative and cytotoxic assays for responsiveness are available to the investigator as tissues from the mice expressing the transgenic class I molecule can be used to provoke immune responses from genetically matched, non-transgenic, responder mice (e.g. non-transgenic littermates when the transgene is inserted into an inbred strain of mice). Thus, skin grafts from H-2 class I transgenic mice provoke a strong T cell response in matched non-transgenic responder mice which results in graft rejection. T cell responses causing graft rejection can be studied in detail by culturing T cells from the responder mice *in vitro* in the presence of appropriate stimulator cells from the transgenic mice (MLC, mixed lymphocyte culture) and testing for antigen-specific cytotoxicity or proliferation. Non-classical murine MHC class I molecules, such as Qa-2 molecules, also behave as strong transplantation antigens when expressed in transgenic mice and their tissues are grafted onto non-transgenic responder mice (Mellor *et al.*, 1991). This is surprising since the same molecules provoke only very weak responses when grafted between normal mice. The reason for this distinction seems to be connected with higher levels of expression of the molecules in transgenic mice which may also affect the pattern of expression of the molecules in, for example, skin tissue so that specialized antigen-presenting cells may express raised levels of MHC molecules. Studies with HLA class I transgenic mice have revealed that HLA-A, -B and -C class I molecules behave as strong transplantation antigens and function as restriction elements for virus-specific cytotoxic responses (e.g. see Dill *et al.*, 1988). Despite this, HLA molecules fail to affect positive selection of murine T cells in the thymus of HLA transgenic mice. When the $\alpha 3$ domain of the human HLA class I molecule is replaced by a murine $\alpha 3$ domain, the human/mouse hybrid class I molecules are able to affect positive selection of T cells in transgenic mice. Further mapping studies suggest that a murine CD8-binding site located in the $\alpha 3$ domain is required for positive selection and that the human $\alpha 3$ domain cannot interact with murine CD8 (G. Hämmerling, personal communication).

B cell responses to MHC class I molecules can also be studied in transgenic mice using the immunization procedures described above. Potentially, HLA transgenic mice can be used to generate very specific B cell (antibody) responses to particular HLA antigens and so result in the production of new and better monoclonal antibodies for research and tissue typing.

3.2.2. Responses to MHC class II molecules

Immune responses in mice carrying murine or human class II transgenes have been studied in a similar way in order to investigate functional effects due to particular class II molecules on immune responses. Murine I-E, I-A and human HLA-DR, -DQ and -DP transgenic mice have been created by various groups for such studies. MHC class II genes can be expressed in an appropriate tissue-specific manner and induction of class II gene expression in the presence of certain lymphokines is maintained by the transgene if appropriate DNA sequences containing transcriptional control elements are retained in the transgene construct. Deletion of DNA sequences containing control elements for MHC class II gene transcription has not only helped in defining the elements themselves but has also resulted in the generation of transgenic mice with restricted expression of the transgenic MHC class II molecule. These mice have proved to be extremely useful in determining which cell types function as presenting cells during selection events in the thymus (Marrack *et al.*, 1988; Benoist and Mathis, 1989). In addition, specific modifications to immune responsiveness can be engineered in transgenic mice by this method. In some cases this can be correlated with an alteration to disease susceptibility phenotype generating a useful system for autoimmune disease study.

3.3. *Induction of tolerance*

One of the most perplexing problems in contemporary immunological research concerns tolerance, not least because failure of tolerance to self is thought to be the mechanism by which pathology develops in several autoimmune diseases. Tolerance is manifested as a state of specific unresponsiveness to structures displayed on an individual's own cells, thus avoiding damaging autoimmune responses (Schwartz, 1989). In some cases, self-tolerance can be explained because the self antigen is sequestered in a specific compartment which is not accessible systemically (e.g. eye lens proteins, thyroglobulin and sperm proteins) or when B cells with appropriate self-reactive receptors exist but, in the absence of T cell help, do not respond (i.e. T cells only are tolerant). However, tolerance to most self molecules, including MHC molecules, is thought to arise by alternative mechanisms. Until recently, arguments revolved around three models invoked to explain how a state of unresponsiveness to self could be generated; clonal deletion, clonal anergy and clonal suppression of self-reactive lymphocytes. In the last 3 years, with the introduction of new reagents and techniques, including transgenesis, experimental support for the clonal deletion model has accumulated. Nevertheless, evidence for additional mechanisms controlling tolerance status has also been obtained, notably from studies involving the creation of transgenic mice which express MHC or viral antigens in a restricted, tissue-specific pattern.

3.3.1. *Clonal deletion*

T cells with receptor specificity for self MHC molecules cannot be detected in the extrathymic compartment by limiting dilution analysis, despite the fact that they are detected amongst immature thymocytes in the thymus. Direct evidence that the complement of self molecules present in the thymus influences selection of the T cell repertoire has been obtained by following the fate of thymocytes bearing particular TCR V_β chains which confer receptor specificity for murine I-E MHC class II or *Mls* molecules (Marrack *et al.*, 1988; Benoist and Mathis, 1989) or by following the fate of T cells expressing transgenes encoding TCR

$\alpha\beta$ chains with specificity for the male-specific H-Y antigen in association with H-2D^b in male and female mice (Kisielow *et al.*, 1988). As discussed above (Section 3.1.2), there is evidence that a clonal deletion mechanism operates in bone marrow, causing immature B cells expressing transgenic H-2K^k-specific Ig molecules to be eliminated before they emerge from the bone marrow of H-2^k haplotype mice (Nemazee and Burki, 1989).

3.3.2. Peripheral tolerance

The concept that T cell tolerance is acquired entirely by clonal deletion mechanisms is inconsistent with results from a series of experiments involving studies of transgenic mice carrying recombinant MHC class I or class II genes designed to be expressed only in the extrathymic environment (results are summarized in Table 2; for reviews, see, J. F. A. P. Miller *et al.*, 1989, 1990). As indicated in Table 2, several different transcriptional promoters have been used to drive transcription of either MHC class I or class II genes. In all studies reported to date, mice expressing transgenic MHC molecules in the extrathymic environment only were immunologically tolerant, as revealed by absence of pathological damage caused by lymphocyte infiltration in the tissue concerned and, when MHC class I transgenes are involved, failure to reject skin grafts and/or inability to mount cytotoxic responses against transgenic self MHC molecules. Nevertheless, tolerance to transgenic self MHC molecules could be broken down, in all but two transgenic mouse systems, by culturing T cells taken from spleen and/or thymus and testing them for specific responsiveness (sometimes T cell growth factor, IL-2, was required) to appropriate target cells *in vitro* (see Table 2). Thus, tolerance of self MHC is acquired in the peripheral, extrathymic environment in these specially engineered transgenic mice since the presence of T cells able to recognize the self MHC molecule could be demonstrated in the periphery of such mice. It is not known how a state of functional unresponsiveness (anergy) is induced in mature T cells but it may result from cell interactions which take place in the absence of appropriate co-stimulatory signals (perhaps from an accessory cell) and which lead to a failure to T cell proliferation, as opposed to activation, in the presence of specific ligand. A subtly different mechanism of tolerance is revealed by transgenic mice which express a viral polypeptide from lymphocytic choriomeningitis virus (LCMV) exclusively in pancreatic β islet cells (Oldstone *et al.*, 1991). No pathology or autoimmunity develops in these mice except when the mice are infected with LCMV, whereupon the mice rapidly become diabetic as a result of lymphocyte infiltration into the β islet cells. More recently, this type of model system to detect peripherally induced tolerance has been extended by generating double transgenic mice which carry a second transgene coding for a TCR idiotype capable

Table 2. Summary of results from studies of transgenic mice carrying recombinant MHC genes

Transgene		Immunological status (tolerance)		Transgene specific T cell proliferation (in vitro)[2]	Reference
MHC gene	Promoter	Pathology	Graft rejection (in vivo)[1]		
I-A^d	Elastase	No	NT	Yes	Murphy et al. (1989)
I-E	Insulin	No	NT	No	Lo et al. (1988)
I-A^k	Insulin	No	NT	Yes	Boehme et al. (1989)
I-E	Elastase	No	NT	Yes (weak)	Lo et al. (1989)
I-A^d	Insulin	No	NT	Yes	J. Miller et al. (1990)
H-2K^b	Insulin	No[3]	NT	No (−IL-2) / Yes (+IL-2)	Morahan et al. (1989a)
H-2K^b	Metallothionein	No	No	Yes	Morahan et al. (1989b)
H-2K^k	Soluble H-2K^k	No	NT	Yes	Arnold et al. (1988)
H-2K^b	Milk protein	No	No	No (−IL-2) / Yes (+IL-2)	A. L. Mellor et al. (unpublished)
H-2K^b	β-Globin	No	No	Yes (−IL-2)	Yeoman and Mellor (1992)
Q10/L^d	Q10 (liver)	No	NT	No	Jones-Youngblood et al. (1990)

NT, not tested.

[1] Tested by grafting skin or tumour cells expressing the MHC antigen encoded by the transgene.

[2] Tested by mixed lymphocyte culture in presence (+) or absence (−) of IL-2.

[3] These mice develop diabetes but this is not due to lymphocyte infiltration.

of recognizing the transgenic self antigen encoded in the extrathymic environment. Thus, Schönrich *et al.* (1991) reported that mice expressing H-2K^b molecules exclusively in neural cells under the control of a promoter from the glial fibrillary acidic protein gene are tolerant of H-2K^b, even when the mice also carry a TCR transgene able to recognize H-2K^b molecules. By following the fate of thymocytes and T cells in these double transgenic mice it is clear that tolerance is maintained, not by clonal deletion, but by anergy, and that this functionally unresponsive state is correlated with down-modulation of TCR molecules on the surface of self-reactive T cells. A similar model system has also been used by Ohashi *et al.* (1991) but in this case tolerance seems to be maintained passively since self-reactive T cells could be activated by infecting mice with a virus expressing the target self-antigen. This raises the possibility that some potentially self-reactive T cells fail to cause autoimmunity because the cognate self-antigen is expressed only on cells which are unable to activate T cells.

Similarly, peripheral mechanisms of tolerance induction have been demonstrated for B cells using transgenic mice expressing hen egg lysozyme (HEL) under the control of an inducible (metallothionein) promoter which was crossed to an Ig (IgM) transgenic mouse lineage expressing a high proportion of B cells with receptor specificity for HEL (Goodnow *et al.*, 1988). Tolerance was rapidly induced in the B cells when HEL levels were up-regulated and this effect correlated with down-regulation of IgM levels on the surface of B cells. Clearly, the biochemical, cellular and molecular biological events underlying these mechanisms of tolerance induction are the next set of goals for research. The model systems afforded by the transgenic mice lineages described in this chapter will be of great use in the next stage of this research. The goal of this research is to arrive at a better understanding of tolerance so that it will become easier to determine why, in certain clinical autoimmune disorders, the state of tolerance to self structures which normally exists seems to break down, leading, in extreme cases, to debilitating or even fatal consequences.

3.4. Disease models

As alluded to in Section 2.5, transgenic mice should provide new ways of studying disease aetiologies, particularly when susceptibility has a clearly defined genetic component. Susceptibility to infectious agents, such as viruses, is partly controlled by the MHC haplotype of an individual. Simply by introducing a new MHC molecule onto a 'non-responder' phenotypic background for a particular virus or protein antigen, the immune response phenotype can be changed into a 'responder' phenotype. This profound change is often due to the ability of the new MHC molecule to complex with a processed peptide derived from the foreign agent and present it to helper or effector T cells. However, subtle effects on the selection of the T cell repertoire can also account for changes to the immune response phenotype. Because it is easy to introduce precise structural changes to proteins via site-directed mutagenesis of cloned DNA, this type of system is a potentially very powerful way of dissecting the effects of minor structural changes on complex events leading to altered immune responses. This approach has already been used in studies of murine autoimmune disease models such as NOD mice (e.g. see Lund *et al.*, 1990) and it seems likely that other autoimmune disease models in mice such as experimental autoimmune encephalitis (EAE, a model for multiple sclerosis) will soon benefit from the application of transgenic mouse technology. Thus, introduction of I-Eα or modified I-Aβ molecules into the NOD mouse genetic background prevents diabetes developing in transgenic mice. Of course, it may not be possible to extrapolate directly from studies using transgenic mice to human diseases. Indeed, there are unconfirmed reports that whereas transgenic HLA-B27 mice fail to develop any signs of the HLA-B27 linked disease called ankylosing spondylitis, HLA-B27 transgenic rats do develop pathological conditions which, in some respects, are reminiscent of the human disease. Nevertheless, results from well-defined model situations engineered using transgenic mice should be of assistance in understanding the underlying principles and processes involved in autoimmune and infectious diseases and may lead to some surprises concerning the basis of or predisposition to disease. Indeed, the double transgenic model system engineered by Ohashi *et al.* (1991) is remarkable, not only for the elegant demonstration that tolerance can arise passively through failure of self-reactive T cells to be activated in the periphery, but also because it emphasizes an old idea that autoimmunity could arise following an immune response activated by an exogenous agent which causes activation of self-reactive T cells.

3.5. Miscellaneous

The list of transgenic mice carrying genes of interest to immunologists is rapidly growing. For example, human T cell markers including CD2, CD4 and CD8 have been successfully expressed in transgenic mice. Possession of these mice is very useful for determining whether such surface molecules are necessary or sufficient in immunological processes. For example, T cells from transgenic mice expressing human CD2 molecules can be activated by anti-CD2 monoclonal antibodies. In addition, the thymus of such mice becomes smaller at earlier times than in normal mice. Preliminary results suggest that this may arise due to increased CD2-mediated cell adhesion between differentiating T cells and other cells in the thymus, which accelerates the rate of thymus degeneration or 'ageing' (D. Kioussis, personal communication). The possibility that serendipitous observations such as this will lead to new insights into immunological processes is another important reason for using transgenic mice. Although it has proved difficult to obtain correctly regulated expression of CD4 and CD8 molecules, transgenic mice expressing these human T cell surface markers will be extremely useful in studies on the role of these molecules in T cell differentiation and selection in the thymus. Moreover, CD4 transgenic mice could be useful in research on human immunodeficiency virus which uses CD4 molecules as receptors mediating entry into target cells. A major focus of attention in current immunological research is the nature and characterization of factors affecting lymphocyte activation, differentiation, anergization and adhesion during cell interaction events. Mice carrying transgenes coding for lymphocyte-associated protein kinase (lck; R. Perlmutter, personal communication), human CD2 (D. Kioussis, personal communication) and murine Qa-2 molecules (Robinson *et al.*, 1989) have been used to investigate structural, functional and developmental aspects of lymphocyte activation. Transgenes coding for several lymphokines have also been introduced into mice. These studies are at an early stage and suffer from the problem that the resultant transgenic mice often have profound, sometimes lethal, alterations in the way their immune systems develop as a consequence of the increased amounts of lymphokines present throughout development. Nevertheless, it has been possible to show unequivocally that human IL-5, or eosinophil differentiation factor, causes massive increases in the numbers of eosinophils when expressed in transgenic mice under the control of a T cell-specific (CD2) transcriptional promoter element (Dent L. A. *et al.*, 1990). Transgenic mice have also been used to further studies into the phenomenon of resistance to tumour growth which seems to involve a natural killer (NK) cell activity directed at cells failing to express self MHC antigens (Höglund *et al.*, 1988; Öhlen *et al.*, 1989). Thus, tumours that grow well in one strain of mice grow very poorly in transgenic mice carrying a novel MHC class I gene on the same genetic background, presumably because the tumour cell now fails to express the full complement of self MHC determinants and becomes a target for NK cells sensitive to the MHC complement of target cells.

The ability to target and/or to delete particular cell surface molecules and/or cell types involved in these events is one way in which transgenic mice are proving useful for this aspect of immunology. A remarkable example of this approach is the creation of transgenic mice which fail to express any surface MHC class I molecules as a result of deletion of the β_2-microglobulin gene by insertional mutagenesis in ES cells (Zijlstra *et al.*, 1990). These mice develop normally except that $CD4^-8^+$ T cells cannot be detected, elegantly demonstrating the requirement for MHC class I molecules during T cell development in order for cytotoxic $CD8^+$ T cells to mature. It will be of interest to learn how these mice cope with infectious agents and tissue grafts so that the relative contribution of cytotoxic effector T cells to various immune responses can be assessed. At present, there is much interest in 'knocking out' a variety of genes, such as CD4,

CD8, lymphocyte adhesion molecules and lymphokine genes, which code for molecules implicated in regulation of immune responses. If they are viable as adult mice, these transgenic mice will be very useful in unravelling some of the complexity of the immune system and, undoubtedly, there will be a few surprises in store along the way.

4. The future of transgenic mice in immunology

Transgenic mice provide a versatile and accessible experimental system with which to study complex processes and resolve them into simpler, easily understood stages or steps. Because of the nature of immunological research with its reliance on the integrity of the whole organism for complete expression of many of the phenomena under scrutiny, transgenic mice have already played a useful role in resolving many aspects of contemporary immunological problems. Improvements in the technique of transgenesis as well as wider use of this technique by researchers interested in immunological problems will no doubt continue be a valuable aid to progress in immunology.

Regulation of immune responses is a very important subject that has barely been touched by new molecular and cell biological approaches which have contributed largely to an increase in knowledge concerning structural, rather than regulatory, aspects of immune responses. A very important goal in many areas of immunological research is the establishment of well-defined *in vitro* systems which mimic events underlying complex phenomena such as repertoire selection, lymphocyte differentiation, activation, anergy and suppression. In the near future, improved analysis of the cellular events underlying these immunological phenomena will provide new information concerning cellular and, ultimately, molecular mechanisms. The ability to target specific cell lineages and cause ablation or oncogenesis in transgenic mice will be a very useful way of studying the cell biology of immunological processes. In addition, the availability of almost homogeneous cell populations from such mice will be a useful source of material for further biochemical and molecular studies once the cell biology is better understood.

Study of disease will also be a beneficiary of wider application of this technology because transgenic mice can be manipulated in very subtle ways without causing major disruption to development or integrity of the mice. Increasingly, immune responses to various pathogens, including parasitic organisms, will be studied using model systems tailor made in transgenic mice. Of course, mice and humans are very different but it is to be hoped that general biological principles not accessible by any other route of study will be forthcoming from studies in mice. Susceptibility to diseases with more complex aetiologies such as tumour growth (cancer) and autoimmune diseases will also be subjected to more detailed scrutiny using transgenic mice. Promising results with transgenic mice have already been obtained in several model systems and further studies should soon give new insight into genetic and other predisposing factors promoting such diseases.

Acknowledgements

I thank all my colleagues who allowed me to quote some of their most recent, sometimes unpublished, experimental results. In particular, I thank Dr Dimitris Kioussis, Dr Elaine Dzierzak, Dr Parmjit Jat, Dr David Nemazee and Professor Gunther Hämmerling. In addition, I thank Dr Elizabeth Simpson and Dr Helen Yeoman for their constructive comments.

References

Adams T. E., Alpert S. and Hanahan D. (1987). Non-tolerance and autoantibodies to a transgenic self antigen expressed in pancreatic β cells. *Nature* **325**, 223–228.

Arnold , B., Dill O., Küblbeck G., Jatsch L., Simon M. M., Tucker J. and Hämmerling G. J. (1988). Alloreactive immune responses of transgenic mice expressing a foreign transplantation antigen in a soluble form. *Proc. Natl Acad. Sci. USA* **85**, 2269–2273.

Benoist C. and Mathis D. (1989). Positive selection of the T cell repertoire: where and when does it occur? *Cell* **58**, 1027–1033.

Boehme J., Haskins K., Stecha P., van Ewijk W., LeMeur M., Gerlinger P., Benoist C. and Mathis D. (1989). Transgenic mice with I-A on islet cells are normoglycemic but immunologically tolerant. *Science* **244**, 1179–1183.

Dent A. L., Matis L. A., Hooshmand F., Widacki S. M., Bluestone J. A. and Hedrick S. M. (1990). Self-reactive $\gamma\delta$ T cells are eliminated in the thymus. *Nature* **343**, 714–719.

Dent L. A., Strath M., Mellor A. L. and Sanderson C. (1990). Eosinophilia in transgenic mice expressing interleukin-5. *J. Exp. Med.* **172**, 1425–1431.

Dill O., Kievits F., Koch S., Ivanyi P. and Hämmerling G. (1988). Immunological function of HLA-C antigens in HLA-Cw3 transgenic mice. *Proc. Natl Acad. Sci. USA* **85**, 5664–5668.

Goodnow C. C., Crosbie J., Adelstein S., Lavoie T. B., Smith-Gill, S. J., Brink R. A., Pritchard-Briscoe H., Wotherspoon J. S., Loblay R. H., Raphael K., Trent R. J. and Basten A. (1988). Altered immunoglobulin expression and functional silencing of self-reactive B lymphocytes in transgenic mice. *Nature* **334**, 676–682.

Hanahan D. (1985). Heritable formation of pancreatic β-cell tumours in transgenic mice expressing recombinant insulin/simian virus 40 oncogenes. *Nature* **315**, 115–122.

Heyman R. A., Borrelli E., Lesley J., Anderson D., Richman D. D., Baird S. M., Hyman R. and Evans R. M. (1989). Thymidine kinase obliteration: creation of transgenic mice with controlled immune deficiency. *Proc. Natl Acad. Sci. USA* **86**, 2698–2702.

Höglund P., Ljunggren H-G., Öhlen C., Ährlund-Richter L., Scangos G., Bieberich C., Jay G., Klein G. and Kärre K. (1988). Natural resistance against lymphoma grafts conveyed by H-2D^d transgene to C57BL mice. *J. Exp. Med.* **168**, 1469–1474.

Jones-Youngblood, S. L., Weities, K., Forman J. and Hammer, R. E. (1990). Effect of the expression of a hepatocyte-MHC molecule in transgene mice on T cell tolerance. *J. Immunol.* **144**, 1187–1195.

Kisielow P., Bluthmann H., Staerz U. D., Steinmetz M. and von Boehmer H. (1988). Tolerance in T-cell receptor transgenic mice involves the deletion of nonmature CD4$^+$8$^+$ thymocytes. *Nature* **333**, 742–746.

Lo D., Burkly L. C., Widera G., Cowing C., Flavell R. A., Palmiter R. D. and Brinster R. L. (1988). Diabetes and tolerance in transgenic mice expressing class II MHC molecules in pancreatic beta cells. *Cell* **53**, 159–165.

Lo D., Burkly L. C., Flavell R. A., Palmiter R. D. and Brinster R. L. (1989). Tolerance in transgenic mice expressing class II major histocompatibility complex on pancreatic acinar cells. *J. Exp. Med.* **170**, 87–104.

Lund T., O'Reilly L., Hutchings P., Kanagawa O., Simpson E., Gravely R., Chandler P., Dyson J., Picard J. K., Edwards A., Kioussis D. and Cooke A. (1990). Prevention of murine insulin dependent diabetes mellitus by transgenes encoding a modified Aβ or a normal Eα. *Nature* **345**, 727–729.

Marrack P., Lo D., Brinster R., Palmiter R., Burkly L., Flavell R. A., and Kappler J. (1988). The effect of thymus environment on T cell development and tolerance. *Cell* **53**, 627–634.

Mellor A. L., Tomlinson D. P., Antoniou J., Chandler P., Robinson P., Felstein M. Slaon J., Edwards A., O'Reilly L., Cooke A. and Simpson E. (1991). Expression and function of Qa-2 major histocompatibility complex class I molecules in transgenic mice. *Int. Immunol.* **3**, 493–502.

Miller J., Daitch L., Rath S. and Selsing E. (1990). Tissue-specific expression of allogeneic class II MHC molecules induces neither tissue rejection nor clonal inactivation of alloreactive T cells. *J. Immunol.* **144**, 334–341.

Miller J. F. A. P., Morahan G. and Allison J. (1989). Immunological tolerance: new approaches using transgenic mice. *Immunol. Today.* **10**, 53–57.

Miller J. F. A. P., Morahan G., Slattery R. and Allison J. (1990). Transgenic models of T cell self tolerance and autoimmunity. *Immunol. Rev.* **118**, 21–33.

Morahan, G., Allison J. and Miller J. F. A. P. (1989a). Tolerance of class I histocompatibility antigens expressed extrathymically. *Nature* **339**, 662–624.

Morahan G., Brennan F., Brathal P. S., Allison J., Cox K. O. and Miller J. F. A. P. (1989b). Expression in transgenic mice of class I histocompatibility antigens controlled by the metallothionein promoter. *Proc. Natl Acad. Sci. USA* **86**, 3782–3785.

Murphy K. M., Weaver C. T., Elish M., Allen P. M. and Loh D. Y. (1989). Peripheral tolerance to allogeneic class II histocompatibility antigens expressed in transgenic mice: evidence against a clonal deletion mechanism. *Proc. Natl Acad. Sci. USA* **86**, 10034–10038.

Nemazee D. A. and Burki K. (1989). Clonal deletion of B lymphocytes in a transgenic mouse bearing anti-MHC class I antibody genes. *Nature* **337**, 562–566.

Ohashi P. S., Oehen S., Buerki K., Pircher H., Ohashi C. T., Odermatt B., Malissen B., Zinkernagel R. M. and Hengartner H. (1991). Ablation of 'tolerance' and induction of diabetes by virus infection in viral antigen transgenic mice. *Cell* **65**, 305–317.

Öhlen C., Kling G., Höglund P., Hansson M., Scangos G., Bieberich C., Jay G. and Karre K. (1989). Prevention of allogeneic bone marrow graft rejection by H-2 transgene in donor mice. *Science* **246**, 666–668.

Oldstone M. B. A., Nerenberg M., Southern P., Price J. and Lewicki H. (1991). Virus infection triggers insulin-dependent diabetes mellitus in a transgenic model: Role of anti-self (virus) immune response. *Cell* **65**, 319–331.

Robinson P. J., Millrain M., Antoniou J., Simpson E. and Mellor A. L. (1989). A glycophospholipid anchor is required for Qa-2-mediated T cell activation. *Nature* **342**, 85–87.

Schönrich G., Kalinke U., Momburg M., Malissen M., Schmitt-Verhulst A-M., Malissen B., Hämmerling G. J. and Arnold B. (1991). Down-regulation of T cell receptors on self-reactive T cells as a novel mechanism for extrathymic tolerance induction. *Cell* **65**, 293–304.

Schwartz R. H. (1989). Acquisition of immunologic self-tolerance. *Cell* **57**, 1073–1081.

Yeoman, H. and Mellor, A. L. (1992). Tolerance and MHC restriction in transgenic mice expressing a MHC class I gene in erythroid cells. *Int. Immunol.* **4**, 59–65.

Zijlstra M., Bix M., Simister N. E., Loring J. M., Raulet D. H. and Jaenisch R. (1990). β2-Microglobulin deficient mice lack $CD4^-8^+$ cytolytic T cells. *Nature* **344**, 742–746.

8
Transgenic mice in neurobiology

WILLIAM D. RICHARDSON

*Department of Biology (Medawar Building), University College London,
Gower Street, London WC1E 6BT, UK*

1. Introduction

Transgenic technology will have a profound influence on several areas of neurobiology. Developmental neurobiology will especially benefit through our ability to disrupt the function of specific genes, to express genes inappropriately during development, and to ablate particular cell lineages in the developing nervous system. Interfering with the expression of genes active in the adult, e.g. neurotransmitters and their receptors, may help us understand the workings of the mature nervous system. For those heritable diseases of the nervous system

TRANSGENIC ANIMALS
ISBN 0-12-304530-4

for which the cellular or molecular defect is known, transgenic animal models of the corresponding human disease may become available. In this chapter I discuss potential applications of the transgenic technology to the study of central nervous system (CNS) development, with special emphasis on the peripheral visual system, which is particularly suited to experimental manipulation. Much of what follows could, however, equally apply to the development of any neural or non-neural structure.

2.   Transgenic mice versus rats for neurobiology

The preferred mammalian model for developmental neurobiology has been the rat. The reasons for this are partly historical and partly experimental expedience – the rat nervous system is larger than that of the mouse for one thing. On the other hand, mouse genetics is much further advanced and essentially all transgenic work so far has been done with mice. The question arises, therefore; should a major effort be expended to develop transgenic rats for neurobiology, or should neurobiologists adapt to working with mice? Rats require much more space than mice and are correspondingly more expensive to maintain and breed. Although there are reports of success with transgenic rats (Hammer et al., 1990; Mullins et al., 1990), only experience will tell whether transgenic rats are as amenable to genetic manipulation as mice. On balance, the practical benefits of working with mice probably outweigh other considerations, even if some of our neurobiology database would need re-evaluating in mice. In any case, this will not always be necessary – the cellular architecture and development of the mouse retina, for example, is already well documented (Blanks and Bok, 1977; Young, 1984, 1985).

3.   Development of the CNS

The mature mammalian CNS is composed of two broad classes of cells, neurons and glia, which are equally numerous, although there are probably many more subtypes of neurons than glia. The extreme cellular and morphological complexity of the mature CNS belies the simplicity of the neural tube from which it develops. The neural tube starts off as a hollow cylinder of undifferentiated and morphologically indistinguishable neuroepithelial cells. Regional variations in the proliferation rate of cells towards the anterior end of the neural tube result in three bulges that eventually form the forebrain, midbrain and hindbrain, while the more posterior neural tube becomes the spinal cord. Further folding of the walls of the neural tube creates additional subdivisions that later develop

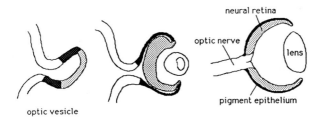

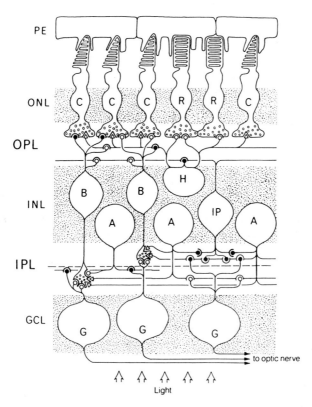

Fig. 1. (a) Development of the optic nerve and retina. The optic nerve and retina are derived from the optic vesicle, an outfolding of the neural tube. The optic nerve, pigment epithelium and neural retina are determined very early, possibly at the optic vesicle stage or before. The eye lens is not derived from the neural tube, but buds off from the overlying skin epithelium. (b) The structure and cell types of the mature vertebrate retina, adapted with permission from Dowling (1979); PE, pigment epithelium; ONL, outer nuclear layer; OPL, outer plexiform layer; INL, inner nuclear layer; IPL, inner plexiform layer; GCL, ganglion cell layer; C, cone photoreceptor cell; R, rod photoreceptor cell; B, bipolar neuron; A, amacrine neuron; H, horizontal cell; IP, interplexiform cell; G, ganglion cell.

into specialized structures such as the cerebellum, the optic stalk and the cortex. The cells of the ventricular zone, which lines the inner cavity of the neural tube, proliferate and migrate outwards into more superficial layers, where they differentiate into the specialized neuronal or glial cell types appropriate for their location. Subsequently, the proliferative zone shifts outwards a few cell diameters to the subventricular layer. In addition, other specialized germinal zones, such as the external granule layer of the cerebellum, become active. The inner fluid-filled canal of the neural tube persists, eventually giving rise to the ventricular system of the adult brain and spinal cord.

At the cellular level, development proceeds in two stages. First, cells are regionalized so that the cells in particular regions of the neural tube become dedicated to the production of only cerebellar cells, only retinal cells and so on. This regional specification of the neural tube is thought to occur very early, prior to any obvious anatomical demarcations, and at a time when there are no overt differences among cells. Secondly, the cells within each region of the neural tube diversify and differentiate into the spectrum of neural cell types peculiar to that region. For example, precursor cells in the developing optic nerve give rise to glial cells but no neurons, while cells in the adjacent section of neuroepithelium differentiate as retinal pigment epithelium, and cells in the next adjacent section give rise to the variety of photoreceptors, neurons and glial cells characteristic of the neural retina (see Fig. 1). During this second phase of cellular development, neurons extend axons and dendrites, sometimes large distances through the brain and spinal cord, and synapse with their specific targets. Also, specialized neuronal–glial interactions develop, the most obvious being the myelin sheaths that are constructed around some central axons by oligodendrocytes. To further complicate matters, these events are accompanied by long-range migrations of some neuronal and glial precursor cells, and by programmed cell deaths. Different parts of the CNS develop at different rates, so that at any one time different regions of the CNS are at different stages of development. For example, most neurons in the brain of the mouse or rat are in place several days before birth, while the majority of cerebellar and retinal neurons differentiate postnatally.

4.   Regional specification of the neural tube: transgenic approaches

Hopes that we may eventually understand the molecular events underlying regional specification of the neural tube have been encouraged by the outstanding success of molecular genetic approaches to *Drosophila* segmentation. The segmented body plan of the fly is established by a hierarchy of genes that are expressed in a defined temporal sequence and in increasingly narrow spatial domains in the *Drosophila* embryo. The segmentation genes subdivide the *Drosophila* embryo into a fixed and reproducible number of compartments; these are independent structural units in the sense that cells in neighbouring compartments do not mix with each other. A different class of genes, the homeotic genes, defines the individual identities of compartments by specifying what structures – leg parts or wing parts, for example – should be elaborated in each. The homeotic genes are expressed in the embryo in restricted anterior-posterior domains whose limits are defined by compartment boundaries. They encode proteins that contain a sequence-specific DNA-binding domain – the homeodomain – and are thought to control the transcription of many subordinate genes that collectively define cell phenotype (for a review of *Drosophila* development see Ingham (1988)).

Recent evidence suggests that regionalization of the vertebrate neural tube may be achieved in an analogous way, by sets of genes that are close relatives of the *Drosophila* homeotic and segmentation genes (for reviews see Keynes and Lumsden, 1990; Lumsden, 1990; Wilkinson and Krumlauf, 1990). First, it has been shown that the visible bulges (rhombomeres) in the developing hindbrain constitute true compartments, in that cells do not cross the rhombomere boundaries once they are established (Fraser *et al.*, 1990). Secondly, rhombomeres are domains of gene expression as well as anatomical structures: several genes that are expressed in restricted rostrocaudal domains of the neural tube have their anterior and/or posterior limits of expression at a rhombomere boundary. One such gene is *Krox-20*, a transcription factor of the 'zinc-finger' class, which is expressed only in rhombomeres r3 and r5 of the mouse (Wilkinson *et al.*, 1989a). Possibly, the most striking example is provided by the *Hox* genes, the murine equivalents of the homeotic genes of *Drosophila*. Each of the four *Hox* gene complexes contains several homeobox-containing genes that correspond almost one-for-one with particular genes in the *Drosophila* homeotic gene complexes, and are arranged in the same order within each gene cluster as their *Drosophila* counterparts (Duboule and Dollé, 1989; Graham *et al.*, 1989). Furthermore, in mice as in *Drosophila*, the order of the genes along the chromosome correlates with the anterior limits of their expression in the embryo; a *Hox* gene that is located further towards the 3 ' end of its gene cluster is expressed more anteriorly in the embryo, in some cases ceasing abruptly at a rhombomere boundary (Graham *et al.*, 1989; Murphy *et al.*, 1989; Wilkinson *et al.*, 1989b). These striking parallels between segmentation in the *Drosophila* embryo and in the vertebrate hindbrain suggest strongly that similar mechanisms are at work in both situations.

The possibility that the *Hox* genes play a role in controlling regional diversification of the neural tube has been tested in transgenic mice. The *Hox-1.1* gene, for example, has been expressed ectopically in transgenic mice under the control of the β-actin promoter (Balling *et al.*, 1989). These mice had abnormalities of neural crest-derived craniofacial structures (Balling *et al.*, 1989) and anterior–posterior transformations of the cervical vertebrae (Kessel *et al.*, 1990). These alterations are consistent with the proposed role of the *Hox* genes in specifying the anterior–posterior axis of the embryo. However, a potential pitfall of this approach is that ectopic over-expression of a gene in inappropriate places could have unpredictable effects that might tell us little about the normal function of the gene. An alternative approach is to create loss-of-function mutations by homologous recombination in embryonic stem (ES) cells and subsequently to incorporate the mutated cells into mice (Mansour *et al.*, 1988). A line of transgenic mice with a disrupted *Hox-1.5* gene has been created in this way (Chisaka *et al.*, 1991). The normal expression domain of *Hox-1.5* extends to a point in the hindbrain somewhere anterior to the otic vesicle (possibly the r4/r5 boundary). Mice homozygous for the *Hox-1.5* mutation have many defects in structures, such as the throat, thyroid and thymus, that are formed in part from neural crest cells that migrate from the hindbrain region. Surprisingly, the neural tube, including the hindbrain rhombomeres, develops apparently normally. In *Drosophila*, deletion of a particular homeotic gene causes a corresponding body segment to lose its own identity and acquire the character of its anterior neighbour. For example, certain mutations in the Bithorax Complex of homeotic genes can cause the third thoracic segment to carry a second pair of wings, instead of the halteres that it normally bears. It is not possible to interpret the phenotypes of the *Hox* mutant mice in such unequivocal terms, because large-scale cell movements and mixing during development make it difficult to assign a definite segmental origin to cells. However, the results obtained so far strongly support the idea that *Hox* genes are important for determining the fates of cells in restricted anterior–posterior regions of the embryo, including the neural plate.

Homologous recombination in ES cells has also been used to create mice that lack both copies of the *Wnt-1(int-1)* proto-oncogene (McMahon and Bradley, 1990; Thomas and Capecchi, 1990). *Wnt-1* is related to the *Drosophila* segment polarity gene *wingless*, and is normally expressed in the developing midbrain and anterior hindbrain. In the mice lacking *Wnt-1*, a large part of the midbrain and part of the hindbrain are completely absent (McMahon and Bradley, 1990; Thomas and Capecchi, 1990), supporting a role for *Wnt-1* in specifying the corresponding regions of the neural tube. A similar approach has been used to inactivate *En-2*, a mammalian homologue of engrailed, another *Drosophila* segment polarity gene (Joyner *et al.*, 1991). Mice homozygous for the *En-2* mutation develop apparently normally, perhaps because there is functional redundancy between the closely similar *En-1* and *En-2* genes of mice. The only obvious defect in the *En-2* mutant mice was in the architecture of the cerebellum where *En-2*, but not *En-1*, is normally expressed. Functional redundancy between related mammalian segmentation/homeotic genes may be a more general effect that may further complicate interpretation of targeted 'knockout' experiments in transgenic mice.

An obvious way forward is to use region-specific enhancer/promoter elements from genes such as *Hox* to direct expression of other mammalian segmentation/homeotic genes to inappropriate rostrocaudal domains of the neural tube. Such region-specific enhancer elements have now been isolated from the murine *Hox-1.1* (Puschel *et al.*, 1990) and *Hox-1.3* (Zakany *et al.*, 1988) genes and the human *Hox-5.1* gene (Tuggle *et al.*, 1990). These elements direct expression of reporter transgenes to defined regions of the brachial and upper cervical spinal cord, respectively, but not to the full range of expression sites characteristic of the corresponding endogenous genes (Zakany *et al.*, 1988; Tuggle *et al.*, 1990), suggesting that there may be additional enhancers associated with these genes that target other regions of the neural tube, including the developing hindbrain. Eventually, as more region-specific enhancers are identified, it will become possible to target expression of particular *Hox* genes, e.g. to regions of the neural tube where they are not normally expressed. Conversely, a region-specific enhancer could be used in conjunction with one of several types of *trans*-dominant inhibitory molecules (see Section 6) to block the activity of a particular *Hox* gene in a defined region of the neural tube.

5. Growth control in the developing optic nerve

Cell–cell signalling is a crucial aspect of development in all vertebrate organ systems, including the CNS. To study these events it is as well to concentrate on a region of the CNS with as few types of cells and cell–cell interactions as possible. For this reason several groups, including my own, have focused on the developing rat optic nerve. The optic nerve is typical of many white matter tracts throughout the brain and spinal cord; it contains glial cells and axons but no neuronal cell bodies, making it one of the simplest parts of the CNS and particularly suitable for studies of glial cell development, both for its own sake and as a model for more general developmental processes (for a review see Richardson *et al.*, 1990). The mature rat optic nerve contains two classes of postmitotic glial cells – oligodendrocytes, the myelin-forming cells of the CNS, and astrocytes, of unknown function. During development, oligodendrocytes are derived from one type of progenitor cell (O-2A progenitor), while the vast majority of astrocytes (type 1) are derived from a different precursor cell (Raff *et al.*, 1983). The newborn rat optic nerve contains rapidly dividing O-2A progenitors but no oligodendrocytes; starting on the day of birth and continuing through the first few postnatal weeks, some of the O-2A progenitors stop dividing and differentiate into oligodendrocytes while others continue to proliferate. Several lines of evidence indicate that platelet-derived growth factor (PDGF) drives O-2A progenitor proliferation during development (reviewed by Richardson *et al.*, 1990), but it is not known what triggers oligodendrocyte differentiation. It seems that, with time, O-2A progenitors lose the ability to divide in response to PDGF, and oligodendrocyte differentiation follows automatically as a result. However, we do not know what causes O-2A progenitors to stop responding to PDGF in the first place.

Other factors that have been proposed to play a role in the development of the O-2A lineage are the fibroblast growth factors (Eccleston and Silberberg, 1985; Saneto and deVellis, 1985; Bögler *et al.*, 1990; McKinnon *et al.*, 1990), insulin and/or insulin-like growth factor I (IGF-I; McMorris and Dubois-Dalcq, 1988) and transforming growth factor β (TGF-β; Van Obberghen-Schilling *et al.*, 1987). Naturally, one would like to know what roles, if any, these polypeptides play in the development of the O-2A lineage *in vivo*. To determine this it will be necessary to interfere with their expression or activities in a developing animal. Introducing homozygous deletions into the relevant growth factor genes by homologous recombination in ES cells would probably not be very useful, because mice carrying such deletions would probably die very early *in utero*. PDGF, for example, is expressed in the preimplantation mouse embryo (Mercola *et al.*, 1990) and almost certainly has pleotropic effects during development, being used as a signalling molecule at different times and in a variety of tissues (Ross *et al.*, 1986); the same is undoubtedly true for many, perhaps most, growth factors. The answer may be to use *trans*-acting inhibitory molecules whose effects can be targeted to specific cells using cell type-specific promoters. Several strategies have been devised for generating such *trans*-dominant inhibitory molecules; these are reviewed below (Section 6). This *trans*-dominant approach requires that we know the cellular source(s) of the growth factor in question. For example, possible sources of PDGF in the optic nerve are type 1 astrocytes (Richardson *et al.*, 1990) or the axons of retinal ganglion neurons (Sasahara *et al.*, 1991; Yeh *et al.*, 1991). We might wish, therefore, to target expression of a PDGF-inhibitory transgene to astrocytes using the glial fibrillary acidic protein gene promoter (Miura *et al.*, 1990), or to neurons using the neurofilament (Monteiro *et al.*, 1990) or neuron-specific enolase (Forss-Petter *et al.*, 1990) gene promoters. Alternatively, if we could identify a promoter that is specifically active in the PDGF-responsive O-2A progenitor cells, we could direct expression of a PDGF receptor inhibitor to O-2A progenitors.

6. *Trans*-dominant strategies for inhibiting gene function

6.1 *Antisense RNA*

An antisense RNA is an RNA that is complementary to, and capable of hybridizing with, another RNA species, usually an mRNA. By forming a double-stranded region in otherwise single-stranded mRNA, the antisense RNA can inhibit translation of the mRNA or, in eukaryotes, post-transcriptional events such as splicing. Antisense RNAs occur naturally in prokaryotic cells, where they are involved in the control of plasmid and phage replication, Tn 10 transposition, and gene expression (reviewed by Green *et al.*, 1986). Antisense RNAs have also been identified in normal mammalian cells, but it is not yet certain whether these have a regulatory function. The potential use of artificially created antisense RNAs as inhibitory agents has attracted much interest (Weintraub *et al.*, 1985; Green *et al.*, 1986). Antisense RNAs generated by *in vitro* transcription have been used to inhibit expression of specific genes by direct microinjection into *Xenopus* oocytes or *Drosophila* embryos, and antisense RNAs transcribed *in vivo* from integrated plasmids have been used to inhibit gene expression in cultured animal cells (Holt *et al.*, 1986), transgenic plants (Ecker and Davis, 1986; Powell *et al.*, 1989) and the slime mould *Dictyostelium* (Crowley *et al.*, 1985; Knecht and Loomis, 1987). However, the antisense approach has so far not been very successful in transgenic mice. In one study, a dysmyelinating phenotype was induced by an antisense myelin basic protein (MBP) transgene (Katsuki *et al.*, 1988). The phenotype was more severe in mice heterozygous for the shiverer mutation, which already lacked one MBP allele, underlining what appears to be the main drawback of the antisense approach, namely that a high molar excess of antisense RNA is required to effectively inhibit target gene expression. Therefore, this approach may become more generally useful if it becomes possible to generate high levels of stable antisense RNAs in mammalian cells, by using a cloned bacteriophage RNA polymerase, for example (reviewed by Moss *et al.*, 1990; see Fig. 2).

6.2. *Ribozymes*

Ribozymes are RNA enzymes that possess highly sequence-specific endoribonuclease activity. The realization that RNA can sometimes possess nuclease activity came with the discovery of introns in ribosomal RNA transcripts of *Tetrahymena* that are capable of self-excision (Cech, 1983). Self-splicing reactions have also been observed in certain small circular RNAs that replicate in plants, either alone (viroid RNAs) or with the aid of a helper virus (satellite RNAs). The catalytic domains of these RNA molecules are regions of conserved sequence and secondary structure containing a stem loop and two single-stranded regions. The susceptible phosphodiester bond is brought in close proximity to the catalytic domain by base pairing between the sequence across the cleavage point and the

178   W. D. RICHARDSON

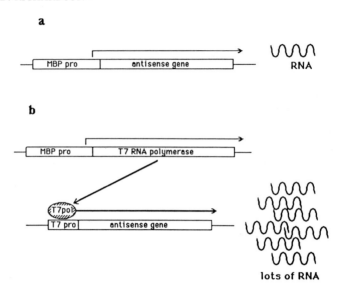

Fig. 2. Using T7 bacteriophage RNA polymerase (T7pol) to amplify the cell type-specific expression of an antisense RNA or an mRNA. (a) Using a mammalian cell type-specific promoter (the MBP promoter in this example) to drive expression of an antisense transgene might result in inadequate levels of antisense RNA. (b) This potential problem might be overcome by using the MBP promoter to drive expression of the gene encoding T7pol, which in turn drives high-level transcription of the antisense gene from a T7 late promoter (26 bp element). The two cooperating transgenes could be introduced separately into two transgenic mouse strains which could then be crossed to combine both transgenes in one animal.

sequences on either side of the catalytic domain. These flanking sequences define the specificity of the cleavage reaction, and altering these sequences provides the means by which RNAs other than the natural substrates can be cleaved (Haseloff and Gerlach, 1988). For example, recombinant ribozymes have been produced that specifically cleave mRNA of the human immunodeficiency virus (HIV) in infected cells in culture, thereby inhibiting growth of the virus (Sarver *et al.*, 1990). It is still too early to know whether ribozymes will be generally useful for targeting the destruction of specific mRNAs in transgenic animals, but these early indications *in vitro* are encouraging. Because of the enzymatic nature of ribozymes, they might be effective inhibitors at relatively low intracellular concentrations.

6.3. Dominant-negative mutations

A mutated protein that is altered in such a way that it inhibits the function of the corresponding wild-type protein is known as an 'antimorph', or 'dominant-negative' variant (Herskowitz, 1987). For a protein that normally functions as a dimer or multimer, a loss-of-function mutation that inactivates any hetero-dimer or -multimer into which it is incorporated would be expected to have a dominant-negative effect. This latter type of mutation in $\alpha 1(I)$ collagen is respon-sible for the lethal human disease osteogenesis imperfecta; a transgenic mouse model of this disease has been generated (Stacey *et al.*, 1988) using an engineered form of the mouse pro-$\alpha 1(I)$ collagen gene. In mice expressing the mutant gene, incorporation of only one mutant collagen subunit in 10 was sufficient to cause the severe skeletal abnormalities characteristic of the human disease.

Our own interest in the molecules that regulate glial cell development has led us to consider ways of inhibiting the activity of particular polypeptide growth factors, or their receptors, in the developing CNS. Several growth factors, including PDGF and TGF-β, are dimeric molecules, and the active forms of many growth factor receptors, including those for PDGF, epidermal growth factor (EGF) and insulin/IGF are ligand-induced dimers (see Fig. 3). Accor-dingly, a dominant-negative approach seems appropriate for inactivating a range of growth factors and their receptors *in vivo*. Figure 3 illustrates (using PDGF and its receptors as examples) some altered types of signalling molecules that have been shown, or would be expected, to exhibit dominant-negative effects.

6.4.   *Targeted deletions using a site-specific recombinase*

The most thorough way of inactivating a gene is to delete it from the genome. It is now possible to delete parts of specific genes in transgenic animals by homologous recombination in ES cells (Mansour *et al.*, 1988). However, dele-tions introduced in ES cells are transmitted indiscriminately to all cells in transgenic mouse lines established from them and, as explained above, this could be a serious drawback. A possible way round this problem is to make use of a site-specific recombinase, such as the Cre recombinase (Sauer and Henderson, 1990) from phage P1 or the FLP recombinase from yeast (Golic and Lindquist, 1989). The Cre enzyme is a well-characterized protein that catalyses reciprocal recombination between 34 bp *lox* recognition sequences; it can perform exci-sions, integrations or inversions depending on the relative orientations of the *lox* sites and whether they are carried on the same or different DNA molecules. Cre has been shown to perform excisive and integrative recombination in cultured mammalian cells (Sauer and Henderson, 1990), and promises to become a valuable part of the molecular geneticist's toolkit. It could be used to make a disabling deletion or rearrangement within a particular gene in a specific cell type, by prelabelling the target gene with *lox* sites and expressing Cre under the control of a cell type-specific promoter. Of course, the *lox* sites would first have to be inserted next to a crucial part of the gene in question, without affecting

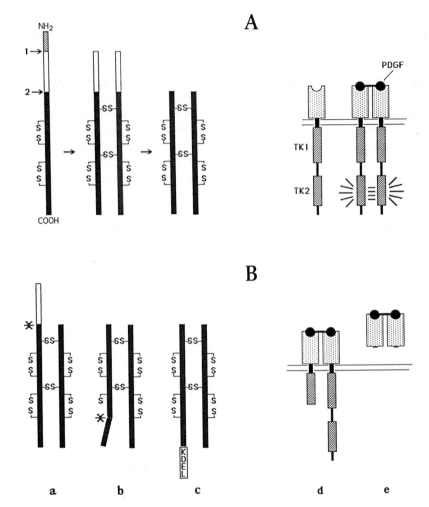

Fig. 3. Dominant-negative variants of PDGF and its receptors. (A, left) PDGF consists of dimers of A and B chains, with the structure AA, AB or BB depending on its source. Each chain of PDGF is synthesized as a pre-propeptide. Covalent dimerization occurs soon after insertion into the lumen of the endoplasmic reticulum and removal of the signal peptide. Later in the secretory pathway, the dimer of propeptides is cleaved further to form an active dimeric PDGF molecule which is then released from the cell. (A, right) Binding of PDGF to its receptors crosslinks two receptor subunits in the plasma membrane, activating their tyrosine kinase domains (for a review see Heldin and Westermark, 1989). Many other growth factor/receptor systems share some of these general features. (B) Some dominant-negative molecules. (a) Mutation of a proteolytic processing site in the PDGF propeptide prevents normal maturation, leading to secretion of inactive PDGF dimers (Mercola *et al.*, 1991). (b) Mutation of a cysteine that participates in a vital disulphide bridge leads to unstable PDGF dimers that are degraded within the cell (Mercola *et al.*, 1991). (c) Carboxy terminal addition of the amino acid sequence KDEL, a signal for retention within the lumen of the endoplasmic reticulum,

the normal expression or activity of the gene, by host-directed homologous recombination in ES cells. The *lox* sites could be placed on either side of an exon that is essential for activity of the encoded protein or whose removal would cause a downstream frameshift. Alternatively, they could be placed flanking essential elements of the gene promoter, so long as this did not interfere with normal transcription. For a small gene, the entire locus could be excised. The Cre recombination system could also be used to activate transgenes in particular cells by excising a *cis*-acting negative regulatory element (see below, Section 10).

7. Cell–cell interactions and cell diversification in the developing CNS: the retina as a model system

Unlike the optic nerve, the neural retina contains several classes of neurons and photoreceptors in addition to glia. Nevertheless, the retina is still fairly simple by comparison to much of the CNS. It is experimentally accessible and, in rodents, much of its development occurs postnatally. The cells of the retina, being organized in well-defined lamellae, are easy to locate *in situ*. Moreover, their lineage relationships are known: all the cells (except for the astrocytes) originate from the same pluripotential retinal neuroepithelial cells (Holt *et al.*, 1988; Wetts and Fraser, 1988; Turner *et al.*, 1990). The neuroepithelial cells reside at the outer surface of the developing retina, where they divide and give rise to postmitotic differentiated cells that migrate inwards (i.e. towards the centre of the eye), before coming to rest at the correct retinal layer. In this respect, morphogenesis in the retina is typical of many other regions of the CNS. For all these reasons the retina is an attractive experimental model for studies of CNS development and is an ideal target for genetic manipulation in transgenic animals. Furthermore, since it is not essential for survival or reproduction, genetic lesions that affect only retinal function can be propagated in the homozygous state.

What developmental questions can we address in the retina using transgenics? Cell–cell communication is likely to be a central issue in retinal development as elsewhere but, being a more complex system of interacting cells than the optic nerve, it is a more difficult task to identify the signalling molecules involved,

---

prevents variant PDGF dimers from leaving the cell (A. Calver and W. Richardson, unpublished). Note that this approach could be adapted easily to other dimeric or multimeric growth factors. (d) Truncation of the PDGF receptor subunit on the cytoplasmic side of the plasma membrane results in variant receptors that can form ligand-induced dimers with wild-type subunits but cannot activate DNA synthesis. This approach has been adopted for the insulin/IGF receptors (Treadway *et al.*, 1991). (e) Removal of the transmembrane and cytoplasmic domains of the PDGF receptor subunit generates a secreted form of the PDGF receptor that may be able to compete with normal receptors for ligand binding, and sequester PDGF in the extracellular space.

and their cellular sources and targets. Consequently, we are only beginning to gather information about growth factors and receptors in the developing retina. Eventually, when we have more precise knowledge about cell–cell interactions in the retina, it will be appropriate to inhibit the activities of the signalling ligands or their receptors by means of *trans*-dominant inhibitors, as described above (Section 6). In the meantime transgenic mice may help us identify some of the signal-transmitting and -receiving cells. In the following two sections I describe approaches that may be useful in this regard.

8.   Cell lineage ablation in the retina

In lower eukaryotes such as *Caenorhabditis elegans*, the ability to obliterate particular cells at defined times during development by laser ablation or micro-surgery has been instrumental in identifying developmental interactions among cells. In mice, genetic manipulation promises to substitute for surgery. By using appropriate transcriptional control elements to drive expression of a toxic gene product, it is possible to induce self-destruction of a chosen population of cells at a defined time during development; this has been called 'cell lineage ablation' (for reviews see Evans, 1989; Chapter 6). Transgenic mice that harbour the diptheria toxin A chain (DT-A) gene under genetic control of the elastase I promoter eliminate those cells that normally express elastase – exocrine cells of the pancreas – soon after the elastase promoter becomes active in those cells (Palmiter *et al.*, 1987). In these animals there is also a large reduction in the number of pancreatic duct and islet cells, which do not normally express the elastase I gene, suggesting either a lineage relationship among these and the exocrine cells or a dependence of duct and islet cells on signals from exocrine cells for differentiation, survival or proliferation. Both DT-A and the A subunit of ricin, a toxic lectin, have been used to ablate cells of the developing eye lens by driving expression of these 'toxigenes' with the lens-specific αA- or γ_2-crystallin promoters (Landel *et al.*, 1988; Breitman *et al.*, 1989; Kaur *et al.*, 1989). Some of the mice resulting from these experiments totally lacked an eye lens; in these animals other ocular tissues such as schlera, cornea and ciliary epithelium were reduced in size, suggesting that proportional growth of these tissues depends in some way on the presence of a lens. These examples illustrate an inherent problem with genetic ablation experiments: pleotropic effects are quite likely to be observed when cells are ablated during development, either because of lineage relationships between ablated and non-ablated populations of cells or through the loss of crucial cell–cell interactions involving ablated and non-ablated cells. In the absence of independent information about cell lineage, such as that available for the retina (Holt *et al.*, 1988; Wetts and Fraser, 1988; Turner *et al.*, 1990), these effects would be difficult to sort out.

What sort of information could we expect to gain from cell lineage ablation in the retina? Ablating a particular population of postmitotic cells in the mature retina, such as Müller glia or bipolar neurons, could provide clues to the structural or functional roles of those cells in the adult. Such an experiment would require a gene promoter that was activated only after retinal development was complete. Alternatively, eliminating cells as they are formed in the developing retina might tell us something about the mechanisms involved in controlling the time and rate of production of each specialized cell type from the pluripotent retinal stem cells. It is possible that during normal retinal development each cell type as it is born signals back to the stem cells, causing them to switch fates and generate the next cell type in their repertoire. Alternatively, the stem cells might possess an innate genetic programme that causes them to switch fates according to a predetermined schedule that is unaffected by other cells in their vicinity. The former model predicts that ablation of a cell type formed early in retinal development, such as amacrine neurons, would throw the stem cells into continual futile production of replacement amacrine cells. Consequently, differentiation of later forming cell types such as rod photoreceptors would be prevented. The latter model predicts that ablation of amacrine cells would allow the development of a full complement of retinal cell types apart from amacrine cells. Cell ablation experiments might therefore help to distinguish between these two possibilities or, more likely, to determine where, between these two extremes, the real situation lies.

9. Transgenic mice as sources of modified cells for *in vitro* studies of retinal development

Progress in developmental neurobiology will always demand an integrated approach and, as transgenic technology develops, it will complement rather than displace other experimental disciplines. In particular, 'designer' transgenic mice may be useful as sources of primary cells for *in vitro* studies. A common problem with *in vitro* studies is the difficulty of identifying cells in culture. Diagnostic antibodies are the usual means of distinguishing cells, but specific antibodies are not always available. Surface-directed antibodies are preferable to those directed against intracellular antigens as they allow living cells to be identified and manipulated *in vitro* – either killed by antibody-mediated complement lysis, or purified by immunoselection. Thus, cell type-specific promoters could be used to drive expression of a marker protein (for which antibodies are readily available) on the surface of defined cells in transgenic mice. A set of such transgenics could be assembled for studies of retinal development. Each mouse strain in this set would provide retinal cell cultures that could be enriched for or depleted of a different cell type, in order to determine how each cell type influences the development of other cells in the culture. In addition, the same cell type could be labelled with different molecules in two separate transgenic mouse strains, to enable cells derived from mice of different ages to be distinguished in mixed primary cultures. Such mixed cell cultures can be used to assess the relative importance of cell-intrinsic genetic programmes versus environmental signals in the timing of cell differentiation (Watanabe and Raff, 1990).

One of the reasons why the cell biology of neural cells is still relatively primitive is the lack of continuously growing cell lines that represent specific types of neurons or glia. Studies of specific neural cells in culture have been restricted to primary cells that can be purified in reasonable numbers, or tumour cells derived from gliomas and neuroblastomas. Recently, immortalized cell lines with the properties of particular neuronal subtypes have been established from transgenic mice in which expression of the simian virus 40 (SV40) large T antigen was targeted to those neurons using specific gene promoters. In one study, the promoter region of the human phenylethenolamine N-methyltransferase (PNMT) gene was used to direct expression of SV40 large-T to catecholaminergic neurons of the adrenal gland and retina, and a cell line with some characteristics of amacrine neurons was subsequently established from an eye tumour (Hammang et al., 1990). In another study the gonadotropin-releasing hormone gene promoter was used to target SV40 large-T induced tumours to neurons in the anterior hypothalamus, and a cell line was established from one such tumour (Mellon et al., 1990). These and other similar cell lines will be invaluable for biochemical studies of defined types of neural cells.

10.   Targeting transgene expression to specific cells in the CNS

It is not known how many cell types make up the CNS, but there may well be thousands of neuronal and dozens of glial cell types. Most of these have not been classified yet, let alone characterized at a molecular level. These cells often have similar morphologies and there are relatively few antibodies that distinguish between them. The consequent difficulty in identifying cells in situ or in vitro has severely hindered progress in developmental neurobiology. The paucity of well-characterized neural cell type-specific gene control elements also now limits the use of transgenic mice for neurobiology. One common approach in the search for cell-specific markers has been to raise monoclonal antibodies to whole cells or subcellular fractions derived from selected regions of the CNS. As a result of several efforts to raise antibodies to retinal cells and cell components, there is now a panel of monoclonal antibodies against specific retinal cell types (Barnstable, 1987; Fry and Lam, 1988; Dreher et al., 1991); some of these antibodies are directed against polypeptide antigens and may eventually allow the cognate genetic control elements to be isolated by conventional cDNA and genomic DNA library screening procedures.

In theory, cell type-specific mRNA sequences can be isolated directly from subtractive cDNA libraries or by screening conventional libraries with substractive cDNA probes. With one or two notable exceptions these approaches have so far been disappointing, partly because of inherent technical difficulties, but mainly because a relatively large and pure population of the cells of interest is required. This can rarely be achieved with CNS cells. However, new procedures for library construction (Hla and Maciag, 1990; Duguid and Dinauer, 1990; Timblin *et al.*, 1990) that make use of the polymerase chain reaction (PCR) to amplify minute amounts of cDNA may allow libraries to be generated from much smaller numbers of cells, e.g. CNS cells purified from mixed cell suspensions by 'panning' with surface-directed antibodies. This does not require the antibody to be directed against a unique gene product or even to be absolutely specific for the cells of interest, only to display specificity within a restricted region of the CNS that can be cleanly removed by dissection. For example, O-2A glial progenitor cells have been purified from rat cerebellum by panning dissociated cells with antibody NG2 (Stallcup and Beasley, 1987), and from rat cerebral cortex (Behar *et al.*, 1988) and optic nerve (B. Barres, personal communication) with antibody A2B5. Rat retinal ganglion neurons have also been purified using anti-thy-1 antibodies (Barres *et al.*, 1988), and it seems likely that other retinal cells such as Müller glia and amacrine neurons could also be purified with specific surface-directed antibodies (Barnstable, 1987). Thus, it is now possible to consider making specific cDNA libraries from these and other CNS cell types.

In general, the character of a cell is determined not by any one gene product but by the combination of gene products that the cell contains, and it may not always be possible to find a single gene that is expressed uniquely in one cell type and no other. This may be especially true for neurons, many of which are morphologically indistinguishable and whose functional diversity presumably is a reflection of the kind of cells with which they synapse, and differences in their pre- and postsynaptic structures. These differences could be relatively subtle and could depend, for example, on particular combinations of neurotransmitter receptors or ion channels, and their precise subunit compositions. Thus, to express a transgene specifically in some neuronal subtypes may require the combinatorial use of two or more gene promoters. To coordinate the activities of these promoters would require an additional level of gene regulation that could be provided by well-characterized *cis*- and *trans*-acting control elements from prokaryotes or lower eukaryotes. For example, the Lac repressor protein from *Escherichia coli* has been shown to bind its operator sequence and regulate gene activity in mammalian cells (Brown *et al.*, 1987; Hu and Davidson, 1987; Figge *et al.*, 1988). The Cre recombinase from bacteriophage P1 is also active in mammalian cells, as discussed above.

(a)

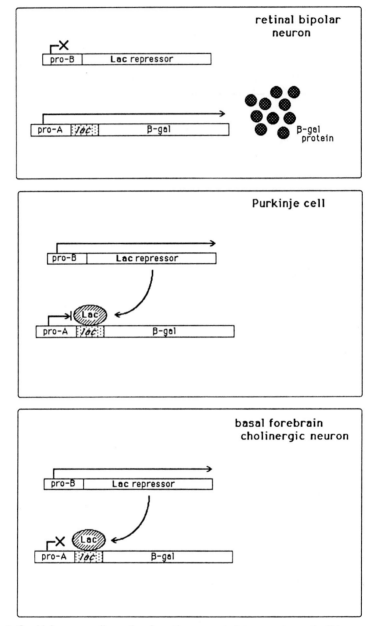

Fig. 4. Combining mammalian and prokaryotic gene controls to generate novel cell-type specificities in transgenic mice. (a) Pro-A is a promoter that is active in retinal bipolar neurons and cerebellar Purkinje cells, while pro-B is active in Purkinje cells and forebrain cholinergic neurons. Combining these two promoters with the Lac repressor/operator system of *E. coli* in the way depicted here might

(b)

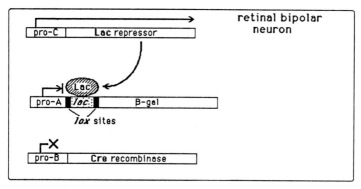

retinal bipolar neuron

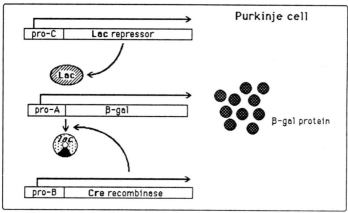

Purkinje cell

β-gal protein

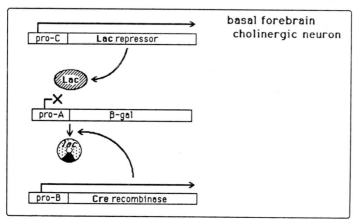

basal forebrain cholinergic neuron

lead to expression of a reporter transgene (β-gal) only in retinal bipolar cells. Interchanging pro-A and pro-B in this scenario would lead to expression of β-gal only in forebrain cholinergic neurons. (b) Combining pro-A and pro-B with Lac repressor and the Cre recombinase of phage Pl might allow expression of β-gal only in Purkinje cells (see Section 10).

Figure 4 illustrates some ways that mammalian and prokaryotic regulatory elements might be superimposed to create novel transcriptional specificities. Recently, a gene promoter was described (denoted pro-A in Fig. 4) that drives transgene expression both in cerebellar Purkinje neurons and in retinal bipolar neurons (Oberdick et al., 1990). The promoter for the nerve growth factor receptor gene (pro-B) drives transgene expression in Purkinje neurons and cholinergic neurons of the basal forebrain (Patil et al., 1990). These two promoters might therefore be used in combination to direct expression of a transgene (β-galactosidase, say) in any one of the three cell types. Specific expression in retinal bipolar neurons could be obtained by creating two transgenic mouse strains, one carrying a transgene consisting of pro-A linked to β-galactosidase (β-gal) coding sequences via the Lac repressor binding site and the other containing the coding sequence for the Lac repressor protein under the transcriptional control of pro-B. Crossing these two mouse strains would generate progeny with both transgenes; β-gal would be expressed only in the retinal bipolar cells of these animals because transcription in Purkinje neurons would be repressed by Lac. An analogous procedure, with pro-A and pro-B interchanged, would yield mice that express β-gal only in cholinergic basal forebrain neurons.

Obtaining mice that express β-gal only in Purkinje cells is more complex. First, a transgenic mouse strain could be created that expresses the Lac repressor in all its cells under the transcriptional control of a ubiquitously active promoter. Into this mouse strain could be introduced another transgene consisting of pro-A linked to β-gal coding sequences via a cassette comprising the Lac repressor-binding site flanked by lox recognition sequences for the Cre site-specific recombinase. This transgene would be active in Purkinje cells and bipolar cells, were it not kept in a repressed state by Lac. A separate transgenic mouse strain could be produced by introduction of a transgene consisting of the coding sequences for the Cre recombinase itself under the transcriptional control of pro-B. Crossing the two transgenic strains would yield progeny that expressed β-gal only in their Purkinje cells, because only these cells would contain both the cellular transcription factors required to use pro-A and the Cre recombinase that can excise the Lac-binding site and permit transcription to proceed. Of course these examples are at present somewhat fanciful and there are many technical problems to be addressed before this sort of multicomponent strategy could be put into practice.

11.   Conclusion

We are now entering an exciting new phase of the transgenic technology. The pioneering, exploratory period is drawing to a close and the technology is becoming part of the developmental biologist's routine experimental repertoire. In future, apart from witnessing rapid growth in the number of groups practising these techniques, we shall see more subtle genetic manipulations directed at very specific questions. The precision of genetic control that this demands will rely on combining simple and well-understood gene regulatory systems from prokaryotes or lower eukaryotes with the cell-type specificity of complex mammalian gene promoters. At present, what perhaps most limits the application of transgenic technology to neurobiology is the lack of available mammalian gene promoters that are active in restricted subsets of neurons and glia. The search for cell type-specific promoters will stimulate the molecular characterization of neural cell types, which in turn may provide useful clues to cell function and lineage. Thus, transgenic technology will catalyse progress at several levels. There clearly are exciting times ahead for neurobiology.

Acknowledgments

I thank my colleagues, past and present, for discussions and ideas. I also thank Ellen Collarini, Leon Nawrocki, Ian Hart, Barbara Barres and Martin Raff for useful comments on the manuscript. I acknowledge the excellent review by Rossant (1990) for drawing my attention to the potential uses of site-specific recombinases.

References

Balling, R., Mutter, G., Grüss, P. and Kessel, M. (1989). Craniofacial abnormalities induced by ectopic expression of the homeobox gene Hox-1.1 in transgenic mice. Cell 58, 337–347.

Barnstable C. J. (1987). Immunological studies of the diversity and development of the mammalian visual system. Immunol. Rev. 100, 47–78.

Barres, B. A., Silverstein, B. E., Corey, D. P. and Chun, L. L. Y. (1988). Immunological, morphological, and electrophysiological variation among retinal ganglion cells purified by panning. Neuron 1, 791–803.

Behar, T., McMorris, F. A., Novotny, E. A., Barker, J. L. and Dubois-Dalcq, M. (1988). Growth and differentiation properties of O-2A progenitors purified from rat cerebral hemispheres. J. Neurosci. Res. 21, 168–180.

Blanks, J. C. and Bok, D. (1977). An autoradiographic analysis of postnatal cell proliferation in the normal and degenerative mouse retina. J. Comp. Neurol. 174, 317–328.

Breitman, M. L., Bryce, D. M., Giddens, E., Clapoff, S., Goring, D., Tsui, L.-C., Klintworth, G. K. and Bernstein, A. (1989). Analysis of lens fate and eye morphogenesis in transgenic mice ablated for cells of the lens lineage. Development 106, 457–463.

Brown, M., Figge, J., Hansen, U., Wright, C., Jeang, K.-T., Khoury, G., Livingstone,

D. M. and Roberts, T. M. (1987). *lac* repressor can regulate expression from a hybrid SV40 early promoter containing the *lac* operator in animal cells. *Cell* **49**, 603–612.

Bögler, O., Wren, D., Barnett, S. C., Land, H. and Noble, M. (1990). Cooperation between two growth factors promotes extended self-renewal and inhibits differentiation of O-2A progenitor cells. *Proc. Natl Acad. Sci. USA* **87**, 6368–6372.

Cech, T. R. (1983). RNA splicing: three themes with variations. *Cell* **34**, 713–716.

Chisaka, O. and Capecchi, M. R. (1991). Regionally restricted developmental defects resulting from targeted disruption of the mouse homeobox gene *hox-1.5*. *Nature* **350**, 473–479.

Crowley, T. E., Nellen, W., Gomer, R. H. and Firtel, R. A. (1985). Phenocopy of discoidin I-minus mutants by antisense tranformation in *Dictyostelium*. *Cell* **43**, 633–641.

Dowling, J. E. (1979). Information processing by local circuits: the vertebrate retina as a model system. In: Schmitt, F. O. and Worden, F. G. (eds) *The Neurosciences: Fourth Study Program*, pp. 213–216. MIT Press, Cambridge, MA.

Dreher, Z., Mitrofanis, J., Dreher, B. and Stone, J. (1991). 4B2: a monoclonal antibody to ganglion and bipolar cells in the retina of the cat, generated by intrasplenic immunization. *J. Neurocytol.* **20**, 39–50.

Duboule, D. and Dollé, P. (1989). The structural and functional organization of the murine HOX gene family resembles that of Drosophila homeotic genes. *EMBO J.* **8**, 1497–1506.

Duguid, J. R. and Dinauer, M. C. (1990). Library subtraction of in vitro cDNA libraries to identify differentially expressed genes in scrapie infection. *Nucleic Acids Res.* **18**, 2789–2792.

Eccleston, P. A. and Silberberg, D. H. (1985). Fibroblast growth factor is a mitogen for oligodendrocytes in vitro. *Dev. Brain Res.* **21**, 315–318.

Ecker. J. R. and Davis, R. W. (1986). Inhibition of gene expression in plant cells by expression of antisense RNA. *Proc. Natl Acad. Sci. USA* **83**, 5372–5376.

Evans, G. A. (1989). Dissecting mouse development with toxigenics. *Genes Dev.* **3**, 259–263.

Figge, J., Wright, C., Collins, C. J., Roberts, T. M. and Livingstone, D. M. (1988). Stringent regulation of stably integrated chloramphenicol acetyltransferase genes by E. coli *lac* repressor in monkey cells. *Cell* **52**, 713–722.

Forss-Petter, S., Danielson, P. E., Catsicas, S., Battenberg, E., Price, J., Nerenberg, M. and Sutcliffe, J. G. (1990). Transgenic mice expressing beta-galactosidase in mature neurons under neuron-specific enolase promoter control. *Neuron* **5**, 187–197.

Fraser, S. E., Keynes, R. J. and Lumsden, A. (1990). Segmentation in the chick embryo hindbrain is defined by cell lineage restrictions. *Nature* **344**, 431–435.

Fry, K. R. and Lam, D. M.-K. (1988). Cell-specific monoclonal antibodies: probes for studying retinal organisation and development. In: Chandler, N. and Chader, G. (eds) *Progress in Retinal Research*, vol. 8 pp. 1–21. Pergamon Press, Oxford.

Golic, K. G. and Lindquist, S. (1989). The FLP recombinase of yeast catalyses site-specific recombination in the *Drosophila* genome. *Cell* **59**, 499–509.

Graham, A., Papalopulu, N. and Krumlauf, R. (1989). The murine and *Drosophila* homeobox gene complexes have common features of organization and expression. *Cell* **57**, 367–378.

Green, P. J., Pines, O. and Inouye, M. (1986). The role of antisense RNA in gene regulation. *An. Rev. Biochem.* **55**, 569–597.

Hammang, J. P., Baetge, E. E., Behringer, R. R., Brinster, R. L., Palmiter, R. D. and Messing, A. (1990). Immortalized retinal neurons derived from SV40 T-antigen-induced tumors in transgenic mice. *Neuron* **4**, 775–782.

Hammer, R. E., Maika, S. D., Richardson, J. A., Tang, J.-P. and Taurog, J. D. (1990). Spontaneous inflammatory disease in transgenic rats expressing HLA-B27 and human β_2m: an animal model of HLA-B27-associated human disorders. *Cell* **63**, 1099–1112.

Haseloff, J. and Gerlach, W. L. (1988). Simple RNA enzymes with new and highly specific endoribonuclease activities. *Nature* **334**, 585–591.

Heldin, C.-H. and Westermark, B. (1989). Platelet-derived growth factor: three isoforms and two receptor types. *Trends Genet.* **5**, 108–111.

Herskowitz, I. (1987). Functional inactivation of genes by dominant negative mutations. *Nature* **329**, 219–222.

Hla, T. and Maciag, T. (1990). Isolation of immediate-early differentiation mRNAs by enzymatic amplification of subtracted cDNA from human endothelial cells. *Biochem. Biophys. Res. Comm.* **167**, 637–643.

Holt, C. E., Bertsch, T. W., Ellis, H. M. and Harris, W. A. (1988). Cellular determination in the *Xenopus* retina is independent of lineage and birth date. *Neuron* **1**, 15–26.

Holt, J. T., Gopal, T. V., Moulton, A. D. and Nienhuis, A. W. (1986). Inducible production of c-*fos* antisense RNA inhibits 3T3 cell proliferation. *Proc. Natl Acad. Sci. USA* **83**, 4794–4798.

Hu, M. C.-T. and Davidson, N. (1987). The inducible *lac* operator-repressor system is functional in mammalian cells. *Cell* **48**, 555–566.

Ingham, P. W. (1988). The molecular genetics of embryonic pattern formation in *Drosophila*. *Nature* **335**, 25–34.

Joyner, A. L., Herrup, K., Auerbach, B. A., Davis, C. A. and Rossant, J. (1991). Subtle cerebellar phenotype in mice homozygous for a targeted deletion of the *En-2* homeobox. *Science* **251**, 1239–1243.

Katsuki, M., Sato, M., Kimura, M., Yokoyama, M., Kobayashi, K. and Nomura, T. (1988). Conversion of normal behaviour to shiverer by myelin basic protein antisense cDNA in transgenic mice. *Science* **241**, 593–595.

Kaur, S., Key, B., Stock, J., McNeish, J. D., Akeson, R. and Potter, S. S. (1989). Targeted ablation of α-crystallin-synthesising cells produces lens-deficient eyes in transgenic mice. *Development* **105**, 613–619.

Kessel, M., Balling, R. and Grüss, P. (1990). Variations of cervical vertebrae after expression of a *Hox-1.1* transgene in mice. *Cell* **61**, 301–308.

Keynes, R. and Lumsden, A. (1990). Segmentation and the origin of regional diversity in the vertebrate central nervous system. *Neuron* **2**, 1–9.

Knecht, D. A. and Loomis, W. F. (1987). Antisense inactivation of myosin heavy chain gene expression in *Dictyostelium discoideum*. *Science* **236**, 1081–1086.

Landel, C. P., Zhao, J., Bok, D. and Evans, G. A. (1988). Lens-specific expression of recombinant ricin induces developmental defects in the eyes of transgenic mice. *Genes Dev.* **2**, 1168–1178.

Lumsden, A. (1990). The cellular basis of segmentation in the developing hindbrain. *Trends Neurosci.* **13**, 329–339.

McKinnon, R. D., Matsui, T., Dubois-Dalcq, M. and Aaronson, S. A. (1990). FGF modulates the PDGF-driven pathway of oligodendrocytes development. *Neuron* **5**, 603–614.

McMahon, A. P. and Bradley, A. (1990). The *Wnt-1* (*int-1*) proto-oncogene is required for development of a large region of the mouse brain. *Cell* **62**, 1073–1085.

McMorris, F. A. and Dubois-Dalcq, M. (1988). Insulin-like growth factor I promotes cell proliferation and oligodendroglial commitment in rat glial progenitor cells developing in vitro. *J. Neurosci. Res.* **21**, 199–209.

Mansour, S. L., Thomas, K. R. and Capecchi, M. R. (1988). Disruption of the proto-oncogene *int-2* in mouse embryo-derived stem cells: a general strategy for targeting mutations to non-selectable genes. *Nature* **336**, 348–352.

Mellon, P. L., Windle, J. J., Goldsmith, P. C., Padula, C. A., Roberts, J. L. and Weiner, R. I. (1990). Immortalization of hypothalamic GnRH neurons by genetically targeted tumorigenesis. *Neuron* **5**, 1–10.

Mercola, M. Wang, C., Kelly, J., Brownlee, C. L., Jackson-Grusby, L., Stiles, C. D. and Bowen-Pope, D. (1990). The selective expression of PDGF A and its receptor during early mouse embryogenesis. *Dev. Biol.* **138**, 114–122.

Mercola, M., Deininger, P. L., Shamah, S. M., Porter, J., Wang, C. and Stiles, C. D. (1991). Dominant-negative mutants of a platelet-derived growth factor gene. *Genes Dev.* **4**, 2333–2341.

Miura, M., Tamura, T. and Mikoshiba, K. (1990). Cell-specific expression of the mouse glial fibrillary acidic protein gene: identification of the *cis*- and *trans*-acting promoter elements for astrocyte-specific expression. *J. Neurochem.* **55**, 1180–1188.

Monteiro, M. J., Hoffman, P. N., Gearhart, J. D. and Cleveland, D. W. (1990). Expression of NF-L in both neuronal and nonneuronal cells of transgenic mice: increased neurofilament density in axons without affecting caliber. *J. Cell Biol.* **111**, 1543–1557.

Moss, B., Elroy-Stein, O., Mizukami, T., Alexander, W. A. and Fuerst, T. R. (1990). New mammalian expression vectors. *Nature* **348**, 91–92.

Mullins, J. J., Peters, J. and Ganten, D. (1990). Fulminant hypertension in transgenic rats harbouring the mouse *Ren-2* gene. *Nature* **344**, 541–544.

Murphy, P., Davidson, D. R. and Hill, R. E. (1989). Segment-specific expression of a homeobox-containing gene in the mouse hindbrain. *Nature* **341**, 156–159.

Oberdick, J., Smeyne, R. J., Mann, J. R., Zackson, S. and Morgan, J. I. (1990). A promoter that drives transgene expression in cerebellar Purkinje and retinal bipolar neurons. *Science* **248**, 223–226.

Palmiter, R. D., Behringer, R. R., Quaife, C. J., Maxwell, F., Maxwell, I. H. and Brinster, R. L. (1987). Cell lineage ablation in transgenic mice by cell-specific expression of a toxin gene. *Cell* **50**, 435–443.

Patil, N., Lacy, E. and Chao, M. V. (1990). Specific neuronal expression of human NGF receptors in the basal forebrain and cerebellum of transgenic mice. *Neuron* **2**, 437–447.

Powell, P. A., Stark, D. M., Sanders, P. R. and Beachy, R. N. (1989). Protection against tobacco mosaic virus in transgenic plants that express tobacco mosaic virus antisense RNA. *Proc. Natl Acad. Sci. USA* **86**, 6949–6952.

Puschel, A. W., Balling, R. and Grüss, P. (1990). Position-specific activity of the *Hox-1.1* promoter in transgenic mice. *Development* **108**, 435–442.

Raff, M. C., Miller, R. H. and Noble, M. (1983). A glial progenitor cell that develops in vitro into an astrocyte or an oligodendrocyte depending on the culture medium. *Nature* **303**, 390–396.

Richardson, W. D., Raff, M. and Noble, M. (1990). The oligodendrocyte-type-2 astrocyte lineage. *Semin. Neurosci.* **2**, 445–454.

Ross, R., Raines, E. W. and Bowen-Pope, D. F. (1986). The biology of platelet-derived growth factor. *Cell* **46**, 155–159.

Rossant, J. (1990). Manipulating the mouse genome: implications for neurobiology. *Neuron* **4**, 323–334.

Saneto, R. P. and deVellis, J. (1985). Characterization of cultured rat oligodendrocytes proliferating in a serum-free, chemically defined medium. *Proc. Natl Acad. Sci. USA* **82**, 3509–3513.

Sarver, N., Cantin, E. M., Chang, P. S., Zaia, J. A., Ladne, P. A., Stephens, D. A. and Rossi, J. J. (1990). Ribozymes as potential anti-HIV-1 therapeutic agents. *Science* **247**, 1222–1225.

Sasahara, M., Fries, J. W. U., Raines, E. W., Gown, A. M., Westrum, L. E., Frosch, M. P., Bonthron, D. T., Ross, R. and Collins, T. (1991). PDGF B-chain in neurons of the central nervous system, posterior pituitary, and in a transgenic model. *Cell* **64**, 217–227.

Sauer, B. and Henderson, N. (1990). Targeted insertion of exogenous DNA into the eukaryotic genome by the Cre recombinase. *New Biologist* **2**, 441–449.

Stacey, A., Bateman, J., Choi, T., Mascara, T., Cole, W. and Jaenisch, R. (1988). Perinatal lethal osteogenesis imperfecta in transgenic mice bearing an engineered mutant pro-α1(I) collagen gene. *Nature* **332**, 131–136.

Stallcup, W. B. and Beasley, L. (1987). Bipotential glial progenitor cells of the optic nerve express the NG2 proteoglycan. *J. Neurosci.* **7**, 2737–2744.

Thomas, K. R. and Capecchi, M. R. (1990). Targeted disruption of the murine *int-1* proto-oncogene resulting in severe abnormalities in midbrain and cerebellar development. *Nature* **346**, 847–850.

Timblin, C., Battey, J. and Kuehl, W. M. (1990). Application for PCR technology to subtractive cDNA cloning: identification of genes expressed specifically in murine plasmacytoma cells. *Nucleic Acids Res.* **18**, 1587–1593.

Treadway, J. L., Morrison, B. D., Soos, M. A., Siddle, K., Olefsky, J., Ullrich, A., McClain, D. A. and Pessin, J. E. (1991). Transdominant inhibition of tyrosine kinase activity in mutant insulin/insulin-like growth factor I hybrid receptors. *Proc. Natl Acad Sci. USA* **88**, 214–218.

Tuggle, C. K., Zakany, J., Cianetti, L., Peschle, C. and Nguyen-Huu, M. C. (1990). Region-specific enhancers near two mammalian homeo box genes define adjacent rostrocaudal domains in the central nervous system. *Genes Dev.* **4**, 180–189.

Turner, D. L., Snyder, E. Y. and Cepko, C. L. (1990). Lineage-independent determination of cell type in the embryonic mouse retina. *Neuron* **4**, 833–845.

Van Obberghen-Schilling, E., Behar, T., Sporn, M. B. and Dubois-Dalcq, M. (1987). Signaling between type-1 astrocytes and glial O-2A progenitors: modulation by transforming growth factor-β (TGFβ). *J. Cell Biol.* **105**, 318a.

Watanabe, T. and Raff, M. C. (1990). Rod photoreceptor development in vitro: intrinsic properties of proliferating neuroepithelial cells change as development proceeds in the rat retina. *Neuron* **2**, 461–467.

Weintraub, H., Izant, J. G. and Harland, R. M. (1985). Antisense RNA as a molecular tool for genetic analysis. *Trends Genet.* **1**, 22–25.

Wetts, R. and Fraser, S. E. (1988). Multipotent precursors can give rise to all major cell types of the frog retina. *Science* **239**, 1142–1145.

Wilkinson, D. G. and Krumlauf, R. (1990). Molecular approaches to the segmentation of the hindbrain. *Trends Neurosci.* **13**, 335–339.

Wilkinson, D. G., Bhatt, S., Chavrier, P., Bravo, R. and Charnay, P. (1989a). Segment-specific expression of a zinc-finger gene in the developing nervous system of the mouse. *Nature* **337**, 461–464.

Wilkinson, D. G., Bhatt, S., Cook, M., Boncinelli, E. and Krumlauf, R. (1989b). Segmental expression of *Hox-2* homeobox-containing genes in the developing mouse hindbrain. *Nature* **341**, 405–409.

Yeh, H.-J., Ruit, K. G., Wang, Y.-X., Parks, W. C., Snider, W. D. and Deuel, T. F. (1991). PDGF A-chain gene is expressed by mammalian neurons during development and in maturity. *Cell* **64**, 209–216.

Young, R. W. (1984). Cell death during differentiation of the retina in the mouse. *J. Comp. Neurol.* **229**, 362–373.

Young, R. W. (1985). Cell differentiation in the retina of the mouse. *Anat. Rec.* **212**, 199–205.

Zakany, J., Tuggle, C. K., Patel, M. D. and Nguyen-Huu, M. C. (1988). Spatial regulation of homeobox gene fusions in the embryonic central nervous system of transgenic mice. *Nueron* **1**, 679–691.

9
Oncogenesis and transgenic mice

D. KIOUSSIS

Division of Molecular Immunology, National Institute for Medical Research, The Ridgeway, Mill Hill, London NW7 1AA, UK

1. Introduction

Oncogenes have been the focus of intense studies during the last 15 years. Initially, it was thought that the only biological activity of the viral oncogenes and their cellular homologues was to immortalize and transform cells which subsequently would develop into malignant or benign tumours. As research proceeded, however, it became evident that several of the cellular proto-oncogenes were implicated in cell differentiation (Westphal and Gruss, 1989), cell cycle (Coppola *et al.*, 1989) and embryonic development (Shackleford and Varmus, 1987; Wilkinson *et al.*, 1987; Mutter *et al.*, 1988). In recent years some of them were identified as transcription factors, such as *fos* and *jun* (Angel *et al.*, 1988). Studies on expression of such oncogenes in transfected cells has elucidated several aspects of the regulation of expression of these genes. However, their exact role in the processes referred to above remains unknown. Therefore, the deliberate introduction of such genes in the germ line of animals would provide more precise information.

Transgenic mice have been exceedingly useful in this respect. Generation of mice carrying oncogenes under the control of their own transcriptional elements has revealed tissue specificity, both in terms of their expression as well as their oncogenic action (Compere *et al.*, 1988; Cory and Adams, 1988; Hanahan, 1988,

1989; Adams and Cory, 1991). It was however, the use of artificial hybrid genes which shed more light on their action and developed into useful tools in the study of cell differentiation and development. Such hybrid genes usually place the expression of the oncogene (activated, cellular, viral) under the control of an element which is either widely active, or of a restricted tissue specificity, i.e. a promoter with or without enhancer sequences. These constructs are introduced in the germ line of mice, and animals that carry these genes usually express the transoncogene in the tissue determined by the regulatory elements of the hybrid gene. This tissue often suffers a developmental disturbance and in most cases a tumour develops from the cells that express the oncogene.

Many such attempts have led to the expression of an activated cellular or viral oncogene in a variety of cell lineages or specific organs. For example, the simian virus 40 (SV40) large T antigen gene has been introduced in the germ line of mice under the control of many promoters, e.g insulin (Teitelman et al., 1988), elastase (Ornitz et al., 1987), α-crystallin (Mahon et al., 1987), glucagon (Efrat et al., 1988), metallothioneine (Messing et al., 1985), or atrial natriuretic factor (Field, 1988) promoters among others. Transgenic animals were also generated which carried the myc gene under the control of: the immunoglobulin enhancer (Adams et al., 1985; Suda et al., 1987; Schmidt et al., 1988), the mouse mammary tumour virus (MMTV) long terminal repeat (LTR) (Leder et al., 1986), the H2 (Morello et al., 1989) or Thy-1 (Spanopoulou et al., 1989) promoters; a ras-activated gene under the MMTV, whey acidic protein (WAP) (Andres et al., 1987), elastase (Quaif et al., 1987) or immunoglobulin transcription elements (Suda et al., 1987; Harris et al., 1988). More recently, other oncogenes have been studied in a similar fashion. Thus, expression of neu (Muller et al., 1988; Bouchard et al., 1989) and fos (Ruther et al., 1987, 1988) has been targeted to specific tissues, almost invariably leading to tumours in the expressing organs.

In several of these animals the tumours were shown to be of monoclonal origin (Adams et al., 1985; Leder et al., 1986; Harris et al., 1988; Teitelman et al., 1988; Spanopoulou et al., 1989). The transgene appears to be expressed in most cells of the organ where the tumours developed, yet not all cells proceed to malignancy. Thus, it became evident that additional events are needed in order to achieve the final transformed phenotype. These secondary events are most likely the expression of additional (onco) genes. More recently, some of the additional events were shown to be the inactivation of anti-oncogenes, such as the retinoblastoma gene (Bernards et al., 1989; Hong et al., 1989; Horowitz et al., 1988). The concept of synergism (cooperativity) between oncogenes had already been demonstrated to be true in experiments with fibroblasts transfected with two oncogenes simultaneously (Land et al., 1983). Transgenic animals expressing different oncogenes in the same tissue have proved instrumental in showing cooperativity between the oncogenes in question. Thus, transgenic animals harbouring the MMTV–myc hybrid gene were crossed with MMTV–ras transgenic animals; the resulting double transgenics developed mammary tumours much faster than either of the single transgenics (Sinn *et al.*, 1987).

Excellent reviews on the results of transgenic experiments with oncogenes already exist in the literature and the reader is directed to some of those for additional information (Compere *et al.*, 1988; Cory and Adams, 1988; Hanahan, 1988, 1989; Adams and Cory, 1991).

In this chapter, three aspects of research with transgenic mice harbouring oncogenes will be considered. One aspect concerns the use of such mice to generate cell lines that have been difficult to obtain by other means. The second concerns the use of these mice to trap other cellular oncogenes which can cause cancer in cooperation with the transoncogene. The final aspect concerns the problem of breeding and maintaining mice which carry harmful oncogenes. These difficulties can be circumvented using carefully designed gene expression systems which lead to regulated expression of the oncogene product.

2. Generation of *in vitro* cell lines using organ-directed oncogenesis

Research on biological systems has been greatly enhanced by the availability of tissue-specific cell lines which can grow indefinitely *in vitro*. So far, such lines have been derived from tumours occurring spontaneously or induced by mutagens. Alternatively, retroviruses carrying oncogenes have been used to transform primary cell cultures. The advent of transgenic mice and the possibility to target the expression of genes in different organs using tissue-specific promoters made it feasible to induce tumours in mice in the tissue of choice. Thus, cell lines have been isolated from several of the transgenic mice which carry oncogenes under the control of different tissue-specific promoters (see references in Section 1). Below, the use of such animals to isolate cell lines from the thymus will be described.

Development of T cells takes place mainly in the thymus (Scollay *et al.*, 1984). Precursor cells arriving from fetal liver or bone marrow undergo a complex differentiation process resulting in mature functional T cells which are released in the periphery (Adkins *et al.*, 1987). During this development the surface phenotype of the thymocytes changes dramatically (Fowlkes and Pardoll, 1989). Thus, early cells express *Thy-1* and CD2 molecules on their membranes, but lack some of the mature markers found in circulating T lymphocytes, such as CD4 and CD8. Therefore, they are known as double-negative and this population (approximately 2–4% of thymocytes) appears to contain most of the precursor cells that can repopulate a depleted thymus. Later in development these cells acquire CD4 and CD8 simultaneously on their surface and start expressing low levels of the T cell receptor complex (TcR$\alpha\beta$ and CD3$\gamma\delta\epsilon\zeta$n chains). This population, by far the most abundant in the thymus (80–90%), are known as double positives and represent an intermediate step between precursors and mature T cells. Many of these cells are destined to die within the thymus. A minority (approximately 10–20%) of them mature into single-positive CD4$^+$CD8$^-$ or CD4$^-$CD8$^+$ mature T cells with high levels of functional T cell receptor. Another smaller subset of T cells appears to have a different receptor on the cell surface, comprised of γ and δ chains.

Mature T cells are capable of recognizing foreign antigens only in the context of their own MHC products (class I or class II), a phenomenon known as MHC restriction (Alison and Lanier, 1987). MHC restriction is the result of two major mechanisms: positive selection, during which functional cells are allowed to mature, and negative selection, during which harmful cells are eliminated or rendered functionally disabled (anergy) (Ramsdell et al., 1989).

During the described thymic maturation process, it has become clear that the thymic microenvironment plays an important role. The stroma of the thymus is composed of a heterogeneous population of epithelial cells of unknown origin which are found throughout the cortex and the medulla of the thymus. It is widely thought that these stromal cells are not just a supporting structural element, but actively participate in the development of T cells, playing a central role in positive and negative selection. The heterogeneity of the stromal cells is manifested both in morphology and in expression of specific markers. Thus, there are monoclonal antibodies which can distinguish between cortical and medullary epithelial cells (Haynes, 1984; Lampert and Ritter, 1988).

The study of interaction of thymic stroma with the developing thymocytes in whole animals has been hampered by the heterogeneity of the microenvironment cells as well as by the contemporaneous existence of lymphocytes at different stages of development. Such in vivo studies, however, have elucidated the precursor–progeny relationship of some of the thymocyte subpopulations (Scollay et al., 1984; Bluestone et al., 1987), but the factors and conditions determining the transition from precursor to product are still unknown. This is probably accomplished by cell surface interactions between the developing thymocytes and stromal cells, which either trigger responses by themselves or result in the production of cytokines that determine the subsequent events in development.

Experimental systems which can reproduce in vitro some of the steps occurring in the thymocyte development, therefore, would be of great assistance in understanding the underlying mechanisms. Immortalized cells from the thymus, including both stromal cells as well as precursor or immature thymocytes, can be derived by immortalization/transformation after introduction into their genome of an oncogene. This theoretically could be accomplished by transfection or retroviral infection of primary cultures of the organs. However, such an approach becomes very difficult when the cells of interest represent a small minority or when they have a limited life span in primary cultures. Transgenic mice harbouring an oncogene under the control of tissue-specific transcriptional elements provide an alternative pathway to the development of such cell lines. This has proved a successful approach in the past and cell lines have been isolated from mice carrying trans-oncogenes under the control of tissue-specific promoters (see references above)

This is illustrated by the experimental system in which transgenic mice were generated which carry the c-*myc* gene under the control of the *Thy-1* gene (Spanopoulou *et al.*, 1989). Such transgenic mice express the *myc* gene in high levels in the thymus and develop thymic tumours between 6 and 16 weeks of age (depending on the levels of *myc* expression). Immunohistochemical analysis of these tumours showed that both stromal and lymphoid components had expanded in the tumour. Fluorescence activated cell sorter (FACS) analysis of the lymphoid component revealed that most of the cells were CD4$^+$8$^+$ or CD4$^+$8$^-$ (Fig. 1(c)). The transformed phenotype of the cells in such tumours was manifested by the ability of homogenates of such thymuses to cause tumours in histocompatible mice and by their ability to establish in culture both adherent and suspension cell lines.

The adherent cell lines morphologically appeared as 'cobblestone pavement', which is typical of epithelial cells, and stained with antibodies which stain cortical epithelial cells *in vivo* (Lampert and Ritter, 1988). The suspension cell lines exhibit markers of the lymphoid lineage. Thus, the majority of the lines are double-positive (CD4$^+$CD8$^+$) and express on their surface high levels of heat-stable antigen (HSA, Jlld) which is a marker for immature thymocytes. They also express low levels of T cell receptor complex, in accordance with their classification as precursor-type thymocytes. Southern blot analysis of DNA from the cell lines, as well as from thymic tumours using a TcRβ constant-region probe, revealed that they are of mono- or oligoclonal origin, testifying further to the transformed phenotype of these cells. In some tumours which were composed mostly of CD4$^+$CD8$^+$ and CD4$^+$8$^-$ cells it was found that the same β chain rearrangement was present in the DNA of these two subpopulations. This is strong evidence that the double-positive cells isolated from these tumours have the capability to mature into single-positive cells, given the right environment.

Interestingly, the adherent lines, apart from their epithelial phenotype, exhibited an additional property which makes them a good candidate for an *in vitro* differentiation system. When co-cultured with established double-positive cell lines from a *Thy–myc* tumour, they would rosette the lymphocytes in a strong and specific manner. In the same culture lymphocytes would ignore colonies of fibroblastic morphology (Fig. 2). This specific interaction between the epithelial cell lines and the lymphoid precursor lines can be exploited in differentiation experiments involving the appropriate MHC haplotype on the epithelial cells and the appropriate T cell receptor in the precursor thymocytes.

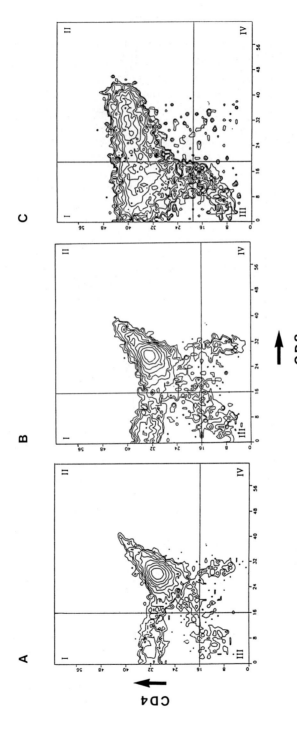

Fig. 1. Contour plots of CD4 and CD8 expression on thymocytes from (A) normal mouse, (B) H2–tsA58 and (C) *Thy–myc* transgenic mice.

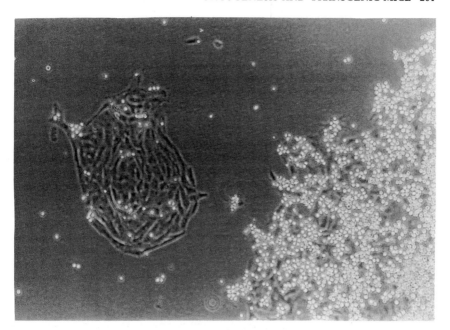

Fig. 2. Specific interaction of double positive (CD4$^+$8$^+$) thymocytes with thymic epithelial cell lines (right). In the same co-culture the lymphocytes do not adhere to the fibroblastic colony (left).

In an extension of this principle of obtaining cell lines through targeted oncogenesis, another construct was devised which ensured the expression of a thermolabile mutant of the SV40 T antigen in a wide variety of tissues in transgenic mice (Jat *et al.*, 1991). The SV40 large T antigen (Tag) gene is one of the most potent immortalizing genes. Several transgenic mice have been generated which carry the wild type of the gene under a variety of promoters (for reviews see Compere *et al.*, 1988; Cory and Adams, 1988; Hanahan, 1989). In almost all these cases, tumours developed in the transgenic mice, usually in the tissues where the expression of the Tag was targeted. In order to avoid the lethal effects due to tumours caused by the Tag, a mutant gene *TsA58* was used which gives rise to protein which is unstable at 39°C and stable at 33°C (Tegtmeyer, 1975). Since the body temperature of the mouse is 37.5°C, the majority of the Tag molecules made in the transgenic mouse would be degraded. If the tissues were dissected, however, and placed in culture conditions which stabilize the protein, i.e. 33°C, then the Tag would be allowed to immortalize the cells. In order to ensure the expression of the transgene in as wide a variety of tissues as possible, use of the MHC class I H2^k promoter was made (Weiss *et al.*, 1983; Kimura *et al.*, 1986; David-Watinne *et al.*, 1990). This promoter has the additional advantage that it can be induced by γ-interferon, thus offering one more level of control (Wallach *et al.*, 1982; Israel *et al.*, 1986; David-Watinne *et al.*, 1990). Indeed, transgenic mice harbouring this hybrid H2-tsA58 construct expressed Tag mRNA in a wide variety of tissues, with thymus and liver ranking amongst the highest expressing tissues (Jat *et al.*, 1991). Skin fibroblasts from these mice were placed at 33°C in the presence of low levels of γ-interferon (1 U ml^{-1}), and yielded immortalized cell lines, which were conditional in their growth. Thus, they would grow continuously without undergoing crisis in the permissive conditions, but would stop dividing when placed at 39°C or in cultures devoid of γ-interferon.

Whereas the rest of the tissues were normal, these mice invariably suffered from thymic hyperplasia. The likely absence of transformation from these thymuses was shown by the following observations: (1) both lobes of the thymus were equally enlarged; (2) unlike the *Thy–myc* thymic tumours, homogenates of these thymuses were unable to cause tumours in histocompatible animals, even when 10^7 cells were injected subcutaneously or intraperitoneally; (3) another difference from the *Thy–myc* tumours was the fact that the enlarged thymuses had polyclonal populations of thymocytes as judged by Southern blot analysis of their β chain gene loci; (4) furthermore, fluocytometry using antibodies against lymphocyte surface markers showed normal distribution of the major thymus subpopulations (Fig. 1(a) and (b)).

All this evidence taken together indicates that the lymphocytic expansion in these enlarged thymuses is probably of a hyperplastic nature rather than neoplastic. The mono- or polyclonality of the epithelial cells which also expand in these thymuses is not easy to determine. However, as mentioned above, homogenates of the enlarged thymuses cannot cause transplantable tumours; therefore, the epithelial cells included in the injection are unlikely to be transformed. The homogenates of these thymuses were put in culture at 33°C in the presence of interferon and they yielded cultures of cells which are keratin-positive and exhibit the typical morphology of thymic epithelial cells. These cultures are also conditional in their growth, in that they stop dividing if they are placed in 39°C. Given the wide tissue distribution of the expression of this transgene in these mice, it is very interesting to see which other tissues can yield conditional cell lines and which cannot.

3.   Identifying cooperating oncogenes

Expression of a transgene in a tissue usually does not guarantee the generation of tumours from all cells in this tissue. Thus, Eμ–*myc* mice develop certain pre-B cell tumours, but not other B cell malignancies (Suda *et al.*, 1987; Harris *et al.*, 1988). Similarly, *Thy–myc* mice develop thymic tumours of predominantly double-positive (CD4⁺CD8⁺) phenotype (Spanopoulou *et al.*, 1989). It is possible that in both of these cases the *myc* gene causes an increase of these populations of cells and thus increasing the probability that secondary events which lead to the final transformation.

In the above cases the secondary event is of an unknown nature. However, sometimes cooperation between oncogenes has been directly shown by crossing different transgenic mice; thus, when mice which express the *myc* transgene in the mammary gland are crossed with transgenic mice expressing the *ras* gene in the same tissue, they develop mammary tumours faster than either parental strain (Teitelman *et al.*, 1988). Similarly, transgenic mice expressing both *myc* and *ras* in their B cells develop pre-B lymphomas at a faster rate than mice expressing either of the genes alone in this compartment (Cory and Adams, 1988). A similar situation occurs when mice carrying the *Thy–myc* hybrid gene are crossed to transgenic mice with the activated H-*ras* gene under the control of the hCD2 gene. The double-transgenic offspring develop thymic tumours much earlier than either of the single transgenic parents (D. Greenberg and D. Kioussis, unpublished observations). Interestingly, the lymphomas developed in this case were monoclonal in origin, indicating that additional events have to take place in order to transform the cells.

In order to be able to identify such additional genetic events, Anton Berns and his colleagues devised a protocol which uses transgenic animals carrying an oncogene and infecting them with a slow-transforming retrovirus devoid of oncogenes (Berns *et al.*, 1989). In this protocol the resident oncogene in the transgenic mice was *pim-1* under the control of its own promoter, the Eμ enhancer and the Moloney murine leukaemia virus (Mo-MuLV) LTR. Such mice express high levels of the transgene in their haemopoietic tissues. However, these mice show no abnormalities in their haemopoietic or lymphoid organs in the early stages (van Lohuizen *et al.*, 1988). Only after a latency period, which can last up to 8 months, do 10% of the transgenic mice develop T cell lymphomas. Thus, the tumour incidence in the Eμ–*pim-1* transgenic mice is very low. Such mice are ideal to use for identification of other genes whose expression or extinction will synergize with *pim* expression and lead to tumour formation. In order to tag such genes, Berns and his colleagues infected newborn normal or *pim* transgenic mice with Mo-MuLV. All the Eμ–*pim-1* transgenic animals developed lymphomas much earlier than their non-transgenic littermates. These lymphomas were oligoclonal and had T cell characteristics. The authors concluded that the Eμ–*pim-1* presence in the genome of these mice rendered them highly susceptible to lymphoma induction by Mo-MuLV. Further investigation of these tumours established that the majority of the proviruses were integrated near the c-*myc* gene, causing transcriptional activation of this oncogene. The activation of the *myc* gene in this case is (one of) the necessary event(s) that leads to the tumour formation. Thus, the synergism between the *pim-1* and *myc* genes is documented and verified by this system. To further document the synergy between these two oncogenes, the authors proceeded to infect Eμ–myc transgenic mice with Mo-MuLV. The non-infected mice develop spontaneous pre-B cell lymphomas and infection by Mo-MuLV accelerated the incidence of such lymphomas. When analysed, it was found that, in a significant fraction of the lymphomas, the *pim-1* gene was activated by the proviral insertion. Such experiments indicate that the Mo-MuLV infection of transgenic mice bearing oncogenes can be exploited to identify other oncogenes which cooperate with the resident trans-oncogene in tumorigenesis. It will be highly interesting to see whether inactivation of anti-oncogenes by retroviral insertion can be detected by this protocol.

4.   Circumventing difficulties in breeding and maintaining mice carrying harmful oncogenes

In the examples quoted above, the animals expressing a trans-oncogene in their tissues almost invariably develop tumours. The latency of such tumours sometimes is short and the tumour can cause physiological disturbances which make maintenance and breeding of these animals problematic. Therefore, it would be of advantage to introduce a gene in the germ line which remains inactive during the life of the animal, but can be activated at will. This could be achieved by the use of temperature-sensitive variants of oncogenes or by placing the oncogene under the control of an element which can be transactivated. In the first case, the inactive transgene (the thermolabile protein) in the animal can be activated by placing the tissues in permissive culture conditions. Such an experimental animal was described above and harbours in its genome a temperature-sensitive SV40 T-antigen gene, *TsA58* (Jat *et al.*, 1991).

The second type of transgenic mouse was attempted in several laboratories using mammalian virus transactivating systems. Thus, transgenic mice were generated which carry the reporter gene chloramphenicol acetyltransferase (CAT) under the control of the human immunodifficiency virus (HIV) LTR, or the HIV *tat* gene under a regulatory element from the αA crystallin gene. When these two strains were crossed, transactivation of the reporter gene by the *tat* product caused expression of CAT in the eyes of double-transgenic mice (Khillan *et al.*, 1988). In another set of experiments, the promoter of an immediate early gene of herpes simplex virus (HSV-1) was transactivated by the VP16 gene product in double-transgenic mice (Byrne and Ruddle, 1989). However, the regulatory elements of mammalian viruses are subject to effects from the transcriptional machinery of the mouse cells. In order to avoid problems of unwanted expression, Leder and his colleagues decided to use regulatory elements from distantly related species (Ornitz *et al.*, 1990). Thus, firstly, a construct is made which places the oncogene of interest under the control of regulatory elements that only respond to the yeast transcriptional activator GAL4 (Ptashne, 1986; Kakidani and Ptashne, 1988; Byrne and Ruddle, 1989). The promoter of such a construct is engineered in such a way that the transgene remains silent in mammalian cells or transgenic mice harbouring it. This construct constitutes the target gene and it can only be expressed in the presence of the transcriptional activator GAL4.

In a second step in this protocol, another transgenic mouse is generated (the transactivator animal). This mouse harbours the yeast *GAL4* gene under the control of specific viral or mammalian transcriptional regulatory elements. The choice of the promoter and the regulatory elements can be designed so that expression of the *GAL4* gene is restricted to a few tissues or at certain times during development. On the other hand, it can be designed to be expressed in a wide variety of tissues. If one mates the two animals described above (the target and transactivator mice), their offspring containing the two transgenes will express the oncogene only in those tissues expressing the *GAL4* gene.

The above protocol was tested using as the target animal a transgenic mouse carrying the *Int-2* gene under the control of GAL4 responsive transcriptional elements. *Int-2* has been implicated in mammary tumorigenesis (Dickson *et al.*, 1984; Ali *et al.*, 1989), and recently in mouse development (Jakobovitz *et al.*, 1986; Wilkinson *et al.*, 1988). When this gene is over-expressed in breast or prostate tissues of transgenic mice it causes an enlargement of the mammary or prostate gland, characterized by hyperplasia of their epithelium (Muller *et al.*, 1990). Mice which carry the GAL4-inducible *Int-2* gene are normal and they do not exhibit any expression of the transgene in their tissues.

The transactivator mouse carries the mouse mammary tumour virus LTR fused to a mutant version of *GAL4* (*GAL4/236*). The presence of the LTR in this construct ensures the expression of fairly high levels of GAL4 in several tissues of the transgenic mouse (Leder *et al.*, 1986).

Offspring of a cross between these two transgenic mice carrying both transgenes express the target gene (*Int-2*) in all tissues that express GAL4. This causes hyperplasia in salivary glands, breast tissue, epididymis and prostate. Such double-transgenic mice have problems reproducing or rearing their offspring, testifying to the usefulness of this binary system.

5. Concluding remarks

Transgenic animals have been proved to be an excellent tool in studying gene regulation and function. Particularly useful applications were employed in studies using oncogenes. In this chapter we describe how such transgenic animal models can be used as tools to answer questions or facilitate procedures which have been problematic in the past, such as isolation of immortalized cell lines. Regulated expression of oncogenes allows the development of normal tissues in mice until the chosen time. Excision of tissues and placement under permissive conditions yields immortalized lines from *Hs-TsA58* transgenic mice. Alternatively, mating of carefully designed target-transactivating mice results in tumours only in the tissues of interest.

Finally, transgenic animals carrying oncogenes can be used to track down additional oncogenes that cooperate with the transgene or to identify loss of function mutants in anti-oncogene loci.

References

Adams, J. M. and Cory, S. (1991). Transgenic models for haemopoietic malignancies. *Biochim. Biophys. Acta* **1072**, 9–31.

Adams, J. M., Harris, A. W., Pinkert, C. A., Corcoran, L. M., Alexander, W. S., Cory, S., Palmiter, R. D. and Brinster, R. L. (1985). The c-myc oncogene driven by immunoglobulin enhancers induces lymphoid malignancy in transgenic mice. *Nature* **318**, 533–538.

Adkins, B., Mueller, C., Okada, C. Y., Reichert, R. A., Weissman, I. L. and Spangrude, G. J. (1987). Early events in T-cell maturation. *An. Rev. Immunol.* **5**, 325–365.

Ali, I. U., Merlo, G., Callahan, R. and Lidereau, R. (1989). The amplification unit on chromosome 11q13 in aggressive primary human breast tumors entails the bcl-1, int-2 and hst loci. *Oncogene* **4**, 88–92.

Alison, J. P. and Lanier, L. L. (1987). Structure, function and serology of the T-cell antigen receptor complex. *An. Rev. Immunol.* **5**, 503–540.

Andres, A. C., Schonenberger, C. A., Groner, B., Henninghausen, L., LeMeur, M. and Gerlinger, P. (1987). Ha-ras oncogene expression directed by a milk protein gene promoter: tissue specificity, hormonal regulation, and tumor induction in transgenic mice. *Proc. Natl Acad. Sci. USA* **84**, 1299–1303.

Angel, P., Allegretto, E., Okino, S. T., Hattori, K., Boyle, W. J., Hunter, T. and Karin, M. (1988). Oncogene jun encodes a sequence-specific trans-activator similar to AP-1. *Nature* **332**, 166–171.

Bernards, R., Schackleford, G. M., Geiber, M. R., Horowitz, J. M., Friend, S. H., Schartl, M., Bogenmann, E., Rapaport, J. M., McGee, T., Dryja, T. P. and Weinberg, R. A. (1989). Structure and expression of the murine retinoblastoma gene and characterization of its encoded protein. *Proc. Natl Acad. Sci. USA* **86**, 6474–6478.

Berns, A., Breuer, M., Verbeek, S. and van Lohuizen, M. (1989). Transgenic mice as a means to study synergism between oncogenes. *Int. J. Cancer* Supp. **4**, 22–25.

Blackman, M. A., Gerhard-Burgert, H., Woodland, D. L., Palmer, E., Kappler, J. W. and Marrack, P. (1990). A role for clonal inactivation in T cell tolerance to Mls-la. *Nature* **345**, 540–542.

Blackman, M. A., Kappler, J. W. and Marrack, P. (1988). T-cell specificity and repertoire *Immunol. Rev.* **101**, 5–19.

Bluestone, T. A., Pardoll, D. M., Sharrow, S. O., and Fowlkes, B. J. (1987). Characterization of murine thymocytes with CD3-associated T-cell receptor structures. *Nature* **326**, 82–84.

Bouchard, L., Lammarre, L., Tremblay, P. J. and Jolicoeur, P. (1989). Stochastic appearance of mammary tumors in transgenic mice carrying the MTV/c-neu oncogene. *Cell* **57**, 931–936.

Byrne, G. W. and Ruddle, F. H. (1989). Multiplex gene regulation: a two-tiered approach to transgene regulation in transgenic mice. *Proc. Natl Acad. Sci. USA* **86**, 5473–5477.

Compere, S. J., Baldacci, P. and Jaenisch, R. (1988). Oncogenes in transgenic mice. *Biochim. Biophys. Acta* **948**, 129–149.

Coppola, J. A., Parker, J. M., Schuler, G. D. and Cole, M. (1989). Continued withdrawal from the cell cycle and regulation of cellular genes in mouse erythroleukemia cells blocked in differentiation by the c-myc oncogene. *Mol. Cell. Biol.* **9**, 1714–1720.

Cory, S. and Adams, J. M. (1988). Transgenic mice and oncogenesis. *An. Rev. Immunol.* **6**, 25–48.

David-Watinne, B., Israel, A. and Kourilsky, P. (1990). The regulation and expression of MHC class I genes. *Immunol. Today* **11**, 286–292.

Dickson, C., Smith, R., Brookes, S. and Peters, G. (1984). Tumorigenesis by mouse mammary tumor virus: proviral activation of a cellular gene in the common integration region *int-2*. *Cell* **37**, 529–536.

Efrat, S., Teitelman, G., Anwar, M., Ruggiero, D. and Hanahan, D. (1988). Glucagon gene regulatory region directs oncoprotein expression to neurons and pancreatic alpha cells. *Neuron* **1**, 605–613.

Field, L. J. (1988). Atrial natriuretic factor-SV40 T antigen transgenes produce tumors and cardiac arrhythmias in mice. *Science* **239**, 1029–1033.

Fowlkes, B. J. and Pardoll, D. M. (1989). Molecular and cellular events of T cell development. *Adv. Immunol.* **44**, 207–264.

Hanahan, D. (1988). Dissecting multistep tumorigenesis in transgenic mice. *An. Rev. Genet.* **22**, 479–519.

Hanahan, D. (1989). Transgenic mice as probes into complex systems. *Science* **246**, 1265–1275.

Harris, A. W., Langdon, W. Y., Alexander, W. S., Hariharan, I. K., Rosenbaum, H., Vaux, D., Webb, E., Bernard, O., Crawford, M., Abud, H., Adams, J. M. and Cory, S. (1988). Transgenic mouse models for hematopoietic tumorigenesis. *Curr. Top. Microbiol. Immunol.* **141**, 82–93.

Haynes, B. F. (1984). The human thymic microenvironment. *Adv. Immunol.* **36**, 87–142.

Hong, F. D., Huang, H. J., To, H., Young, L. J., Oro, A., Bookstein, R., Lee, E. Y. and Lee, W. H. (1989). Structure of the human retinoblastoma gene. *Proc. Natl Acad. Sci. USA* **86**, 5502–5506.

Horowitz, J. M., Friend, S. H., Weinberg, R. A., Whyte, P., Buchkovich, K. and Harlow, E. (1988). Anti-oncogenes and the negative regulation of cell growth. *Cold Spring Harbor Symp. Quant. Biol.* **53** (part 2), 843–847.

Israel, A., Kimura, A., Fournier, A., Fellous, M. and Kourilsky, P. (1986). Interferon response sequence potentiates activity of an enhancer in the promoter region of a mouse H-2 gene. *Nature* **322**, 743–746.

Jakobovitz, A., Shackleford, G. M., Varmus, H. E. and Martin, G. R. (1986). Two proto-oncogenes implicated in mammary carcinogenesis, int-1 and int-2, are independently regulated during mouse development. *Proc. Natl Acad. Sci. USA* **83**, 7806–7810.

Jat, P. S., Noble, M. D., Ataliotis, P., Tanaka, Y., Yannoutsos, N., Larsen L. and Kioussis, D. (1991). Direct derivation of conditionally immortal cell lines from an H-2K^b tsA58 transgenic mouse. *Proc. Natl Acad. Sci. USA* **88**, 5096–5100.

Kakidani, H. and Ptashne, M. (1988). GAL4 activates gene expression in mammalian cells. *Cell* **52**, 161–167.

Khillan, J. S., Deen, K. C., Yu, S., Sweet, R. W., Rosenberg, M. and Westphal, H. (1988). Gene transactivation mediated by the TAT gene of human immunodeficiency virus in transgenic mice. *Nucl. Acids Res.* **16**, 1423–1430.

Kimura, A., Israel, A., LeBail, O. and Kourilsky, P. (1986). Detailed analysis of the mouse H-2Kb promoter: enhancer-like sequences and their role in the regulation of class I gene expression. *Cell* **44**, 261–272.

Lampert, I. A. and Ritter, M. A. (1988). The origin of the diverse epithelial cells of the thymus; is there a common stem cell? *Thymus Update* **1**, 5–25.

Land. H., Parada, L. F. and Weinberg, R. A. (1983). Tumorigenic conversion of primary embryo fibroblasts requires at least two co-operating oncogenes. *Nature* **304**, 596–602.

Leder, A., Pattengale, P. K., Kuo, A., Stewart, T. A. and Leder, P. (1986). Consequences of widespread deregulation of the c-myc gene in transgenic mice: multiple neoplasms and normal development. *Cell* **45**, 485–495.

Mahon, K. A., Chepelinsky, A. B., Khillan, J. S., Overbeek, P. A., Piatigorsky, J. and Westphal, H. (1987). Oncogenesis of the lens in transgenic mice. *Science* **235**, 1622–1628.

Messing, A., Chen, H. Y., Palmiter, R. D. and Brinster, R. L. (1985). Peripheral neuropathies, hepatocellular carcinomas and islet cell adenomas in transgenic mice. *Nature* **316**, 461–463.

Morello, D., Lavenu, A., Bandeira, A., Portnoi, D., Gaillard, J. and Babinet, C. (1989). Lymphoproliferative syndrome associated with c-myc expression driven by a class I gene promoter in transgenic mice. *Oncogene Res.* **4**, 111–125.

Muller, W. J., Lee, F. S., Dickson, C., Peters, G., Pattengale, P. and Leder, P. (1990). The int-2 gene product acts as an epithelial growth factor in transgenic mice. *EMBO J.* **9**, 907–913.

Muller, W. J., Sinn, E., Pattengale, P. K., Wallace, R. and Leder, P. (1988). Single-step induction of mammary adenocarcinoma in transgenic mice bearing the activated c-neu oncogene. *Cell* **54**, 105–115.

Mutter, G. L., Grills, G. S. and Wolgemuth, D. J. (1988). Evidence for the involvement of the proto-oncogene c-mos in mammalian meiotic maturation and possibly very early embryogenesis. *EMBO J.* **7**, 683–689.

Ornitz, D. M., Hammer, R. E., Messing, A., Palmiter, R. D. and Brinster, R. L. (1987). Pancreatic neoplasia induced by SV40 T-antigen expression in acinar cells of transgenic mice. *Science* **238**, 188–193.

Ornitz, D. M., Moreadith, R. W. and Leder, P. (1990). Binary system for regulating transgene expression in mice: targeting int-2 gene expression with yeast GAL4/UAS control elements. *Proc. Natl Acad. Sci. USA* **88**, 698–702.

Ptashne, M. (1986). Gene regulation by proteins acting nearby and at a distance. *Nature* **322**, 697–701.

Quaife, C. J., Pinkert, C. A., Ornitz, D. M., Palmiter, R. D. and Brinster, R. L. (1987). Pancreatic neoplasia induced by ras expression in acinar cells of transgenic mice. *Cell* **48**, 1023–1034.

Ramsdell, F., Lantz, T. and Fowlkes, B. J. (1989). A nondeletional mechanism of thymic self tolerance. *Science* **246**, 1038–1041.

Ruther, U., Garber, C., Komitowski, D., Muller, R. and Wagner, E. F. (1987). Deregulated c-fos expression interferes with normal bone development in transgenic mice. *Nature* **325**, 412–416.

Ruther, U., Muller, W., Sumida, T., Tokuhisa, T., Rajewsky, K. and Wagner, E. F. (1988). c-fos expression interferes with thymus development in transgenic mice. *Cell* **53**, 847–856.

Schmidt, E. V., Pattengale, P. K., Weir, L. and Leder, P. (1988). Transgenic mice bearing the human c-myc gene activated by an immunoglobulin enhancer; a pre-B-cell lymphoma model. *Proc. Natl Acad. Sci. USA* **85**, 6047–6051.

Scollay, R., Bartlett, P. and Shortman, K. (1984). T-cell development in the adult murine thymus: changes in the expression of the surface antigens Ly2, L314, and B242 during development from early precursor cells to emigrants. *Immunol. Rev.* **82**, 79–103.

Shackleford, G. M. and Varmus, H. E. (1987). Expression of the proto-oncogene int-1 is restricted to postmeiotic male germ cells and the neural tube of mid-gestational embryos. *Cell* **50**, 89–95.

Sinn, E., Muller, W., Pattengale, P., Tepler, I., Wallace, R. and Leder, P. (1987). Coexpression of MMTV/v-Ha-ras and MMTV/c-myc genes in transgenic mice: synergistic action of oncogenes in vivo. *Cell*, **49**, 465–475.

Spanopoulou, E., Early, A., Elliott, J., Crispe, N., Ladyman, H., Ritter, M., Watt, S., Grosveld, F. and Kioussis, D. (1989). Complex lymphoid and epithelial thymic tumours in Thy1-myc transgenic mice. *Nature* **342**, 185–189.

Suda, Y., Aizawa, S., Hirai, S., Inoue, T., Furuta, Y., Suzuki, M., Hirohashi, S. and Ikawa, Y. (1987). Driven by the same Ig enhancer and SV40 T promoter ras induced lung adenomatous tumors, myc induced pre-B cell lymphomas and SV40 large T gene a variety of tumors in transgenic mice. *EMBO J.* **6**, 4055–4065.

Tegtmeyer, P. (1975). Function of simian virus 40 gene A in transforming infection. *J. Virol.* **15**, 613–618.

Teitelman, G., Alpert, S. and Hanahan, D. (1988). Proliferation, senescence, and neoplastic progression of beta cells in hyperplasic pancreatic islets. *Cell* **52**, 97–105.

van Lohuizen, M., Verbeek, S., Krimpenfort, P., Domen, J., Saris, C., Radaszkiewicz, T. and Berns, A. (1988). Predisposition to lymphomagenesis in pim-1 transgenic mice: cooperation with c-myc and N-myc in murine leukemia virus-induced tumors. *Cell* **56**, 673–682.

Wallach, D., Fellous, M. and Revel, M. (1982). Preferential effect of γ-interferon on the synthesis of HLA antigens and their mRNAs in human cells. *Nature* **299**, 833–836.

Weiss, E. H., Mellor, A., Golden, L., Fahrner, K., Simpson, E., Hurst, J. and Flavell, R. A. (1983). The structure of a mutant H-2 gene suggests that the generation of polymorphism in H-2 genes may occur by gene conversion-like events. *Nature* **301**, 671–674.

Westphal, H. and Gruss, P. (1989). Molecular genetics of development studied in the transgenic mouse. *An. Rev. Cell. Biol.* **5**, 181–196.

Wilkinson, D. G., Bailes, J. A. and McMahon, A. P. (1987). Expression of the proto-oncogene int-1 is restricted to specific neural cells in the developing mouse embryo. *Cell* **50**, 79–88.

Wilkinson, D. G., Peters, G., Dickson, C. and McMahon, A. P. (1988). Expression of the FGF-related proto-oncogene int-2 during gastrulation and neurulation in the mouse. *EMBO J.* **7**, 691–695.

10
Retrovirus vectors for gene therapy procedures

D. VALERIO

Gene Therapy Department of the Institute of Applied Radiobiology and Immunology – TNO, PO Box 5815, 2280 HV Rijswijk, The Netherlands

1. Introduction

Our increasing knowledge of molecular and cellular biology has enabled us to identify the genetic basis of a large number of inherited diseases. It can be expected that the genes involved in all major congenital disorders will be cloned and characterized before the end of this century. Until now, medicine has

TRANSGENIC ANIMALS
ISBN 0-12-304530-4

benefited from this knowledge through the development of diagnostics and the biotechnological production of pharmaceuticals. However, in the near future we can expect that our greater understanding of molecular genetics will allow us to devise new cures for inborn errors of metabolism by the so-called somatic cell gene therapy. This therapy is based on the introduction of an intact version of a gene into the genetically crippled cells of a patient and thereby restoring their function. Application of gene therapy technology to these patients could decrease morbidity and greatly improve their quality of life. The ethical considerations of gene therapy have been widely discussed and the consensus emerged that the genetic modification of a patient's somatic cells with the purpose of correcting severe genetic disorders is ethically acceptable (Williamson, 1982; Friedmann, 1983; Motulsky, 1983; Anderson, 1984; Belmont and Caskey, 1986; Walters, 1986).

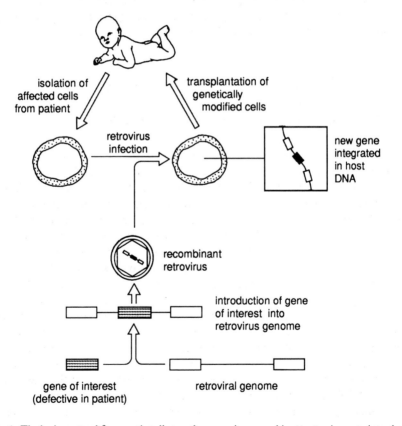

Fig. 1. The basic protocol for somatic cell gene therapy using recombinant retroviruses to introduce a correct version of a gene whose defect underlies the disease of a patient.

Successful gene therapy is dependent on efficient gene transfer systems. The requirements to be met by the gene delivery system include:

1. Efficient transfer into relatively large cell populations.
2. Stability of the introduced genes, which should preferably be integrated in the host cell genome.
3. Appropriate expression patterns of the newly introduced genes.
4. Acceptable safety margins for both the patient and the environment, to be evaluated in risk/benefit assessments.

Currently, retrovirus-mediated gene transfer is the only technique that fulfills requirements (1) through (3). Moreover, improvement in the design of recombinant retrovirus vectors and packaging cells has resulted in a broad applicability and high safety of retrovirus-mediated gene transfer. Most of the current approaches towards gene therapy therefore make use of retrovirus vectors and this chapter will deal with their design as useful tools in gene therapy procedures. A general outline of the envisaged therapy is depicted in Fig. 1.

2. Retrovirus-mediated gene transfer

2.1. Basic principles of retrovirus vector technology

The natural life cycle of retroviruses has prompted many scientists to use recombinant retroviruses as vehicles for foreign gene transfer. Clearly, the existing general understanding of viral replication (Weiss *et al.*, 1984) has been helpful in the design, generation and application of retrovirus vectors. The cycle can be divided into two distinct phases (see Fig. 2): first, viral entry and integration of a DNA copy of the viral RNA genome; and, secondly, steps involved in the expression of viral genes leading to the formation of new virus particles. The specific properties of the replication that render retroviruses suited for foreign gene transfer are: (1) retroviruses can infect a wide variety of cell types with high efficiency; (2) a large number of cells can be infected at once; (3) after adhesion and penetration of the host cells, viral RNA is reverse transcribed into a double-stranded DNA molecule which stably integrates into the cellular DNA in an orderly fashion; and, lastly, (4) the sequences required for viral replication can be physically separated into *cis*- and *trans*-acting elements (see also Fig. 3). The latter characteristic allowed the development of vector systems that generate replication-defective recombinant retroviruses free of wild-type viruses.

In principle, all retroviruses could be used as vector systems. However, for most purposes, including human gene therapy, the vectors should have a mammalian host range. Here we will limit ourselves to vectors based on murine

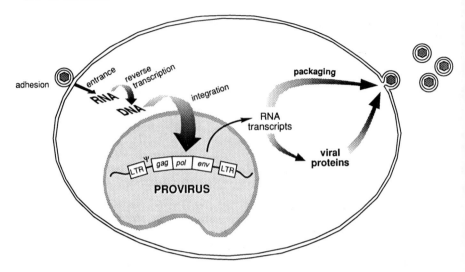

Fig. 2. The retroviral life cycle. Following adhesion, the virus RNA enters a cell and is reverse transcribed into a DNA molecule which is subsequently integrated into the cellular chromosome to form the provirus. Viral RNA is transcribed and translated into viral proteins which encapsidate another subset of viral RNA. The encapsidated viral RNA buds from the cell to give progeny virus particles.

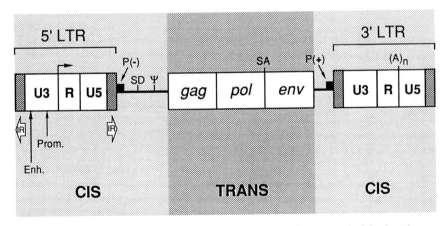

Fig. 3. Viral genome of MuLV and its division into sequences that are required *in cis* or *in trans* for virus replication. The *in cis* requirements consist of: (1) two LTRs which are subdivided into U3, R and U5 and carry inverted repeats (IRs) as well as the transcriptional enhancer, promoter and polyA signal $(A)_n$; (2) the primer binding sites $(P(-)$ and $P(+))$; (3) the packaging signal (Ψ). The *in trans* requirements are indicated as the coding regions of the viral genes *gag*, *pol* and *env*. Also shown in the figure are the splice donor (SD) and splice acceptor (SA) required for the splicing of the *env* mRNA. The transcription initiation site is indicated by a horizontal arrow and the position where a poly(A) tract is added to the mRNA is indicated by $(A)_n$.

leukaemia viruses (MuLV). Such systems, as first described by Mann *et al.* (1983), consist of two building blocks: a *retroviral vector* and a *packaging cell line*. These will be discussed separately in Sections 2.2 and 2.3.

2.2. Vector construction

Retroviral vectors are recombinant DNA molecules that carry the foreign gene(s) to be transduced as well as all the *cis*-acting sequences required for packaging and integration of the viral genome. These *cis*-acting elements and their location in the virus genome are depicted in Fig. 3. At each end, the viral DNA contains a long terminal repeat (LTR) that is divided into three regions, U3, R and U5. The viral promoter and enhancer are located in U3. R is a short sequence repeated at both ends of the RNA-form of the viral genome which (in the case of MuLVs) carries the poly(A) signal. The region that separates R from the 3′ end of the LTR is defined as U5. At the borders of an LTR small inverted repeats (IRs) are located; these are essential for provirus integration. Juxtaposing the LTRs are the primer binding sites required for negative (P($-$)) and positive (P($+$)) DNA strand synthesis. Furthermore, the vector should include a recognition signal essential for packaging genomic RNA into virions, the so-called ψ or packaging signal, which is located downstream of P($-$). These sequences make up the minimal requirements for retrovirus vectors. The space for accommodating foreign DNA is created by deleting the coding sequences for the viral *gag*, *pol* and *env* gene products. In the retroviral vector system these proteins are supplied *in trans* by the packaging cells (see Section 2.3).

2.2.1. Basic configurations

In principle, retroviral vectors can be designed to harbour any type of foreign DNA up to a size of about 8 kb, provided that a viral messenger can be transcribed which carries the elements required *in cis* (i.e. for packaging, reverse transcription and integration). In practice, this usually results in vectors that have retained all retroviral sequences present in the lightly shaded area of Fig. 3. Most often these elements are maintained in their natural position although it should be mentioned that the location of the packaging signal can be manipulated (Mann and Baltimore, 1985). In the last few years a great variety of constructs has been produced, in which one or two genes are present whose transcription relied on the activity of the viral promoter or internal promoter(s) and on different splicing processes. In Fig. 4 some of these configurations are given schematically. In many cases, a dominant marker was introduced in order to simplify the titration of viruses and to allow the detection and selection of

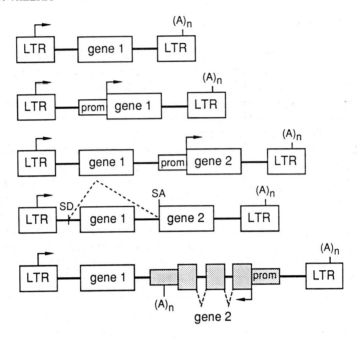

Fig. 4. Basic types of retrovirus vectors. In addition to all of the *in cis* requirements (see Fig. 2) the vectors carry one or two foreign gene(s) whose transcription is driven from the LTR (either directly or through splicing, indicated by the dashed line) or from an additional internal promotor (prom). The transcription initiation site is indicated by a horizontal arrow and the position where a poly(A) tract is added to the mRNA is indicated by $(A)_n$.

infected cells. When foreign genes are to be introduced into retroviral vectors in a 'sense' orientation (i.e. in the same orientation as the viral RNA transcription), two main restraints should be taken into account. (1) The introduction of poly(A) signals should be avoided since the presence of such sequences will lead to the formation of transcripts that lack the 3′ LTR and are therefore incapable of completing the replication cycle. As a result, viral titres of such vectors are drastically reduced. (2) The introduction of cellular genes carrying introns will lead to the deletion of the intron sequences from the viral messenger. This can be of consequence, especially when important regulatory elements are contained within the intron sequences of a gene. As a remedy for this shortcoming, vectors can be constructed that harbour the gene in the opposite orientation. Such genes can than be expressed from internal promoters that drive transcription of RNA molecules complementary to the viral RNA (see also Fig. 4).

2.2.2 *Alterations in the LTR*

The fact that not all of U3 is required for the proper packaging and integration of the viral genome has prompted vector designers to delete or replace parts of the promoters present in U3 of the 3 ′ LTR. By virtue of the fact that, upon viral replication, the U3 regions in both proviral LTRs are derived from the 3 ′ LTR of the vector, the alterations will eventually appear in both the 3 ′ and the 5 ′ LTR (this is schematically depicted in Fig. 5).

Since the enhancer element in U3 is known to be responsible for the tissue specificity of viral expression, several investigators attempted to modulate the expression patterns of recombinant viruses by supplying U3 with an extra enhancer (see Fig. 6, Mo + enh ′ LTR) or by replacing the viral enhancer by other enhancers (see Fig. 6, Δ(enh)Mo + enh ′ LTR). In most cases, these alterations were introduced into replication-competent viruses. The biological effects of these hybrid LTRs manifested themselves in the disease patterns induced in mice by viruses harbouring such modified LTRs. As examples of the insertion of enhancer elements into the LTR of Moloney MuLV (Mo-MuLV; a virus which induces T cell lymphomas in mice) the following cases can be mentioned:

1. Replacement of the Mo-MuLV enhancer with either the enhancer from the Friend murine leukaemia virus or from the Friend mink cell focus-forming virus resulted in viruses which induced erythroid leukaemia (Chatis *et al.*, 1983; Ishimoto *et al.*, 1987).

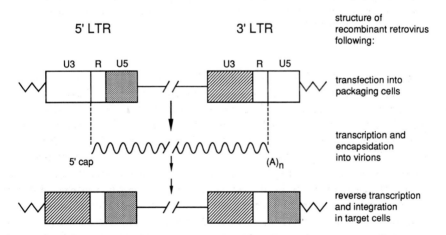

Fig. 5. Alterations in the U3 region of a 3 ′ LTR will end up in both LTRs after one round of replication. Transcription of a retroviral vector with a modification in U3 of the 3 ′ LTR will yield viral genomic messengers carrying this modified U3 region. The cap as well as the poly(A) tract of the mRNA are indicated. Following reverse transcription and integration, the resulting provirus will feature the altered U3 in both the 5 ′ and 3 ′ LTR.

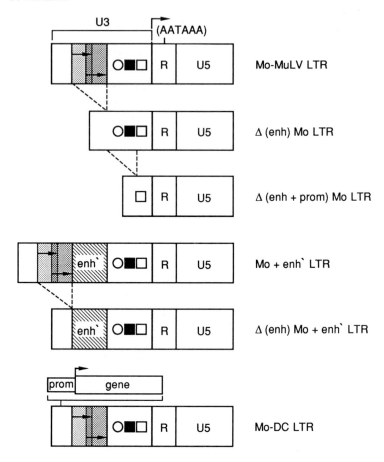

Fig. 6. Schematic representation of the Mo-MuLV LTR and modifications that can be introduced in the U3 region. The positions of the TATA box (open rectangle), CAAT box (solid rectangle) and G/C motif (open circle) that are part of the promoter, the two direct repeats that make up the enhancer element (stippled area) as well as the poly(A) signal (AATAAA) are indicated. Modifications that could be introduced include: a deletion of the enhancer element (Δ(enh)Mo LTR); a deletion of both the enhancer and the majority of the core promoter (Δ(enh + prom)Mo LTR); the addition of an enhancer element next to the Mo-MuLV enhancer (Mo + enh 'LTR); the replacement of the Mo-MuLV LTR by another enhancer (Δ(enh)Mo + enh 'LTR); and lastly the introduction of a mini gene 5 ' of the enhancer element (Mo-DC LTR).

2. Insertion of simian virus 40 (SV40) enhancer sequences into the Mo-MuLV LTR next to the endogenous enhancer yielded infectious virus that, like Mo-MuLV, induced primarily T cell-derived lymphoblastic lymphomas. However, viruses in which the SV40 enhancer replaced the Mo-MuLV enhancer induced pre-B and B cell lymphomas, as well as acute myeloid leukaemia (Hanecak et al., 1988).

3. The insertion of regulatory sequences from murine mammary tumour virus into the Mo-MuLV LTR resulted in glucocorticoid-responsive viruses (Overhauser and Fan, 1985).

4. Interesting LTRs with respect to gene therapy purposes seem to be those in which the enhancer of a polyoma virus host range mutant (PyF101) has been introduced. It was established that these viruses do not induce any form of leukaemia when they are injected subcutaneously into newborn mice, even though the viruses do replicate to high titres in tissue culture fibroblasts and can establish infection in animals (Davis et al., 1985). Two such LTRs were constructed, one in which the PyF101 enhancer was introduced next to the Mo-MuLV enhancer (Mo + PyF101 LTR), and another in which it replaced the retroviral enhancer (ΔMo + PyF101) (Linney et al., 1984). The latter LTR could be used to generate replication-defective retroviral vectors which consequently have a non-pathogenic backbone (Valerio et al., 1989; Van Beusechem et al., 1990a, b). An additional advantage of using the ΔMo + PyF101 LTR in viral vectors is that their expression pattern seems quite favourable for gene therapy purposes (see Section 3.2; Valerio et al., 1989; Van Beusechem et al., 1990b)).

In conclusion, enhancer replacements in the LTR are possible and seem an appropriate strategy to obtain altered virus expression specificities.

By deleting the promoter region in U3, the so-called self-inactivating (SIN) vectors were produced that lack a part of U3 in both LTRs after one round of replication (Yu et al., 1986). LTRs can be made in which only the enhancer or both the enhancer and most of the 'core promoter' is deleted (see Fig. 6, Δ(enh) Mo LTR and Δ(enh + prom) Mo LTR, respectively). As a result, these vectors, once integrated, are fatally crippled in their capacity to produce new viruses. Moreover, the absence of viral enhancers will reduce the risk that retroviral integration might activate a flanking cellular (proto-onco-) gene in the target cell, which is one of the main causes for concern in the application of retrovirus vectors in gene therapy. A drawback of these vectors is that their titres are usually very low.

Lastly, LTRs turned out to be flexible enough to harbour complete minigenes (Mo-DC LTR). Upon integration, such constructs result in a provirus with two copies of a minigene; one of these is present in the 5′ LTR outside of the retrovirus transcriptional unit, hence the name 'double-copy (DC) LTR' (Reik *et al.*, 1985; Hantzopoulos *et al.*, 1989) (see Fig. 6). Such viruses can also exhibit significantly altered expression patterns as compared to constructs driven by the Mo-MuLV LTR (Hantzopoulos *et al.*, 1989).

2.3. Packaging cell lines

Packaging cells provide the viral proteins required in a vector system. They synthesize all of the proteins encoded by the genes *gag*, *pol* and *env*, which are required *in trans* for the production of viable virus particles (see also Fig. 3). Packaging cells themselves, however, are incapable of releasing infectious virus. The functions of a packaging cell are schematically depicted in Fig. 7. The first MuLV-derived packaging line to be described was the Ψ-2 cell line (Mann *et al.*, 1983) that was generated by transfecting NIH/3T3 cells with Mo-MuLV from which the packaging signal (Ψ) had been deleted. Hence, Ψ-2 cells produce all the viral proteins but their viral RNA cannot be packaged into virions. The

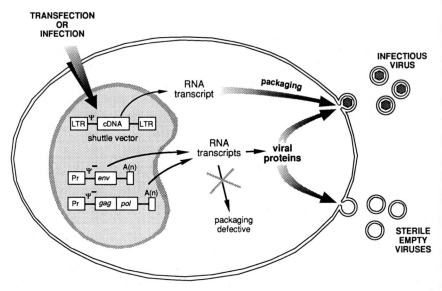

Fig. 7. A packaging cell at work. A retroviral vector carrying all of the *in cis* requirements as well as the gene(s) of interest is introduced into the packaging cell. This can be achieved either by physical transfection procedures or by infection with a recombinant retrovirus whose entrance is not hindered by interference. The packaging cell by itself produces all of the viral proteins, but generates no functional virus particles. Transcripts from the vector, however, can be packaged by the virus proteins to form infectious virus. Virus stocks from such cells are therefore helper-free and the recombinant retroviruses are replication-defective so that they can only undergo the first stages of the virus life cycle (see also Fig. 2).

introduction of a retroviral vector into the packaging cell line will therefore result in cells that can transcribe the recombinant retroviral genome, which is subsequently packaged into infectious but replication-defective particles to be released into the culture medium. With this procedure, virus-producing cell lines can be derived that stably produce recombinant virus stocks, free of helper virus, in titres up to 10^7 infectious particles per millilitre.

2.3.1. The host range of recombinant retroviruses

The host range of retroviruses is to a large extent determined by the proteins of the virus particle. The main denominator in this respect is the *env* gene-encoded envelope protein (see Weiss *et al.*, 1984). This protein is located at the surface of the virions and is required for the recognition of specific cell surface receptors during the adhesion and penetration phases of viral infection. Based on the envelope–receptor interaction, four host range classes of MuLVs can be recognized. These are:

1. *Ecotropic viruses*, which can infect mouse cells and, to a lesser extent, rat cells but not cells of other mammalian species.
2. *Xenotropic viruses*, which are derived from mouse cells but cannot infect that species due to a lack of receptors on the surface of the cells. On the other hand, xenotropic viruses have a wide host range in heterologous species including monkeys and humans.
3. *Amphotropic viruses*, which have a broad host range and can infect both homologous and heterologous cells, including those from monkeys and humans.
4. *Polytropic viruses*, which, like the amphotropic viruses, can infect mice, as well as a variety of other species. They belong, however, to a distinct class since, in contrast to the amphotropic envelope, the polytropic envelope is recognized by antisera against both ecotropic and xenotropic viruses.

Initially, most packaging cell lines were based on the *gag, pol* and *env* genes from the ecotropic Mo-MuLV. By replacing the Mo-MuLV *env* gene by one from an amphotropic, polytropic or xenotropic virus, packaging lines can be derived that produce recombinant viruses with the corresponding host ranges. In general, the *env* gene from the amphotropic virus 4070A (Chattopadhyay *et al.*, 1981) was used to derive packaging cells that produce viruses with a broad host range (see also Table 1). Obviously, the presence and concentration of viral receptors on the target cell is one of the factors that determines the infection efficiency. To date, however, our knowledge in this field is limited. The progress in the cloning and characterization of MuLV receptor genes (Albritton *et al.*, 1989) should be helpful in the elucidation of such questions. In the future, packaging cells might be designed that further define the cell-type specificity of recombinant viruses. One can, for example, envisage replacements or modifications of the *env* gene that will enable the resulting viruses to carry a ligand that interacts specifically with (receptor) molecules present on the target cells of choice (e.g. see Roux *et al.*, 1989). In Table 1, the most commonly used ecotropic and amphotropic packaging cell lines are presented with some of their characteristics.

2.3.2. *Improvements introduced to reduce the risk of helper virus formation*

One of the advantages of the modern retroviral vector systems is that, in principle, they can produce recombinant viruses free of replication-competent helper virus. For safety reasons this is considered to be a prerequisite in gene therapy protocols. To avoid the production of intact viruses, the viral proteins in packaging cells are encoded by retroviral sequences which, by themselves, are not transmissible as helper virus. In its simplest form this is achieved by a single deletion of the Ψ-packaging signal. Packaging cell lines such as Ψ-2, Ψ-Am and PA12 contain such constructs (see Table 1). These cell lines initially, indeed, generate helper-free virus stocks but, eventually, they may start to produce replication-competent viruses. This is presumably due to a recombinational event between the crippled helper virus genome and the vector which, by definition, carries the packaging sequences. To alleviate this problem, packaging lines were designed in which the regions of homology between the viral vectors and the helper genome are drastically reduced (see Table 1, the cell lines PA317 (Miller and Buttimore, 1986) and T19-14X (Sorge *et al.*, 1984)). In the case of PA317 cells this was achieved by introducing deletions in the helper provirus that encompass the 3 ′ LTR, the packaging signal and a portion of the 5 ′ LTR. The PA317 packaging cell line is widely used and has greatly reduced chances of helper-virus formation. Experiences from several laboratories have, however, indicated that helper-virus production can also occur in PA317 cells (Bosselman *et al.*, 1987; Danos and Mulligan, 1988; D. Valerio, unpublished observations). This can only be explained by a minimum of two recombinational events between the genomes of the helper virus and the vector. In view of these alarming reports, other groups approached this safety hazard by devising packaging cells in which deletions in the packaging signal and the LTRs are combined with a physical separation of the *gag* and *pol* genes from the *env* gene (see Fig. 7). This physical separation should be achieved by sequentially transfecting the 'packaging vectors' into the cells rather than by co-transfecting them, since recombination between co-transfected plasmids is known to occur at high frequencies (Small and Scangos, 1983). Indeed, a packaging cell line that was generated by co-transfecting separate *gag/pol* and *env* constructs eventually generated helper

Table 1. Packaging cell lines derived from MuLVs. All lines were generated in NIH/3T3 cells. PA12 and PA317 were derived from NIH/3T3 TK⁻ cells

Packaging line	Tropism	Recombination events for helper[a]	Helper virus detected	Marker gene[b]	Reference
Ψ-2	Ecotropic	1	+	*gpt*	Mann *et al.* (1983)
clone 32	Ecotropic	d	+	*neo* R	Bosselman *et al.* (1987)
Ψ-CRE	Ecotropic	c	−	*gpt, hygro* R	Danos and Mulligan (1988)
GP + E -86	Ecotropic	d	−	*gpt*	Markowitz *et al.* (1988b)
ΩE	Ecotropic	d	−	*gpt*	Morgenstern and Land (1990)
Ψ-AM	Amphotropic	1	+	*neo* R	Cone and Mulligan (1984)
PA12	Amphotropic	1	+	*tk*	Miller *et al.* (1985)
T19-13Z	Amphotropic	1	+	*neo* R	Sorge *et al.* (1984)
T19-14X	Amphotropic	2	−	*neo* R	Sorge *et al.* (1984)
PA317	Amphotropic	2	+	*tk*	Miller and Buttimore (1986)
Ψ-CRIP	Amphotropic	c	−	*gpt, hygro* R	Danos and Mulligan (1988)
GP + *env* Am12	Amphotropic	c	−	*gpt, hygro* R	Markowitz *et al.* (1988a)

[a] Minimum number of recombination events between the retrovirus vector and the helper genome(s) that could lead to the formation of helper virus.

[b] The marker gene indicates which (dominant) selectable marker genes were used to introduce the helper genome(s): *gpt*, *Escherichia coli* xanthine–guanine phosphoribosyltransferase gene; *neo* R, *E. coli* aminoglycoside-3'-phosphotransferase gene; *hygro* R, *E. coli* hygromycin B phosphotransferase gene; *tk*, herpes simplex virus thymidine kinase gene.

[c] In these cell lines the chances of helper formation are drastically reduced since two complementary helper genomes (*gag–pol* and *env*) are present at two separate locations in the cellular genome. Helper formation could only occur through subsequent recombination events at both these sites.

[d] These packaging cell lines were generated by co-transfecting two separate *gag–pol* and *env* constructs. As a consequence, recombination could have occured prior to integration in the cellular genome. The constructs employed in the ΩE cell line, however, had minimal sequence overlap and decreased sequence homology, which reduced the chances for such a recombination.

viruses after some time of *in vitro* culture (Bosselman *et al.*, 1987). The packaging cell lines ΨCRE, ΨCRIP and GP + *env* Am12 were generated by subsequent transfection of the *gag/pol* and *env* helper genomes (Danos and Mulligan, 1988; Markowitz *et al.*, 1988a). The generation of intact retroviruses from these lines would require a complicated sequence of events involving three recombinations, and to date this has not been reported. An added risk of the earlier packaging cells was that of the transfer of the helper virus genome into secondary cells. This phenomenon has been shown to occur with Ψ-2 and Ψ-AM cells (Bosselman *et al.*, 1987; Danos and Mulligan, 1988) which produce low levels of virus containing the Ψ⁻ genome and thereby transfer the 'packaging construct' into recipient cells, albeit at a low efficiency. Such a release of packaging-defective helper viruses has not been reported to take place in any of the redesigned packaging cell lines.

In conclusion, the introduced improvements satisfy the helper-virus safety issue as well as possible and permit the use of retroviral vectors for experiments directed towards human gene therapy.

2.4. Strategies to increase viral titres

In order to achieve cell lines that produce high titres of retroviral vectors several strategies have been described:

1. Armentano *et al.* (1987) discovered that a region of the *gag* gene which extends 418 bp downstream from the *gag* ATG translation initiation codon is responsible for the production of 10–50-fold increased titres. Consequently, most current vectors are designed to carry these sequences.
2. Hwang and Gilboa (1984) showed that the expression of genes introduced through retroviral infection is significantly higher than that of genes transfected with the calcium phosphate precipitate technique (Graham and Van der Eb, 1973). As a consequence, most investigators prefer to generate virus-producing cell lines by infecting the packaging cell line of choice with the recombinant retrovirus vector. Since packaging cell lines continuously produce the *env* gene product, competition for the virus receptor takes place when an infection is attempted with viruses containing the same *env* protein (viral interference). One has to overcome this viral interference block in order to be able to infect such cells. This can be done by (a) treating the cells with tunicamycin or (b) by a switch in tropism such that there is no interference between the infectious particle and the cells to be infected (e.g. by infecting amphotropic packaging cells with viruses derived from an ecotropic packaging cell line). By performing repeated reinfection cycles one can achieve extremely high proviral copy numbers with increased chances of high viral titres (Bestwick *et al.*, 1988). It should be mentioned, however, that this technique is not devoid of risk since it can lead to increased chances of generating helper virus (D. Valerio, unpublished observations).
3. By a stepwise increase in the concentration of a cytotoxic drug one can achieve amplification of a gene whose product detoxifies the culture

medium. Miller *et al.* (1985) have shown that viral constructs which carry such an amplifiable gene are amenable to standard gene amplification techniques which can also lead to increased viral titres. Examples of amplifiable genes that could be used for these purposes are those encoding dihydrofolate reductase (DHFR), adenosine deaminase (ADA) or a P-glycoprotein unidirectional efflux pump encoded by the so-called multidrug resistance (*mdr*) genes (Schimke *et al.*, 1978; Yeung *et al.*, 1983; Kane *et al.*, 1989; Israel and Kaufman, 1990).

3. Gene therapy protocols employing retrovirus vectors

3.1. Targets

In principle, all disorders caused by a gene defect should be curable by modifying the genetic make up of some of the patient's somatic cells. In addition, gene transfer can also be used to manipulate cellular (e.g. immune) responses, to interfere with infectious agents or simply as an advanced drug delivery system. In this, somewhat extended, view of what gene therapy might offer in the future, approaches are not limited solely to the inherited disorders but strategies could also be worked out for the improved treatment of afflictions caused by infectious agents, chronic diseases, neurological disorders and even cancer. Since, currently, the only feasible approach is one of gene addition rather than gene replacement, recessively inherited disorders are generally perceived as appropriate candidates. Scarpa and Caskey (1989) recently published a list of no less than 72 congenital diseases that are amenable to gene therapy. When choosing model disorders for gene therapy research, one should take into account several points of consideration.

1. The severity of the disease. As a rule, every experimental therapy should be preceded by a risk–benefit assessment. Similarly, the possible risks of gene therapy protocols should be studied in preclinical animal models in order to be able to compare them with the anticipated benefits of the procedure. However, the long-term side-effects in humans of any novel treatment might not be completely evident even long after the onset of clinical trials. Due to this inherent and unavoidable uncertainty, initial attempts should be aimed at disorders with severe pathology that can, at least in part, be alleviated by the proposed therapy. Currently, gene therapy research is therefore restricted to lethal diseases or to those that cause a dramatic loss in the quality of life.

2. Lack of effective therapies. Arguments similar to the previous one require that the future gene therapist should be able to indicate that the proposed treatment can be considered an improvement to available therapeutic possibilities. For this reason untreatable lethal disorders like cystic fibrosis or Duchenne's muscular dystrophy can be considered as candidate diseases (Drumm *et al.*, 1990; England *et al.*, 1990; Wolff *et al.*, 1990). Other initial target diseases could be those where severe side-effects are known to be associated with the presently available therapies. For example, bone marrow transplantation can be used to treat severe combined immunodeficiency (SCID) but, in the last 20 years, the over-all failure rate of the procedure due to the conditioning regime, graft-versus-host disease and take failure was about 25% when an HLA-matched donor could be used and even higher (about 50%) when HLA-mismatched donors had to be used (Dooren and Vossen, 1985; Fischer *et al.*, 1990). Recently, better results are obtained in the HLA-matched situation but, with HLA-mismatched donors (two-thirds of all cases), half of the patients still do not survive the treatment (Fischer *et al.*, 1990).

3. Chances of success.   The success of an experimental gene therapy will be influenced by:

(a) *Accessibility of the target tissue(s).* Several target tissues can be considered for genetic manipulation. The main issues here are whether (enough) cells can be isolated, infected and retransplanted or, otherwise, whether an acceptable form of *in vivo* gene transfer can be envisaged. The latter approach has not yet been extensively investigated and, until now, most attention has been focused on such transplantable cells as T lymphocytes, hepatocytes, fibroblasts, keratinocytes and endothelial cells (e.g. see Kantoff *et al.*, 1986; Braakman, *et al.*, 1992; Ledley *et al.*, 1987; Palmer *et al.*, 1987; Miyanohara *et al.*, 1988; Wilson *et al.*, 1988; Anderson *et al.*, 1989; Fenjves *et al.*, 1989; Zwiebel *et al.*, 1989; Kasid *et al.*, 1990). Although progress has been made with the use of all these cell types, this chapter will be limited to the use of bone marrow cells as discussed in Section 3.2.

(b) *Requirements for fine-tuning of foreign gene expression patterns.* The rules that govern expression of retrovirus-mediated genes are not always completely understood (see also Section 3.2.2). Currently, diseases can therefore only be considered as models for gene therapy if some freedom exists in the specificities and levels of gene expression that are expected to have a therapeutic effect. Examples of such diseases are deficiencies of humoral factors whose activities are largely regulated after their release into the serum (e.g. blood-clotting factors, St Louis and Verma, 1988; Palmer *et al.*, 1989; Hoeben *et al.*, 1990; Israel and Kaufman, 1990) or pathologies caused by the absence of a protein whose levels are known to vary greatly between healthy individuals (e.g. adenosine deaminase deficiency; see below).

(c) *Possible selective advantages of genetically cured cells.* When the corrected cells have a selective advantage over affected cells one can expect an enhanced repopulation of the afflicted compartment in a patient. This is especially important in cases where the available techniques do not permit a high efficiency of gene transfer. Examples can again be found among the haematological disorders. Bone marrow transplantation studies showed a clear selective advantage for healthy bone marrow stem cells to repopulate immunodeficient, thalassaemic or osteopetrotic recipients. This usually resulted in a chimeric blood system in which the cell lineages affected by the congenital defect were preferentially reconstituted by the donor cells (Volf *et al.*, 1978; Loutit and Nisbet, 1982; Dooren and Vossen, 1985; Wagemaker *et al.*, 1986).

4. The availability of a preclinical testing model. Proper testing of gene therapy protocols requires an animal model in which both the efficacy and possible toxic side-effects of the gene therapy protocol can be studied. Usually, standard laboratory animals will be used to test the *in vivo* characteristics of genetically modified cells. In some cases, strains of animals exist that harbour the genetic lesion that is under investigation. In those instances the therapeutic effects of a protocol can be evaluated directly in the animal. Several such animal models have been described and many more of them are to be expected in the near future, owing to the availability of embryonic stem cell lines and of techniques to select for homologous recombination events (see elsewhere in this book and Capecchi, 1989).

As a final evaluation before clinical trials, most investigators propose to test their protocols in an animal model more analogous to man than rodents. For this reason, my group and others are currently developing preclinical gene therapy models in non-human primates (Kantoff *et al.*, 1987; Van Beusechem *et al.*, 1990a; Valerio and Van Beusechem, 1991; see also Section 2.2.4).

Lastly, xenografting human cells in animals may be performed since these systems offer the credibility of an *in vivo* model coupled to the realistic situation of working with human (patient) material (Kamel-Reid and Dick, 1988; McCune *et al.*, 1988; Mosier *et al.*, 1988).

Based on the considerations mentioned above, several inherited disorders can be considered as targets for gene therapy. In the next section I will discuss one such disease, adenosine deaminase deficiency, which is the subject of gene therapy research in many laboratories, including my own.

3.2. Preclinical testing in the adenosine deaminase deficiency model

3.2.1. General considerations

It is generally believed that the initial development of techniques required for gene therapy should be directed to the treatment of a few model diseases that might serve to surmount the initial scientific and ethical hurdles. The knowledge so amassed can then be evaluated and applied to other diseases as well. One of the prime model targets for somatic cell gene therapy is the autosomally inherited form of SCID that is caused by malfunctioning of the human adenosine deaminase (hADA) gene. The disease is restricted to the haemopoietic system where the absence of functional ADA results in a defect in T and B cell differentiation (for a review see Thompson and Seegmiller, 1980). The cytotoxicity is probably caused by the accumulation of deoxyadenosine, which results in toxic concentrations of dATP in T cells. The deficiency in humoural immunity is possibly related to the absence or a defective function of helper T cells necessary for an antibody response. Despite the fact that the disease is quite rare (less than 1: 100 000 newborns (Hirschhorn, 1977)), a great number of laboratories have chosen ADA$^-$ SCID as a target for gene therapy (Valerio et al., 1984; Belmont et al., 1986; Kantoff et al., 1986; Williams et al., 1986; Palmer et al., 1987; Bordignon et al., 1989; Wilson et al., 1990). The reasons for this are:

1. The clinical syndrome is profoundly debilitating and, if untreated, lethal. Currently, ADA$^-$ SCID disease is treated with bone marrow transplantation. This treatment is possible if an HLA-identical donor is available (one-third of all cases) but HLA-mismatched parental donors have also been used. However, the major problem for the success of bone marrow transplantation in ADA$^-$ SCID is that the chances of graft failure or rejection are much larger than for other forms of SCID. As a result, the treatment fails in 40–60% of all cases (Fischer, et al., 1990).

2. Since it is a haemopoietic disorder that can be cured by bone marrow transplantation, genetic correction of haemopoietic stem cells (HSCs) should result in a complete and lasting cure and the future gene therapist can benefit from the vast experience and knowledge accumulated in the field of bone marrow transplantation. Moreover, the development of techniques to introduce functional genes into HSCs will be of great use in gene therapy protocols for a wide range of other (inherited) diseases.

3. Precise and organ-specific control of ADA gene expression is probably not required for a curative effect, since large variations of ADA levels have been found between healthy individuals (Thompson and Seegmiller, 1980;

Herbschleb-Voogt, 1983). Based on these observations one can predict that only 10% of the average ADA level will be sufficient to overcome the immunodeficiency.

4. Most importantly, it is to be expected that genetically cured blood cells will have a selective advantage over unaltered cells. This can be deduced from the chimeric reconstitution patterns of ADA⁻ SCID patients following bone marrow transplantation. In most of these cases, donor cell types can only be found in the lymphoid compartment, whereas all other haemopoietic lineages remain of the recipient type (Dooren and Vossen, 1985).

In a desirable gene therapy protocol, hADA coding sequences will be introduced into the HSCs obtained from a patient's bone marrow with the use of retroviral vector technology. Reimplantation and outgrowth of these cells should result in a lasting restoration of the patient's lymphoid system.

Recently, some ADA⁻ SCID patients have been treated experimentally with purified bovine ADA enzyme coupled to polyethylene glycol (PEG) (Hershfield et al., 1987; Levy et al., 1988). Although this form of treatment has been successful in some respects, a major problem was encountered when some of the patients developed ADA-inactivating antibodies (Hershfield and Finkelberg, 1989; A. Fischer, personal communication). One of the advantages of this treatment could be that it offers a new possibility to perform gene therapy, since PEG–ADA-treated patients usually have circulating T cells that can be isolated and infected with recombinant ADA viruses (Braakman et al., 1992). Reinfusion of such corrected cells into the patient can be expected to have a therapeutic effect either directly, when enough (different) T cells have been cured to restore some immunological functions or, more importantly, by detoxifying the patient's serum from the toxic adenine nucleosides as it were in an improved form of enzyme therapy. Recently, such a protocol was approved by the authorities in the USA to be performed by the same group that had already studied the fate of retrovirally marked tumour infiltrating lymphocytes (TILs) in patients (Kasid et al., 1990; Rosenberg et al., 1990). As a consequence, this procedure is now being applied before the more permanent bone marrow gene therapy has become operational. However, in contrast to the genetic modification of HSCs, one cannot expect that ADA gene transfer into mature T cells will result in the outgrowth of a lasting, completely cured and educated immune system. In my view, therefore, the transfer of a functional gene into HSCs that can repopulate the haemopoietic system in vivo remains one of the most important goals to be pursued for gene therapy in general and for the treatment of ADA⁻ SCID in particular.

retroviral vectors

retroviral vectors		LTR	expression of proviral ADA gene in:		reference
			NIH/3T3 cells	CFU-S derived spleen colonies	
(diagram)	ZIP-ADA	Mo-MuLV	+	-	1
(diagram)	DHFR*-ADA	Mo-MuLV	+	-	2
(diagram)	DHFR*-SVADA	Mo-MuLV	+	-	2,3
(diagram)	LHMsAL	Mo-MuLV	+	-	4
(diagram)	LHMiAL	Mo-MuLV	+	-	4,5
(diagram)	LDAAL	Mo-MuLV	+	-	5
(diagram)	LDTAL	Mo-MuLV	+	-	6
(diagram)	LDEAL	Mo-MuLV	+	-	6

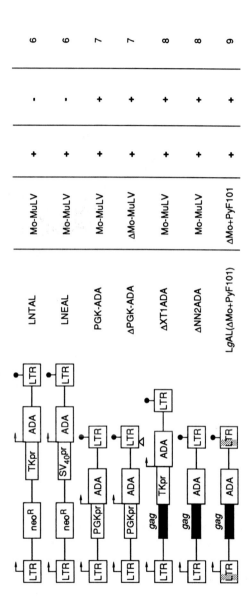

Fig. 8. Recombinant retroviral vectors carrying hADA gene sequences and their expression in cultured NIH/3T3 fibroblasts or CFU-S-derived spleen colonies. Possible transcription initiation sites are indicated by horizontal arrows. The poly(A) signal is depicted as a solid circle. In some vectors, retroviral splicing signals are included, as indicated by a dashed line. Selectable marker genes present in the viruses include: neo^R, encoding the bacterial aminoglycoside 3'-phosphotransferase; DHFR*, encoding methotrexate-resistant dihydrofolate reductase; and HPRT, encoding hypoxanthine phosphoribosyltransferase. In addition to the LTRs, different promoters are used to drive ADA expression, which include: SV_{40}pr, the simian virus 40 early-region promoter; MTpr and MTpr$_1$, a short and long version, respectively, of the mouse metallothioneine promoter; ADApr, the hADA gene promoter; TKpr, the herpes simplex virus thymidine kinase gene promoter; PGKpr, the phosphoglycerate kinase gene promoter. The black horizontal bar indicates a part of the retroviral gag gene that was included in some of the vectors to obtain higher viral titres (see Section 2.4). In addition to the standard Mo-MuLV LTRs, an enhancerless LTR (ΔMo-MuLV), as well as an LTR in which the enhancer sequences were replaced by the enhancer of a polyoma virus selected to multiply in embryonal carcinoma cells (ΔMo + PyF101 LTR). References: 1, Valerio et al. (1984); 2, Williams et al. (1986); 3, Williams et al. (1988); 4, McIvor et al. (1987a); 5, Valerio et al. (1986); 6, McIvor et al. (1987b); 7, Lim et al. (1987); 8, Belmont et al. (1988); 9, Van Beusechem et al. (1990b).

3.2.2.   Obstacles in achieving ADA expression in blood cells of transplanted mice in vivo

Efficient transfer of foreign genes into HSCs can be obtained via retroviral vector technology (e.g. see Williams *et al.*, 1984; Keller *et al.*, 1985). However, a drawback of retrovirus-mediated gene transfer has been that several vectors were found to be incapable of directing sustained expression in haemopoietic cells *in vivo*. It has been reported, for example, that a number of vectors are not expressed in blood cells of animals transplanted with infected HSCs (e.g. see Valerio *et al.*, 1986; Williams *et al.*, 1986; McIvor *et al.*, 1987a, b). In most of the early studies, gene transfer and expression were tested in the class of HSCs that can clonally grow out to form macroscopic colonies on the spleens of lethally irradiated mice 10–14 days after transplantation (CFU-S) (Till and McCulloch, 1961). In Fig. 8, a selection of recombinant ADA viruses is presented that were tested in such an assay by my laboratory and by other groups. As can be seen, a great number of vectors that are potent expressors in NIH/3T3 fibroblasts were silent in the CFU-S-derived spleen colonies. This repression in the haemopoietic system was also observed with other genes and, to overcome it, most investigators introduced additional promoters into the viral transcription unit (Valerio *et al.*, 1986; Lim *et al.*, 1987; Magli *et al.*, 1987; McIvor *et al.*, 1987a, b; Belmont *et al.*, 1988; Bowtell *et al.*, 1988; Guild *et al.*, 1988). Although this has led to some progress in the expression patterns, the outcome of such alterations seems unpredictable, which is witnessed by the fact that a number of strong promoters that act constitutively in various cell types, including blood cells, were inactivated when introduced into HSCs as a part of retrovirus vectors (see the references cited in Fig. 8). The combined data in Fig. 8 also reveal that, ultimately, ADA expression in CFU-S-derived spleen colonies was achieved with simple 'one-gene-only' vectors in which a number of different promoters were shown to be active. Since most Mo-MuLV-derived vectors are transcriptionally inactive in undifferentiated embryo carcinoma (EC) cells, it has been suggested that an analogy exists in the mechanisms responsible for the expression block in EC cells and in haemopoietic cells *in vivo* (McIvor *et al.*, 1987a; Guild *et al.*, 1988; Valerio *et al.*, 1989). The results obtained with the ΔNN2ADA virus (see Fig. 8), however, argue against this hypothesis, since this virus expresses hADA in haemopoietic cells *in vivo*, despite the fact that the gene is driven by a Mo-MuLV LTR, which will most likely render the vector inactive in EC cells.

In an elegant experiment, Williams *et al.* (1988) have shown that the inhibition of expression in the haemopoietic system is not just associated with the differentiation state of the infected cells. When they used DHFR*–SVADA virus (see Fig. 8) to infect murine primary haemopoietic progenitor cells (CFU-S), hADA was not expressed in the mature progeny from these stem cells. In contrast, the hADA enzyme was readily detected in colonies derived from immortalized

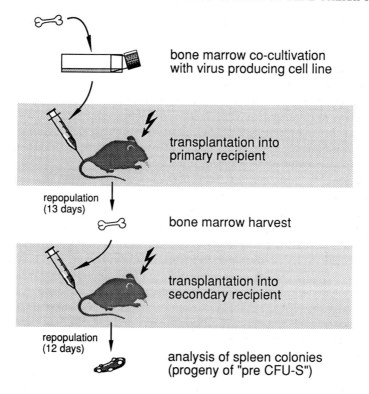

bone marrow co-cultivation
with virus producing cell line

transplantation into
primary recipient

repopulation
(13 days)

bone marrow harvest

transplantation into
secondary recipient

repopulation
(12 days)

analysis of spleen colonies
(progeny of "pre CFU-S")

Fig. 9. Improved method to quantify retroviral gene transfer into primitive haemopoietic stem cells.
See text for details.

multipotent factor-dependent cells. In the light of this study it should be mentioned that extreme caution is necessary when interpreting expression data derived from immortalized cells in culture, especially when the ultimate goal is to achieve expression in primary cells *in vivo*.

Inspired by the above, we have set up an *in vitro* model to monitor retrovirus expression in primary murine HSCs and their mature progeny. For these experiments it is necessary to infect purified HSCs. These can be obtained in relatively large numbers (up to 1×10^6 cells) from the bone marrow by a modification of the purification protocol described by Bertoncello *et al.* (1989) (Einerhand *et al.*, 1992). Purified stem cells are infected with retroviruses (either replication-competent or recombinant, replication-defective). After single-cell-per-well cloning the cells can be stimulated with the appropriate haemopoietic growth factors to obtain cell differentiation into a chosen lineage. Virus integration, expression and/or production can be monitored in the progeny of the stem cells present in the wells and these data will serve to quantify expression phenomena. Initial results obtained in this *in vitro* model gave further evidence for the presence of a retroviral expression block in primary HSCs infected with Mo-MuLV (Einerhand *et al.*, 1992).

Since the enhancer sequences in the LTR of retroviruses are known to be responsible for expression specificity (Weiss *et al.*, 1984), our efforts to obtain expression in haemopoietic cells *in vivo* have been directed toward alterations of this element. We have recently described (Valerio *et al.*, 1989) the construction of a recombinant retrovirus in which a marker gene was placed under the transcriptional control of a hybrid LTR designated ΔMo + PyF101 (Linney *et al.*, 1984). In this construct the enhancer from Mo-MuLV was replaced by the enhancer of a mutant polyoma virus (PyF101) that was selected to grow in EC cells. We showed that this vector directs gene expression in EC cells as well as in HSCs. Moreover, upon transplantation of infected bone marrow cells into lethally irradiated mice, virus expression was sustained in primitive haemopoietic cells of the engrafted recipients (Valerio *et al.*, 1989). These characteristics render the ΔMo + PyF101 LTR quite useful for the design of vectors to be employed in future gene therapy protocols. An additional advantage of such vectors would be that they are derived from a non-pathogenic retrovirus, since Davis *et al.* (1985) have shown that the leukaemogenic potential of Mo-MuLV is suppressed when its LTR is replaced by the ΔMo + PyF101 LTR. We have therefore also constructed a vector in which sequences encoding hADA are under the transcriptional control of the ΔMo + PyF101 LTR (Van Beusechem *et al.*, 1990). In Fig. 8 it can be seen that this virus is also capable of directing hADA expression in spleen colonies derived from infected CFU-S.

3.2.3.   ADA gene transfer into pluripotent HSCs

Apart from the necessity of overcoming the '*in vivo* expression hurdle', the need was also felt to further define the bone marrow stem cell type to be infected with the recombinant viruses. The CFU-S seemed a proper cell type to use as an assay for initial efficiency and toxicity tests. It is quite easy to measure CFU-S in a quantitative way, the assay is relatively fast and enough material can be obtained from the resulting CFU-S-derived spleen colonies to perform both biochemical and DNA analyses on individual colonies. In this light it may be relevant that most successful reports showing long-term persistence and expression of hADA gene sequences *in vivo* were preceded by encouraging results obtained with the recombinant viruses in CFU-S-derived spleen colonies (Lim *et al.*, 1987, 1989; Belmont *et al.*, 1988; Valerio *et al.*, 1989; Van Beusechem *et al.*, 1990b). From several bone marrow transplantation studies it became quite clear, however, that not all CFU-S represent pluripotent HSCs (PHSCs) (e.g. see Magli *et al.*, 1982;

Visser *et al.*, 1984; Bertoncello *et al.*, 1985). Other assays are therefore needed to test the efficiency with which such PHSCs can be infected. The most straightforward way of measuring this is by experiments in which the persistence of the introduced ADA gene and its expression are followed over an extended period of time in the peripheral blood cells of transplanted animals. Such experiments have been performed in mice and indicated that it is possible to infect PHSCs (Lim *et al.*, 1989; Wilson *et al.*, 1990). A limitation of these studies, however, was that they were executed by using viruses with an ecotropic host range. We ourselves recently succeeded in obtaining long-term expression of hADA in mice upon transplantation of bone marrow cells infected with amphotropic viruses which can also infect human cells (Van Beusechem *et al.*, 1990b). To assess the nature of the infected cells responsible for the hADA expression in these mice, we performed retransplantation experiments. Bone marrow was harvested from the primary recipients and used to transplant lethally irradiated secondary recipients. In the analysis, a persistence of hADA expression was observed in lymphoid, myeloid and erythroid blood lineages from the secondarily transplanted mice. DNA analysis revealed that different haemopoietic lineages of these mice harboured a proviral copy of the recombinant ADA virus in an identical chromosomal position. We concluded from these data that the cells were all descendants from one PHSC, which proves that this cell type can be infected with an amphotropic virus. This finding is directly relevant to the application of such viruses in protocols for human gene therapy.

A drawback of such long-term repopulation experiments is that it is difficult to derive quantitative information. Usually, haemopoietic regeneration in transplanted mice is the result of outgrowth from a limited number of PHSCs (Dick *et al.*, 1985; Keller *et al.*, 1985; Lemischka *et al.*, 1986; Snodgrass and Keller, 1987; Van Beusechem *et al.*, 1990b). From the study of such transplanted mice no information is obtained about the actual number of PHSCs that were infected prior to transplantation, nor does it tell us what the toxicity of the procedure was on this cell type. In order to try to solve this problem, we have turned to another experimental design in which cells with extensive self-renewing capacity are identified by an assay which measures marrow-repopulating ability (MRA). In Fig. 9 a schematic outline of the assay is given. Following infection of the bone marrow cells (either by co-cultivating them with the virus-producing cells or by supernatant infections) they are transplanted into lethally irradiated mice. Thirteen days post-transplantation the bone marrow cells of the mice are harvested and subsequently assayed for a primitive cell type (CFU-S in our case). Recent publications on the purification of HSCs indicate that the functional and phenotypic characteristics of this MRA CFU-S (or pre-CFU-S) closely resembles the cell type considered to be the PHSC (Hodgson and Bradley, 1979; Bertoncello *et al.*, 1985; Ploemacher and Brons, 1989). Moreover, MRA CFU-S present in retrovirus-infected bone marrow can be titrated by means of this assay

(Einerhand *et al.*, submitted). The efficiency and/or toxicity of the infection procedures can therefore be quantified on MRA CFU-Ss, quite primitive stem cells with extensive self-renewing capacity. Preliminary data obtained with this assay indicate that the MRA CFU-S can be infected with our amphotropic ADA virus and, in the secondary mice, gives rise to spleen colonies that express hADA. Studies using this assay led to the identification of factors that have an effect on the infection efficiency and the survival of PHSCs in experimental gene therapy protocols (Einerhand *et al.*, submitted).

3.2.4. *Preclinical testing in non-human primates*

Most investigators have performed studies in the mouse as an initial model for the retrovirus-mediated delivery of genes into HSCs (see Sections 3.2.2 and 3.2.3.). The results so far can serve to formulate some of the initial requirements for a gene therapy protocol. However, since mouse and human differ significantly with respect to the size, regulation, distribution and composition of their haemopoietic systems, the current technology cannot be translated directly into a clinical situation. Moreover, since the vectors are based on murine retroviruses, the genetic background of mice disqualifies them for establishing whether the protocols can lead to unwarranted recombinations with endogenous proviruses resulting in the production of recombinant retroviruses. Therefore, studies in larger animals more related to humans (i.e. non-human primates) are required to (1) solve logistic problems concerning the purification, culture, infection and transplantation of primate bone marrow, (2) confirm whether successful integration and expression of genes can also be obtained in primates, and (3) study the occurrence of unwanted recombinational events and possible toxic side-effects due to the protocol. To date, the two studies in which genetically altered bone marrow cells have been used to repopulate irradiated monkeys have had only limited success with respect to the efficiency of gene transfer and expression. Furthermore, the presence of helper virus in these experiments may have influenced their outcome (Anderson *et al.*, 1986; Kantoff *et al.*, 1987; Bodine *et al.*, 1990, Cornetta *et al.*, 1991). In both studies the regenerating capacity of the bone marrow graft was severely affected by the co-culture infection procedure.

Our group has also been involved in preclinical testing in non-human primates. The experimental protocol that we have designed is an autologous bone marrow transplantation in rhesus monkeys using genetically modified haemopoietic stem cells. In short, bone marrow cells obtained by puncturing the femoral shafts are approximately 10-fold enriched for pluripotent stem cells by albumin density gradient centrifugation (Dicke *et al.*, 1969). Genetic modification of the cells is obtained by co-cultivation with a cell line producing recombinant retroviruses, under conditions that stimulate division of repopulating stem cells. The bone marrow is subsequently reimplanted into the irradiated recipient.

Our initial experiments were directed at the establishment of culture conditions that allow the infection of rhesus haemopoietic stem cells. We monitored the effect of the culture conditions by measuring the content of myeloid progenitors (CFU-C), since these were shown to reflect the repopulating capacity of a bone marrow graft (Gerritsen et al., 1988; Van Beusechem et al., 1990a). These studies led to the finding that normal rhesus monkey serum (NMS) supports the formation of CFU-C. During a 3 day culture period the number of CFU-C progenitors doubled when NMS was present in the medium. On the 3rd day of culture we were able to infect approximately 15% of the bone marrow cells that gave rise to CFU-C (Valerio et al., 1989). We concluded that a primitive population of haemopoietic cells was stimulated to replicate, thereby becoming susceptible to retroviral infection.

In the murine model the most efficient gene transfer could be obtained by co-culturing bone marrow cells with irradiated virus-producing cells in the presence of interleukin 3 (IL-3). We therefore investigated whether this approach was also valid in the rhesus monkey model. In contrast to the cytotoxic effects of irradiated fibroblasts to monkey bone marrow that we and others have reported (Kantoff et al., 1987; Valerio et al., 1989; Bodine et al., 1990), a 3 day co-culture with our ADA virus-producing cells in the presence of IL-3 resulted in a complete recovery of CFU-C. Moreover, the non-adherent cells could be harvested and stimulated in vitro with Granulocyte-Macrophage Colony Stimulating Factor to produce mature myeloid cell types. Analysis of these cells revealed the presence of human ADA in quantities equal to the endogenous rhesus monkey ADA, indicating that progenitor cells were efficiently infected with a virus that allows expression in their mature progeny (Van Beusechem et al., 1990a).

The quality of the modified bone marrow grafts can also be measured in vivo by reinfusing the cells into the monkeys following lethal irradiation. To measure the repopulating capacity of the graft we employ a semiquantitative model, developed by Gerritsen et al. (1988), that describes the relation between the number of transplanted bone marrow stem cells and the regeneration time for reticulocytes in the peripheral blood. The data indicated that co-cultivation procedures exist during which the in vivo regenerative capacity of bone marrow cells can be conserved. These conditions also resulted in human ADA expression in peripheral blood cells of the transplanted animals. The foreign gene was observed in the haemopoietic system for as long as the animals were analysed (over the years) (Valerio and Van Beusechem, 1990; Van Beusechem et al., 1990a; Van Beusechem et al., submitted).

4. Summary

After the first successful introduction of DNA into mammalian cells by Graham and Van der Eb, in 1973, gene transfer has become a vital technique in biology and medicine and it is also the basis of a novel therapeutic approach termed 'gene therapy'. Since gene therapy requires a high efficiency of gene transfer, many investigators have turned to the use of retrovirus vectors. By nature, retroviruses are capable of efficiently and stably introducing genes into the chromosomes of vertebrate cells. The advent of recombinant DNA technology has made it possible to construct retrovirus vectors, as well as packaging cells which provide the viral proteins. In combination, these systems allowed the production of high-titre, helper-free, recombinant retroviruses capable of introducing the gene (or genes) of interest into a wide variety of cell types. Currently, the use of retrovirus vectors is widespread and their application in human gene therapy has recently been approved. Several groups have chosen an inherited form of severe combined immunodeficiency caused by a defect in the ADA gene (ADA$^-$ SCID) as their model disease for gene therapy. A desirable protocol that can be envisaged with the present knowledge would be to introduce the ADA gene into the pluripotent haemopoietic stem cells obtained from a patient's bone marrow with the use of retrovirus vectors. Reimplantation and outgrowth of these cells should then result in a lasting restoration of the patients lymphoid system. Preclinical tests in mice proved that it is possible to transfer an ADA gene into pluripotent haemopoietic stem cells with the help of amphotropic retroviruses that can also infect human cells. Initial obstacles in achieving long-term expression of the introduced gene in haemopoietic cells *in vivo* have been overcome in mice and similar results can be obtained in non-human primates. Thus, progress in the field of retrovirus vectors has brought human gene therapy within reach and actual clinical applications of bone marrow gene therapy can be anticipated within the coming years.

Acknowledgements

I thank the investigators in the Gene Therapy Department of the Institute for Applied Radiobiology and Immunology for sharing results and for their comments concerning this manuscript, in particular Victor Van Beusechem and Mark Einerhand, who contributed to most of the experiments performed in my laboratory. I also thank Hans van Ormondt (University of Leiden) for his valuable comments regarding the manuscript. Part of the research described was funded by the Netherlands Organization for Scientific Research (NWO).

References

Albritton, L. M., Tseng, L., Scadden, D. and Cunningham, J. M. (1989). A putative murine ecotropic retrovirus receptor gene encodes a multiple membrane-spanning protein and confers susceptibility to virus infection. *Cell* **57**, 659–666.

Anderson, K. D., Thompson, J. A., DiPietro, J. M., Montgomery, K. T., Reid, L. M. and Anderson, W. F. (1989). Gene expression in implanted rat hepatocytes following retroviral-mediated gene transfer. *Somat. Cell Mol. Genet.* **15**, 215–227.

Anderson, W. F. (1984). Prospects for human gene therapy. *Science* **226**, 401–409.

Anderson, W. F., Kantoff, P., Eglitis, M., McLachlin, J., Karson, E., Zwiebel, J., Nienhuis, A., Karlsson, S., Blaese, R. M., Kohn, D., Gilboa, E., Armentano, D., Zanjani, E. D., Flake, A., Harrison, M. R., Gillio, A., Bordignon, C. and O'Reilly, R. (1986). Gene transfer and expression in nonhuman primates using retroviral vectors. *Cold Spring Harbor Symp. Quant. Biol.* **LI**, 1073–1081.

Armentano, D., Yu, S. F., Kantoff, P. W., Von Ruden, T., Anderson, W. F. and Gilboa, E. (1987). Effect of internal viral sequences on the utility of retroviral vectors. *J. Virol.* **61**, 1647–1650.

Belmont, J. W. and Caskey, C. T. (1986). Developments leading to human gene therapy. In: Kucherlapati, R. (ed.) *Gene Transfer*, pp. 411–434. Plenum Press, New York.

Belmont, J. W., Henkel-Tigges, J., Chang, S. M. W., Wager-Smith, K., Kellems, R. E., Dick, J. E., Magli, M. C., Phillips, R. A., Bernstein, A. and Caskey, C. T. (1986). Expression of human adenosine deaminase in murine haematopoietic progenitor cells following retroviral transfer. *Nature* **322**, 385–387.

Belmont, J. W., MacGregor, G. R., Wager-Smith, K., Fletcher, F. A., Moore, K. A., Hawkins, D., Villalon, D., Chang, S. M. W. and Caskey, C. T. (1988). Expression of human adenosine deaminase in murine hematopoietic cells. *Mol. Cell. Biol.* **8**, 5116–5125.

Bertoncello, I., Hodgson, G. S. and Bradley, T. R. (1985). Multiparameter analysis of transplantable hemopoietic stem cells homing to marrow and spleen on the basis of rhodamine-123 fluorescene. *Exp. Hematol.* **13**, 999–1006.

Bertoncello, I., Bradley, T. R. and Hodgson, G. S. (1989). The concentration and resolution of primitive hemopoietic cells from normal mouse bone marrow by negative selection using monoclonal antibodies and dynabead monodisperse magnetic microspheres. *Exp. Hematol.* **17**, 171–176.

Bestwick, R. K., Kozak, S. L. and Kabat, D. (1988). Overcoming interference to retroviral superinfection results in amplified expression and transmission of cloned genes. *Proc. Natl Acad. Sci. USA* **85**, 5404–5408.

Bodine, D. M., McDonagh, K. T., Brandt, S. J., Ney, P. A., Agricola, B., Byrne, E. and Nienhuis, A. W. (1990). Development of a high-titer retrovirus producer cell line capable of gene transfer into rhesus monkey hematopoietic stem cells. *Proc. Natl Acad. Sci. USA* **87**, 3738–3742.

Bordignon, C., Yu, S. F., Smith, C. A., Hantzopoulos, P., Ungers, G. E., Keever, C. A., O'Reilly, R. J. and Gilboa, E. (1989). Retroviral vector-mediated high-efficiency expression of adenosine deaminase (ADA) in hematopoietic long-term cultures of ADA-deficient marrow cells. *Proc. Natl Acad. Sci. USA* **86**, 6748–6752.

Bosselman, R. A., Hsu, R., Burszewski, J., Hu, S., Martin, F. and Nicolson, M. (1987). Replication-defective chimeric helper proviruses and factors affecting generation of competent virus: expression of Moloney murine leukemia virus structural genes via the metallothionein promoter. *Mol. Cell. Biol.* **7**, 1797-1806.

Bowtell, D. D. L., Cory, S., Johnson, G. R. and Gonda, T. J. (1988). Comparison of expression in hemopoietic cells by retroviral vectors carrying two genes. *J. Virol.* **62**, 2464-2473.

Braakman, E., Van Beusechem, V. W., Van Krimpen, B. A., Fischer, A., Bolhuis, R. L. H. and Valerio, D. (1992). Genetic correction of cultured T cells from an adenosine deaminase deficient patient: characteristics of non-transduced and transduced cells. *Eur. J. Immunol.* **22**, 63-69.

Capecchi, M. R. (1989). The new mouse genetics: altering the genome by gene targetting. *Trends Genet.* **5**, 70-76.

Chatis, P. A., Holland, C. A., Hartley, J. W., Rowe, W. P. and Hopkins, N. (1983). Role for the 3' end of the genome in determining disease specificity of Friend and Moloney murine leukemia viruses. *Proc. Natl Acad. Sci. USA* **80**, 4408-4411.

Chattopadhyay, S. K., Oliff, A. I., Linemeyer, D. L., Lander, M. R. and Lowy, D. R. (1981). Genomes of murine leukemia viruses isolated from wild mice. *J. Virol.* **39**, 777-791.

Cone, R. D. and Mulligan, R. C. (1984). High-efficiency gene transfer into mammalian cells: generation of helper-free recombinant retrovirus with broad mammalian host range. *Proc. Natl Acad. Sci. USA* **81**, 6349-6353.

Cornetta, K., Morgan, R. A., Gillio, A., Sturm, S., Baltrucki, L., O'Reilly, R. and Anderson, W. F. (1991). No retroviremia or pathology in long term follow-up of monkeys exposed to a marine amphotropic retrovirus. *Human Gene Ther.* **2**, 215-219.

Danos, O. and Mulligan, R. C. (1988). Safe and efficient generation of recombinant retroviruses with amphotropic and ecotropic host ranges. *Proc. Natl Acad. Sci. USA* **85**, 6460-6464.

Davis, B., Linney, E. and Fan, H. (1985). Suppression of leukemia virus pathogenicity by polyoma virus enhancers. *Nature* **314**, 550-553.

Dick, J. E., Magli, M. C., Huszar, D., Philips, R. A. and Bernstein, A. (1985). Introduction of a selectable gene into primitive stem cells capable of long-term reconstitution of the hematopoietic system of W/W^v mice. *Cell* **42**, 71-79.

Dicke, K. A., Tridente, G. and Van Bekkum, D. W. (1969). The selective elimination of immunologically competent cells from bone marrow and lymphoctye cell mixtures. III. In vitro test for the detection of immunocompetent cells in fractionated mouse spleen cell suspensions and primate bone marrow suspensions. *Transplantation* **8**, 422-434.

Dooren, L. J. and Vossen, J. M., (1985). Severe combined immunodeficiency: Reconstitution of the immune system following bone marrow transplantation In: Van Bekkum, D. W. and Löwenberg, B. (eds) *Bone Marrow Transplantation*, vol. 3, pp. 351-381. Marcel Dekker, New York .

Drumm, M. L., Pope, H. A., Cliff, W. H., Rommens, J. M., Marvin, S. A., Tsui, L., Collins, F. S., Frizzel, R. A. and Wilson. J. M. (1990). Correction of the cystic fibrosis defect in vitro by retrovirus-mediated gene transfer. *Cell* **62**, 1227–1233.

Einerhand, M. P. W., Bakx, T. A., Visser, J. W. M. and Valerio, D. (1992). Retrovirus infection and culture of purified hemopoietic stem cells. *Bone Marrow Transpl.*, in press.

England, S. B., Nicholson, L. V. B., Johnson, M. A., Forrest, S. M., Love, D. R., Zubrzycka-Gaarn, E. E., Bulman, D. E., Harris, J. B. and Davies, K. E. (1990). Very mild muscular dystrophy associated with the deletion of 46% of distrophin. *Nature* **343**, 180–182.

Fenjves, E. S., Gordon, D. A., Pershing, L. K., Williams, D. L. and Taichman, L. B. (1989). Systemic distribution of apolipoprotein E secreted by grafts of epidermal keratinocytes: implications for epidermal function and gene therapy. *Proc. Natl Acad. Sci. USA* **86**, 8803–8807.

Fischer, A., Griscelli, C., Friedrich, W., Kubanek, B., Levinsky, R., Morgan, G., Vossen, J., Wagemaker, G. and Landais, P. (1980). Bone-marrow transplantation for immunodeficiencies and osteopetrosis: European survey, 1968–1985, *Lancet* **ii**, 1080–1084.

Fischer, A., Lardais, P., Friedrich, W., Morgan, G., Gerritsen, B., Fasth, A., Porta, F., Griscelli, C., Goldman, S. F., Levinsky, R. and Vossen, J. (1990). European experience of bone-marrow transplantation for severe combined immunodeficiency. *Lancet* **336**, 850–854.

Friedmann, T. (1983). *Gene Therapy: Fact and Fiction.* Cold Spring Harbor Laboratory, Cold Spring Harbor.

Gerritsen, W. R., Wagemaker, G., Jonker, M., Kenter, M. J. H., Wielenga, J. J., Hale, G., Waldmann, H. and Van Bekkum, D. W. (1988). The repopulation capacity of bone marrow grafts following pretreatment with monoclonal antibodies against T lymphocytes in rhesus monkeys. *Transplantation* **45**, 301–307.

Graham, F. L. and Van der Eb, A. J. (1973). A new technique for the assay of infectivity of human adenovirus 5 DNA. *Virology* **52**, 456–467.

Guild, B. C., Finer, M. H., Housman, D. E. and Mulligan, R. C. (1988). Development of retrovirus vectors useful for expressing genes in cultured murine embryonal cells and hematopoietic cells in vivo. *J. Virol* **62**, 3795–3801.

Hanecak, R., Pattengale, P. K. and Fan, H. (1988). Addition or substitution of simian virus 40 enhancer sequences into the Moloney murine leukemia virus (M-MuLV) long terminal repeat yields infectious M-MuLV with altered biological properties. *J. Virol.* **62**, 2427–2436.

Hantzopoulos. P. A., Sullenger, B. A., Ungers, G. and Gilboa, E. (1989). Improved gene expression upon transfer of the adenosine deaminase minigene outside the transcriptional unit of a retroviral vector. *Proc. Natl Acad. Sci. USA* **86**, 3519–3523.

Herbschleb-Voogt, E. (1983). Expression of human adenosine deaminase and its complexing protein in health and disease. Ph.D Thesis, State University of Leiden.

Hershfield, M. S., Buckley, R. H., Greenberg, M. L., Melton, A. L., Schiff, R., Hatem, C., Kurtzberg, J., Markert, M. L., Kobayashi, R. H., Kobayashi, A. L., and Abuchowski, A. (1987). Treatment of adenosine deaminase deficiency with polyethylene glycol-modified adenosine deaminase. *N. Engl. J. Med.* **316**, 589–596.

Hershfield, M. S. and Finkelberg, Z. (1989). ADA deficiency treatment. *Science* **246**, 1375.

Hirschhorn, R. (1977). Defects of purine metabolism in immunodeficiency diseases. In: Schwartz, R. S. (ed.) *Progress in Clinical Immunology*, pp. 67–83. Grune and Stratton, New York.

Hodgson, G. S. and Bradley, T. R. (1979). Properties of haematopoietic stem cells surviving 5-fluorouracil treatment: evidence for a pre-CFU-S cell? *Nature* **281**, 381–382.

Hoeben, R. C., Einerhand, M. P. W., Cramer, S. J., Briët, E., Van Ormondt, H., Valerio, D. and Van der Eb, A. J. (1990). Toward gene therapy in hemophilia A: introduction of facor VIII expression vectors into somatic cells. In: Crommelin, D. J. A. and Schellekens, H. (eds) *From Clone to Clinic*, pp. 373–378. Kluwer, Dordrecht.

Hwang, S. L. H. and Gilboa, E. (1984). Expression of genes introduced into cells by retroviral infection is more efficient than that of genes introduced into cells by DNA transfection. *J. Virol* **50**, 417–424.

Ishimoto, A., Takimoto, M., Adachi, A., Kakuyama, M., Kato, S., Kakimi, K., Fukuoka, K., Ogiu, T. and Matsuyama, M. (1987). Sequences responsible for erythroid and lymphoid leukemia in the long terminal repeats of Friend-mink cell focus-forming and Moloney murine leukemia viruses. *J. Virol.* **61**, 1861–1866.

Israel, D. I. and Kaufman, R. J. (1990). Retroviral-mediated transfer and amplification of a functional human factor VIII gene. *Blood* **75**, 1074–1080.

Kamel-Reid, S. and Dick, J. E. (1988). Engraftment of immune-deficient mice with human hematopoietic stem cells. *Science* **242**, 1706–1709.

Kane, S. E., Reinhard, D. H., Fordis, C. M., Pastan, I. and Gottesman, M. (1989). A new vector using the human multidrug resistance gene as as a selectable marker enables overexpression of foreign genes in eukaryotic cells. *Gene* **84**, 439–446.

Kantoff, P. W., Kohn, D. B., Mitsuya, H., Armentano, D., Sieberg, M., Zwiebel, J. A., Eglitis, M. A., McLachlin, J. R., Wiginton, D. A., Hutton, J. J., Horowitz, S. D., Gilboa, E., Blaese, R. M. and Anderson, W. F. (1986). Correction of adenosine deaminase deficiency in cultured human T and B cells by retrovirus-mediated gene transfer. *Proc. Natl Acad. Sci. USA* **83**, 6563–6567.

Kantoff, P. W., Gillio, A. P., McLachlin, J. R., Bordignon, C., Eglitis, M. A., Kernan, N. A., Moen, R. C., Kohn, D. B., Yu, S. , Karson, E., Karlsson, S., Zwiebel, J. A., Gilboa, E., Blaese, R. M., Nienhuis A., O'Reilly, R. J. and Anderson, W. F. (1987). Expression of human adenosine deaminase in nonhuman primates after retrovirus-mediated gene transfer. *J. Exp Med.* **166**, 219–234.

Kasid, A., Morecki, S., Aebersold, P., Cornetta, K., Culver, K., Freeman, S., Director, E., Lotze, M. T., Blaese, R. M., Anderson, W. F. and Rosenberg, S. A. (1990). Human gene transfer: characterization of human tumor-infiltrating lymphocytes as vehicles for retroviral-mediated gene transfer in man. *Proc. Natl Acad. Sci. USA* **87**, 473–477.

Keller, G., Paige, C., Gilboa, E, and Wagner, E. F. (1985). Expression of a foreign gene in myeloid and lymphoid cells derived from multipotent haematopoietic precursors. *Nature* **318**, 149–154.

Ledley, F. D., Darlington, G. J., Hahn, T. and Woo, S. L. C. (1987). Retroviral gene transfer into primary hepatocytes: implications for genetic therapy of liver-specific functions. *Proc. Natl Acad. Sci. USA* **84**, 5335–5339.

Lemischka, I. R., Raulet, D. H. and Mulligan, R. C. (1986). Developmental potential and dynamic behavior of hematopoietic stem cells. *Cell* **45**, 917–927.

Levy, Y., Hershfield, M. S., Fernandez-Meija, C., Polmar, S. H., Scudiery, D., Berger, M. and Sorensen, R. U. (1988). Adenosine deaminase deficiency with late onset of recurrent infections: response to treatment with polyethylene glycol-modified adenosine deaminase. *J.Pediat.* **113**, 312–317.

Lim, B., Williams, D. A. and Orkin, S. H. (1987). Retrovirus-mediated gene transfer of human adenosine deaminase: expression of functional enzyme in murine hematopoietic stem cells in vivo. *Mol. Cell. Biol.* **7**, 3459–3465.

Lim, B., Apperley, J. F., Orkin, S. H. and Williams, D. A. (1989). Long-term expression of human adenosine deaminase in mice transplanted with retrovirus-infected hematopoietic stem cells. *Proc. Natl Acad. Sci. USA* **86**, 8892–8896.

Linney, E., Davis, B., Overhauser, J., Chao, E. and Fan, H. (1984). Non-function of a Moloney murine leukaemia virus regulatory sequence in F9 embryonal carcinoma cells. *Nature* **308**, 470–472.

Loutit, J. F. and Nisbet, K. W. (1982). The origin of osteoclasts. *Immunobiology* **161**, 193–203.

McCune, J. M., Namikawa, R., Kaneshima, H., Shultz, L. D., Lieberman, M. and Weissman, I. L. (1988). The SCID-hu mouse: murine model for the analysis of human hematolymphoid differentiation and function. *Science* **241**, 1632–1639.

McIvor, R. S., Johnson, M. J., Miller, A. D., Pitts, S., Williams, S. R., Valerio, D., Martin Jr., D. W. and Verma, I. M. (1987a). Human purine nucleoside phosphorylase and adenosine deaminase: gene transfer into cultured cells and murine hematopoietic stem cells by using recombinant amphotropic retroviruses. *Mol. Cell. Biol.* **7**, 838–846.

McIvor, R. S., Pitts, S. and Martin, D. W. (1987b). Gene transfer and expression of human purine nucleoside phosphorylase and adenosine deaminase: possibilities for therapeutic application. In: Sasazuki, T. (ed.) *New Approach to Genetic Diseases*, pp. 231–244. Academic Press, London.

Magli, M. C., Iscove, N. N. and Odartchenko, N. (1982). Transient nature of early haematopoietic spleen colonies. *Nature* **295**, 527–529.

Magli, M. C., Dick, J. E., Huszar, D., Bernstein, A. and Phillips, R. A. (1987). Modulation of gene expression in multiple hematopoietic cell lineages following retroviral vector gene transfer. *Proc. Natl Acad. Sci. USA* **84**, 789–793.

Mann, R. and Baltimore, D. (1985). Varying the position of a retrovirus packaging sequence results in the encapsidation of both unspliced and spliced RNAs. *J. Virol.* **54**, 401–407.

Mann, R., Mulligan, R. C. and Baltimore, D. (1983). Construction of a retrovirus packaging mutant and its use to produce helper-free replication defective retrovirus. *Cell* **33**, 153–159.

Markowitz, D., Goff, S. and Bank, A. (1988a). Construction and use of a safe and efficient amphotropic packaging cell line. *Virology* **167**, 400–406.

Markowitz, D., Goff, S. and Bank, A. (1988b). A safe packaging line for gene transfer: separating viral genes on two different plasmids. *J. Virol.* **62**, 1120–1124.

Miller, A. D. and Buttimore, C. (1986). Redesign of retrovirus packaging cell lines to avoid recombination leading to helper virus production. *Mol. Cell. Biol.* **6**, 2895–2902.

Miller, A. D., Law, M. F. and Verma, I. M. (1985). Generation of helper-free amphotropic retroviruses that transduce a dominant-acting methotrexate-resistant dihydrofolate reductase gene. *Mol. Cell. Biol.* **5**, 431–437.

Miyanohara, A., Sharkey, M. F., Witztum, J. L., Steinberg, D. and Friedmann, T. (1988). Efficient expression of retroviral vector-transduced human low density lipoprotein (LDL) receptor in LDL receptor-deficient rabbit fibroblasts in vitro. *Proc. Natl Acad. Sci. USA* **85**, 6538–6542.

Morgenstern, J. P. and Land, H. (1990). Advanced mammalian gene transfer: high titre retroviral vectors with multiple drug selection markers and a complementary helper-free packaging cell line. *Nucl. Acids Res.* **18**, 3587–3596.

Mosier, D. E., Gulizia, R. J., Baird, S. M. and Wilson, D. B. (1988). Transfer of a functional human immune system to mice with severe combined immunodeficiency. *Nature* **335**, 256–259.

Motulsky, A. G. (1983). Impact of genetic manipulation on society and medicine. *Science* **219**, 135–140.

Overhauser, J. and Fan, H. (1985). Generation of glucocorticoid-responsive Moloney murine leukemia virus by insertion of regulatory sequences from murine mammary tumor virus into the long terminal repeat. *J. Virol.* **54**, 133–144.

Palmer, T. D., Hock, R. A., Osborne, W. R. A. and Miller, A. D. (1987). Efficient retrovirus-mediated transfer and expression of a human adenosine deaminase gene in diploid skin fibroblasts from an adenosine deaminase-deficient human. *Proc. Natl Acad. Sci. USA* **84**, 1055–1059.

Palmer, T. D., Thomson, A. R. and Miller, A. D. (1989). Production of human factor IX in animals by genetically modified skin fibroblasts. *Blood* **73**, 438–445.

Ploemacher, R. E. and Brons, N. H. C. (1989). Separation of CFU-S from primitive cells responsible for reconstitution of the bone marrow hemopoietic stem cell compartment following irradiation: evidence for a pre-CFU-S cell. *Exp. Hematol.* **17**, 263–266.

Reik, W., Weiher, H. and Jaenisch, R. (1985). Replication-competent Moloney murine leukemia virus carrying a bacterial suppressor tRNA gene: selective cloning of proviral and flanking host sequences. *Proc. Natl Acad. Sci. USA* **82**, 1141–1145.

Rosenberg, S. A., Aebersold, P., Cornetta, K., Kasid, A., Morgan, R. A., Moen, R., Karson, E. M., Lotze, M. T., Yang, J. C., Topalian, S. L., Merino, M. J., Culver, K., Miller, A. D., Blaese, R. M. and Anderson, W. F. (1990). Gene transfer into humans – immunotherapy of patients with advanced melanoma, using tumor-infiltrating lymphocytes modified by retroviral gene transduction. *N. Engl. J. Med.* **323**, 570–578.

Roux, P., Jeanteur, P. and Piechaczyk, M. (1989). A versatile and potentially general approach to the targeting of specific cell types by retroviruses: application to the infection of human cells by means of major histocompatibility complex class I and class II antigens by mouse ecotropic murine leukemia virus-derived viruses. *Proc. Natl Acad. Sci. USA* **86**, 9079–9083.

Scarpa, M. and Caskey, C. T. (1989). The use of retroviral vectors in human disorders.

In: Baum, S. J., Dicke, K. A., Lotzova, E. and Pluznik, D. H. (eds) *Experimental Hematology Today – 1988*, pp. 81–91. Springer-Verlag, New York.

Schimke, R. T., Kaufman, R. J., Alt, W. and Kellems, R. F. (1978). Gene amplification and drug resistance in cultured murine cells. *Science* 202, 1051–1055.

Small, J. and Scangos, G. (1983). Recombination during gene transfer into mouse cells can restore the function of deleted genes. *Science* 219, 174–176.

Snodgrass, R. and Keller, G. (1987). Clonal fluctuation within the haematopoietic system of mice reconstituted with retrovirus-infected stem cells. *EMBO J.* 6, 3955–3960.

Sorge, J., Wright, D., Erdman, V. D. and Cutting, A. E. (1984). Amphotropic retrovirus vector system for human cell gene transfer. *Mol. Cell. Biol.* 4, 1730–1737.

St Louis, D. and Verma, I. M. (1988). An alternative approach to somatic cell gene therapy. *Proc. Natl Acad. Sci. USA* 85, 3150–3154.

Thompson, L. F. and Seegmiller, J. E. (1980). Adenosine deaminase deficiency and severe combined immunodeficiency disease. *Adv. Enzymol.* 51, 167–210.

Till, J. E. and McCulloch, E. A. (1961). A direct measurement of the radiation sensitivity of normal mouse bone marrow cells. *Radiat. Res.* 14, 213–222.

Valerio, D. and Van Beusechem, V. W. (1990). Expression of human adenosine deaminase in mice and rhesus monkeys following transplantation of bone marrow cells infected with recombinant retroviruses. *J. Cell. Biochem.* Meeting Abstract 14A 375.

Valerio, D. and Van Beusechem, V. W. (1991). Expression of human adenosine deaminase in mice and rhesus monkeys following transplantation of bone marrow cells infected with recombinant retroviruses. In: Chapel, H. M., Levinsky, R. J. and Webster, A. D. B. (eds) *Progress in Immune Deficiency*, vol. III, p. 229. Royal Society of Medicine London.

Valerio, D., Duyvesteyn, M. G. C. and Van der Eb, A. J. (1984). Introduction of sequences encoding functional human adenosine deaminase into mouse cells using a retroviral shuttle system. *Gene* 34, 163–168.

Valerio, D., Visser, T. P., Wagemaker, G., Van der Eb, A. J. and Van Bekkum, D. W. (1986). The introduction of human ADA sequences into mouse hematopoietic stem cells. In: Vossen, J. and Griscelli, C. (eds) *Progress in Immunodeficiency Research and Therapy*, vol II, pp. 335–355. Elsevier, Amsterdam.

Valerio, D., Einerhand, M. P. W., Wamsley, P. M., Bakx, T. A., Li, C. L. and Verma, I. M. (1989a). Retrovirus-mediated gene transfer into embryonal carcinoma cells and hemopoietic stem cells: expression from a hybrid long terminal repeat. *Gene* 84, 419–427.

Valerio, D., Van Beusechem, V. W., Einerhand, M. P. W., Hoogerbrugge, P. M., Van der Putten, H., Wamsley, P. M., Berkvens, T. M., Verma, I. M., Kellems, R. E., Van der Eb, A. J. and Van Bekkum, D. W. (1989b). Towards gene therapy for adenosine deaminase deficiency. In: Baum, S. J., Dicke, K. A., Lotzova, E. and Pluznik, D. H. (eds) *Experimental Hematology Today – 1988*, pp. 92–98. Springer-Verlag, New York.

Van Beusechem, V. W., Bakx, T. A., Kukler, A., Van der Eb, A. J., Van Bekkum, D. W. and Valerio, D. (1990a) Progress in the development of somatic cell gene therapy for adenosine deaminase deficiency. In: Gorin, N. C. and Douay, L. (eds) *Experimental Hematology Today – 1989*, pp. 95–99. Springer-Verlag, New York.

Van Beusechem, V. W., Kukler, A., Einerhand, M. P. W., Bakx, T. A., Van der Eb, A. J. Van Bekkum, D. W. and Valerio, D. (1990b) Expression of human adenosine deaminase in mice transplanted with hemopoietic stem cells infected with amphotropic retroviruses. *J. Exp. Med.* **172**, 729–736.

Visser, J. W. M., Bauman, J. G. J., Mulder, A. H., Eliason, J. F. and De Leeuw, A. M. (1984). Isolation of murine pluripotent hemopoietic stem cells. *J. Exp. Med.* **59**, 1576–1590.

Volf, D., Sensenbrenner, L. L., Sharkis, S. J., Elfenbein, G. J. and Scher, I. (1978). Induction of partial chimerism in nonirradiated B-lymphocyte-deficient CBA/N mice. *J. Exp. Med.* **147**, 940–945.

Wagemaker, G., Visser, T. P. and Van Bekkum, D. W. (1986). Cure of murine thalassemia by bone marrow transplantation without eradication of endogenous stem cells. *Transplantation* **42**, 248–251.

Walters, L. R. (1986). The ethics of human gene therapy. *Nature* **320**, 225–227.

Weiss, R., Teich, N., Varmus, H. and Coffin, J. (1984). *RNA Tumor Viruses.* Cold Spring Harbor Laboratory, Cold Spring Harbor.

Williams, D. A., Lemischka, I. R., Nathan, D. G. and Mulligan, R. C. (1984). Introduction of new genetic material into pluripotent haematopoietic stem cells of the mouse. *Nature* **310**, 476–480.

Williams, D. A., Orkin, S. H. and Mulligan, R. C. (1986). Retrovirus-mediated transfer of human adenosine deaminase gene sequences into cells in culture and into murine hematopoietic cells in vivo. *Proc. Natl Acad. Sci. USA* **83**, 2566–2570.

Williams, D. A., Lim, B., Spooncer, E., Longtine, J. and Dexter, T. M. (1988). Restriction of expression of an integrated recombinant retrovirus in primary but not immortalized murine hematopoietic stem cells. *Blood* **71**, 1738–1743.

Williamson, B. (1982). Gene therapy. *Nature* **298**, 416–418.

Wilson, J, M., Jefferson, D. M., Chowdury, J. R., Novikoff, P. M., Johnston, D. E. and Mulligan, R. C. (1988). Retrovirus-mediated transduction of adult hepatocytes. *Proc. Natl Acad. Sci. USA* **85**, 3014–3018.

Wilson, J. M., Danos, O., Grossman, M., Raulet, D. H. and Mulligan, R. C. (1990). Expression of human adenosine deaminase in mice reconstituted with retrovirus-transduced hematopoietic stem cells. *Proc. Natl Acad. Sci. USA* **87**, 439–443.

Wolff, J. A., Malone, R. W., Williams, P., Chong, W., Acsadi, G., Jani, A. and Felgner, P. L. (1990). Direct gene transfer into mouse muscle in vivo. *Science* **247**, 1465–1468.

Yeung, C. Y., Ingolia, D. E., Bobonis, C., Dunbar, B. S., Riser, M. E., Siciliano, M. J. and Kellems, R. E. (1983). Selective overproduction of adenosine deaminase in cultured mouse cells. *J. Biol. Chem.* **258**, 8338–8345.

Yu, S. F., Von Ruden, T., Kantoff, P. W., Garber, C., Seiberg, M., Rüther, U., Anderson, W. F., Wagner, E. F. and Gilboa, E. (1986). Self-inactivating retroviral vectors designed for transfer of whole genes into mammalian cells. *Proc. Natl Acad. Sci. USA* **83**, 3194–3198.

Zwiebel, J. A., Freeman, S. M., Kantoff, P. W., Cornetta, K., Ryan, U. S. and Anderson, W. F. (1989). High-level recombinant gene expression in rabbit endothelial cells transduced by retroviral vectors. *Science* **243**, 220–222.

11
Germ line manipulation: applications in agriculture and biotechnology

A. J. CLARK, J. P. SIMONS and I. WILMUT

AFRC Institute of Animal Physiology and Genetics, Edinburgh Research Station, Roslin, Midlothian EH25 9PS, UK

1. Introduction

During the last decade the techniques for manipulating the germ line of mice were established. This technology has been used to address a variety of fundamental questions in biology in areas such as the control of gene expression, developmental biology, immunology and cancer (for reviews see Palmiter and Brinster, 1986; Jaenisch, 1988; this volume). The techniques developed in the mouse have also been applied to a number of agricultural species. We review the progress that has been made in adapting this technology to domestic

TRANSGENIC ANIMALS
ISBN 0–12–304530–4

(a)

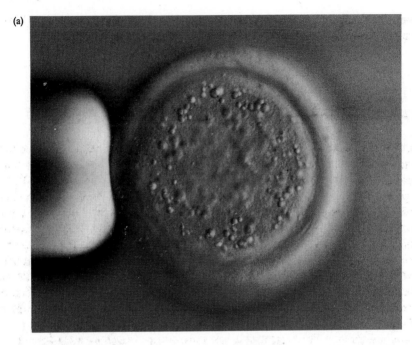

(b)

livestock, poultry and fish, and discuss the potential for its commercial exploitation in agriculture and biotechnology.

2. Production of transgenic livestock

2.1. Microinjection

At the time of writing, pronuclear microinjection is the only proven route for gene transfer in livestock. In principle, the technique is the same as that developed for the mouse in that it involves the recovery of fertilized eggs from superovulated donors, microinjection of DNA and surgical transfer of surviving embryos to hormonally synchronized recipients. In practice, work with farm animals suffers from a number of constraints that makes this technique more difficult than it is in mice. Fewer eggs are available, they are more prone to damage, there is greater variation in the stages of the embryo that are recovered, and the granular nature of the egg cytoplasm can complicate visualization of the pronuclei (Fig. 1; see also Hammer et al., 1985a; Simons et al., 1988). Litter size is limited to one or two in sheep and cattle and, therefore, large numbers of animals have to be employed as recipients for the microinjected eggs.

Despite these difficulties, transgenic pigs, sheep and cattle have been produced. From published data, the average frequency of success (eggs injected and transfered/transgenic animals) is estimated to be 0.59% for pigs and 0.74% for sheep (Table 1). These frequencies compare poorly to the 2–5% that can be obtained routinely in mice (e.g. see Brinster et al., 1985). In cattle there is not sufficient published data to make a reliable assessment, although it would seem unlikely that substantially higher success rates will be obtained.

The low efficiency of generating transgenic livestock is a severe constraint on the adoption of this technology. In sheep and cattle the small number of eggs available for manipulation limits the efficiency of the process. Recent developments, however, have shown that viable embryos can be obtained by the in vitro maturation and fertilization of oocytes removed from the ovaries of slaughtered sheep and cattle (Lu et al., 1987; Fukui et al., 1988). This approach has the potential to provide large numbers of eggs for manipulation. At present the low viability of embryos produced in this manner (e.g. see Lu et al., 1988) limits the utility of this approach. Improvements may be forthcoming, which may

---

Fig. 1. Visualization of pronuclei in fertilized sheep eggs. The eggs are held by gentle suction on the holding pipette (left) and viewed by means of differential interference contrast microscopy. There is considerable variation among the eggs recovered. In some, the extreme granular nature of the cytoplasm prevents visualization of the pronuclei (a) whereas in others both pronuclei are clearly visible (b). Most sheep eggs lie in between these two extremes.

ultimately enable embryos to be cultured *in vitro* to a stage at which their transgenic status could be assessed by the polymerase chain reaction (PCR), and nonsurgical transfer into the recipient is possible.

2.2.   Embryonic stem (ES) cells

In mice, ES cells allow for the targeted mutation of endogenous genes but, in most instances, offer no advantage over pronuclear injection for the introduction of new DNA seque... ~~ into the germ line. However, in farm animals where major limitations include the supply of embryos and the low success rates obtained by pronuclear injection (see above), this route may be advantageous for the introduction of new DNA sequences. The appropriate transgenic ES cell line could be established, characterized and stored prior to introduction into embryos. In cattle particularly, where blastocyst stage embryos can be recovered and transferred non-surgically, ES cells would obviate the need for surgery.

The isolation of ES cells from domestic livestock has not yet been conclusively demonstrated. Isolation and culture of inner cell mass cells from sheep embryos has been accomplished but it has proved difficult to establish undifferentiated cell lines from these (Fig. 2; Handyside *et al.*, 1987). Recently, however, Notarianni *et al.* (1990) have described the derivation of apparently pluripotent, embryonic cell lines from porcine and ovine blastocysts. These cells bear close resemblance to cells in primary outgrowths from inner cell masses and differentiate spontaneously in culture into a range of phenotypes. The pig cells have been transformed to geneticin resistance and used in blastocyst injection experiments. Preliminary results indicate the presence of the neomycin resistance gene sequences in DNA prepared from fetuses, indicating that these cells can contribute to tissues in the developing animal. Demonstration that these cells are able to contribute to the germ line is awaited. The successful isolation of ES cells from farm livestock will open up the possibilities of specific gene targeting in these species.

In cattle and sheep (but not in mice), young have been born following transfer of nuclei from cells of the inner cell mass to enucleated oocytes (Marx, 1988; Smith and Wilmut, 1989). There is, therefore, the possibility that if ES cells can be established from these species it will be possible to transfer nuclei from them, thus avoiding the chimeric generation and ensuring germ line transmission.

2.3.   Retroviral vectors

For domestic livestock, retroviruses have the potential advantage that they could enable the introduction of foreign DNA sequences into the germ line without

(a)

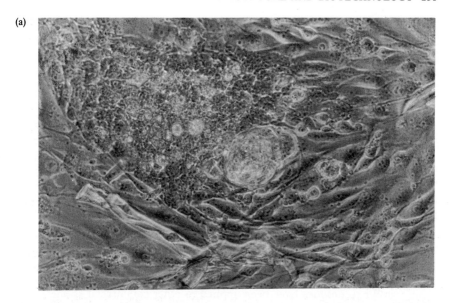

(b)

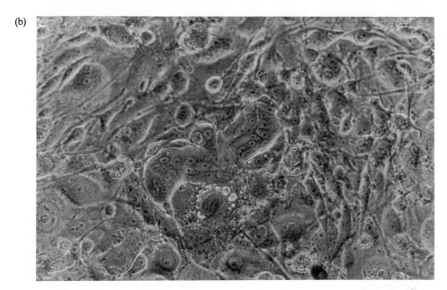

Fig. 2. Embryo-derived cells from sheep. (a) Primary explant. Day 10 embryonic discs were micro-dissected from the trophectoderm and explanted onto inactivated STO fibroblasts. (b) First passage. Primary explants were passaged onto a fresh feeder layer. These small, round cells have a large nucleus and prominent nucleoli and, overall, their morphology is reminiscent of mouse ES cells. With subsequent passages, however, the undifferentiated phenotype was not maintained. (Photographs courtesy of Jim McWhir, IAPGR.)

recourse to the considerable skills and expensive equipment required for pronuclear injection. There are reports of the use of retroviruses in both pigs (Petters *et al.*, 1988) and sheep (Harvey *et al.*, 1990), although it should be noted that neither of these studies provides evidence for integration of the DNA into the germ line and both reports describe the use of wild-type retroviruses rather than recombinants carrying a specific gene of interest .

There are also a number of disadvantages associated with using retroviral vectors and, in general, the approach has not been adopted widely, even in mice. These problems include the difficulty of producing high-titre recombinant retroviruses, the limit to the size of foreign DNA that can be incorporated into retroviral vectors (~ 8 kb) and the fact that this technique usually creates a mosaic animal in which only a fraction of the animal's cells carry the foreign DNA.

3.   Germ line manipulation of other commercial species

Gene transfer techniques have been applied to other commercial species, particularly poultry and fish. There are considerable differences from mammals in the reproductive biology of these organisms, necessitating somewhat different technical approaches.

3.1.   Poultry

In the chicken the yolk-laden ovum is fertilized in the anterior oviduct and, after deposition of albumen and shell, the egg is laid 24 hours later. The embryo or blastoderm of the new-laid egg comprises about 60 000 cells at the surface of the yolk. At this stage the chicken embryo is not amenable to microinjection of DNA as a means of gene transfer, but its ready availability has prompted a number of workers to attempt germ line manipulation by utilizing retroviral vectors. Bosselman *et al.* (1989a, b) have introduced the neomycin phosphotransferase herpes simplex thymidine kinase and chicken growth hormone genes into the chicken germ line by inoculating embryos with a reticuloendotheliosis virus (REV) retroviral vector. In these experiments, 8% of the males that hatched contained the foreign DNA in their sperm and transmitted it to their progeny. It has also been shown that non-defective recombinant avian leukosis (ALV) viruses and REV can be transmitted through the germ line following infection of the embryo or even chicks (Salter *et al.*, 1986, 1987; Salter and Crittenden, 1989).

Strategies for manipulating earlier stages of the chicken embryo are being developed that rely on the use of *in vitro* culture to hatch chicken embryos isolated from the oviduct at the one-cell stage (Perry, 1988). In this case, gene injection is feasible although the foreign DNA (10^5–10^6 copies) is injected into the cytoplasm at the presumed location of paternal and maternal pronuclei. Direct pronuclear injection is not possible because of the extreme opacity of the egg cytoplasm and the fact that there are many defunct male pronuclei present after fertilization (there is no block to polyspermy in the chicken). Injected DNA replicates, persists and is expressed, but transgenic chickens have yet to be produced by this method (Sang and Perry, 1989; Perry and Sang, 1990).

A third approach that is being taken is to produce chimeras by transferring isolated blastodermal cells from the early chick embryo and introducing them into a recipient embryo. Petitte *et al.* (1990) have described the production of both somatic and germ line chimeras using this methodology. These workers have also shown that these isolated blastodermal cells can be efficiently transfected using liposomes (Verrinder-Gibbins *et al.*, 1990), demonstrating that this could be a route for introducing foreign DNA into the germ line.

3.2. Fish

The generation of transgenic fish has been described by a number of workers (reviewed by Chen and Powers, 1990). Fish eggs are large and yolky and, in most species, the pronuclei are not visible. This has necessitated cytoplasmic injection of relatively large amounts of foreign DNA (10^5–10^6 copies). The large amount of DNA used and its extrachromosomal persistence has complicated the analysis of integration into the host genome, particularly when the analysis was carried out using fry. Complex Southern blotting patterns were often obtained between different tissues, indicative of mosaicism. Indeed, in only a few cases has transmission to the next generation been reported (Stuart *et al.*, 1988; Guyomard *et al.*, 1989; Zhang *et al.*, 1990), so confirming integration of foreign DNA into the germ line.

A present limitation to work with transgenic fish is that there is a dearth of fish genes. Many of the constructs that have been used comprise mammalian or viral regulatory elements. Thus, simian virus 40 (SV40), mouse metallothionein (mMT) and RSV promoters have been all been employed in constructs introduced into fish. Surprisingly, expression has been obtained in some cases, although it would appear that such promoters, presumably adapted for working homeothermic species, do not work particularly efficiently in poikilotherms. A few fish genes have been cloned and have been used in gene transfer experiments. For example, the complete flounder antifreeze protein gene has been introduced into the germ line of Atlantic salmon and low levels of expression detected (Davies *et al.*, 1989).

Table 1. Pronuclear injection experiments in domestic livestock. The various constructs used in gene transfer experiments are tabulated. Most are fusion genes. The regulatory DNA sequences are underlined

Construct[a]	Embryos injected and transferred	Transgenics (%)	Reference
Pigs			
MT–hGH	2035	20 (0.98)	Hammer *et al.* (1985a)
MT–hGH	268	1 (0.37)	Brem *et al.* (1985)
MLV–rGH	170	1 (0.59)	Ebert *et al.* (1988)
PEPCK–bGH	1057	2 (0.18)	Wieghart *et al.* (1988)
MT–pGH	423	6 (1.4)	Vize *et al.* (1988)
MT–hGRF	1041	6 (0.58)	Brem *et al.* (1988)
MT–Mx	1083	6 (0.55)	Brem *et al.* (1988)
MT–bGH	2330	9 (0.39)	Miller *et al.* (1989)
MT–hIGF-I	387	4 (1.0)	Pursel *et al.* (1989)
MT–hGRF	2236	7 (0.31)	Pursel *et al.* (1989)
PRL–bGH	289	5 (1.7)	Polge *et al.* (1989)
mWAP	850	5 (0.58)	Wall *et al.* (1991)
Total	12 169	72 (0.59)	
Sheep			
MT–hGH	1032	1 (0.1)	Hammer *et al.* (1985)
MT–bGH	711	2 (0.28)	Pursel *et al.* (1987)
oMT–oGHSV	1079	4 (0.37)	Murray *et al.* (1989)
oMT–oGH	409	3 (0.73)	Murray *et al.* (1989)
TF–bGH	247	7 (2.8)	Rexroad *et al.* (1988)
MT–hGRF	435	9 (2.0)	Rexroad and Pursel (1988)
MT–TK	108	1 (0.9)	Simons *et al.* (1988)
BLG–FIX	245	4 (1.6)	Simons *et al.* (1988)
BLG–AAT	298	3 (1.0)	Clark *et al.* (1988)
Total	4564	34 (0.74)	
Cattle			
αFP–UG	237	4 (1.6)[b]	Church *et al.* (1986)
RSV–CAT	175	4 (2.2)[b]	Biery *et al.* (1988)
pBPV	156	5 (3.2)[b]	Roschlau *et al.* (1989)
MMTV–ADH	130	9 (0.69)[b]	Roschlau *et al.* (1989)
MMTV–bGH	250	2 (0.8)[c]	Roschlau *et al.* (1989)

[a] MT, mouse metallothionein promoter; MLV, Moloney leukaemia virus long terminal repeat (LTR); PEPCK, phosphoenolpyruvate carboxykinase promoter; PRL, prolactin promoter; oMT, ovine metallothionein promoter; SV, simian virus 40 (SV40) origin of replication; TF, transferrin promoter; BLG, β-lactoglobulin sequences; αFP, α-fetoprotein sequences; RSV, Rous sarcoma virus LTR; MMTV, mouse mammary tumour virus LTR. Structural gene sequences: hGH, human growth hormone; rGH, rat growth hormone; bGH, bovine growth hormone; pGH, porcine growth hormone; hGRF, human growth hormone releasing factor; Mx, mouse Mx; hIGF-I, human insulin-

4. Exploiting gene transfer

The ability to manipulate the germ line of animals has stimulated approaches for the direct genetic improvement of commercial species. Some of the approaches that have been adopted have been aimed at improving conventional production traits, particularly growth rate. This suffers from at least two drawbacks. Firstly, such traits are controlled by many different genes, most of which have not been identified and are not obviously amenable to a technology that essentially involves the manipulation of single genes. Secondly, any improvements obtained by this route have to compete directly with the 'slow but sure' genetic gains that are obtainable by conventional selective breeding (Smith *et al.*, 1987). Gene transfer, however, also opens up the possibility of producing animals with quite new properties, thus creating commercial opportunities outside of conventional agriculture.

In the following sections, various opportunities for exploiting germ line manipulation technology are reviewed. Not surprisingly, transgenic mice have figured prominently in assessing these approaches, particularly with regard to transgenic livestock. Although there is an obvious common-sense rationale to this approach (it can take several years to evaluate a gene construct in a farm animal), caution has to be exercised, and results obtained in mice are not necessarily applicable to larger animals. Many of the applications of transgenic technology that have been tried or are being developed for livestock are also applicable to fish and poultry.

4.1. Growth and feed efficiency

Many biological processes influencing growth rate and feed efficiency are regulated by protein hormones, e.g. growth hormone (GH) and the insulin-like growth factors (IGFs). Many studies in transgenic mice have employed constructs which direct the expression of growth hormone from various promoters (Palmiter *et al.*, 1982, 1983). Transgenic mice which express GH ectopically frequently grow significantly faster and achieve greater size than control mice. This appeared to be a clear-cut opportunity to exploit the technology in domestic livestock. Indeed, more than 76% of the transgenic pigs or sheep reported in

---

like growth factor I; oGH, ovine growth hormone; TK, herpes simplex thymidine kinase; FIX, human factor IX; AAT, human α_1-antitrypsin; UG, unknown gene; CAT, chloramphenicol acetyl transferase; ADH, *Drosophila melanogaster* alcohol dehydrogenase. Other constructs: mWAP, mouse whey acidic protein gene and its associated flanking sequences; BPV, 69% of the genome of bovine papilloma virus.
[b] DNA analysis performed on embryos of fetuses.
[c] 1 live animal, 1 dead fetus.

the literature carry transgenes designed to manipulate growth and related traits (Table 1).

Transgenic pigs carrying human and bovine GH under the control of the mouse metallothionein (MT) promoter have been produced. High levels of exogenous GH were present in the serum of a number of the animals generated (Hammer *et al.*, 1985a; Pursel *et al.*, 1989). In some cases a growth increase was observed in founder animals, but variations as much as 30% in daily weight gain were seen in both controls and transgenics.

To gain a more reliable assessment of the effects of a GH transgene, Pursel *et al.* (1989) measured average daily weight gain and feed efficiency in two generations of transgenic and control pigs. Two lines of transgenic pigs were shown to grow 11% and 14% faster than their non-transgenic siblings. This, however, was only when the animals were fed a protein-supplemented diet. A 16–18% increase in feed efficiency was also observed. Most dramatically, expression of the GH transgene resulted in reduced carcass fat content and, for example, mean back fat thickness was reduced from an average of 21 mm to 7.5 mm (Fig. 3). The modest improvements in growth performance were, however, outweighed by a number of considerable deleterious side-effects. The most common problems included lethargy, joint abnormalities and exophthalmus (Pursel *et al.*, 1989).

Experiments with MT–GH genes have been carried out with sheep with similar results. In one set of experiments, four transgenic animals were produced. High levels of GH were present in serum at 20 weeks of age (Murray *et al.*, 1989). No enhancement of growth was seen and, again, these animals exhibited a number of negative side-effects such as joint abnormalities. As was observed with the pigs, these transgenic sheep also exhibited a dramatically reduced carcass fat content.

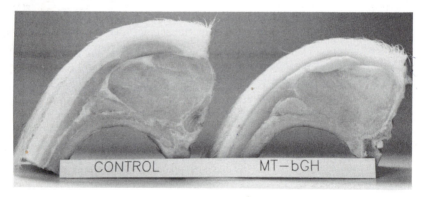

Fig. 3. Back fat reduction in transgenic pigs. Cross-section through the loin of a control and a transgenic pig carrying the MT–bGH construct. The dramatic reduction in back fat is evident in the transgenic animal. (Photograph courtesy of Bob Wall, US Department of Agriculture, Beltsville.)

To circumvent problems associated with chronic expression of MT–GH transgenes, the construction and testing of inducible GH constructs has been attempted. Phosphoenolpyruvate carboxykinase (PEPCK) is a crucial enzyme in the regulation of gluconeogenesis. PEPCK expression may be regulated by alteration of dietary carbohydrate or protein, as well as cyclic AMP administration (McGrane *et al.*, 1988). Transgenic mice have been produced carrying the PEPCK promoter fused to bovine GH (bGH). On a low-carbohydrate/high-protein intake, a 30-fold increase of bGH was observed in serum and, conversely, a high-carbohydrate intake resulted in a 90% decrease. The PEPCK–bGH hybrid gene has also been introduced into pigs. The expression pattern of the transgene and, consequently, the phenotype was, however, very similar to that observed with MT–GH, and dietary manipulation did not control the level of expression (Pinkert *et al.*, 1990).

Growth hormone-releasing factor (GRF) is the hypothalamic hormone which stimulates the synthesis and release of GH from the anterior pituitary. An MT–human GRF (MT–hGRF) transgene has been introduced into transgenic pigs. Two out of the seven founders secreted high concentrations of hGRF into their plasma (Pursel *et al.*, 1989). In these animals, however, production of endogenous GH was not stimulated. The administration of hGRF to pigs does stimulate growth hormone release (Kraft *et al.*, 1985). The failure of the transgene GRF to stimulate GH may be due to the fact that it is synthesized primarily in the liver, a tissue which may not correctly process the GRF precursor peptide (Pursel *et al.*, 1989). This contrasts with the increased growth of mice expressing MT–hGRF transgenes (Hammer *et al.*, 1985b).

In poultry, elevated levels of chicken GH have been detected in embryos injected with an REV vector containing chicken GH. Experiments are underway currently to investigate expression in the progeny of birds that have integrated this transgene in the germ line (Bosselman, 1989b) and assess any effects on growth rate.

In fish, Zhang *et al.* (1990) have shown that progeny inheriting an RSV–rainbow trout GH construct from two mosaic founders grew, on average, 20% and 35% faster than their non-transgenic siblings. Since there is greater plasticity in the growth curve of fish compared to mammals or birds, this may be a viable approach to increase the growth rate of fish reared in commercial farms. Nevertheless, deleterious side-effects may only become apparent when these transgenics are evaluated under commercial aquaculture conditions.

4.2. Production of pharmaceutical proteins

Many mammalian biomedical proteins have post-translational modifications that are essential for their function. These modifications are not performed by microrganisms such as bacteria or yeast and so functional recombinant proteins can only be produced by expression in mammalian cells. In certain circumstances transgenic animals may provide an alternative means of production for recombinant proteins.

Directing expression to the liver and secretion of the product into blood is a possible route for the production of proteins from transgenic animals. Indeed, the high levels of GH present in the plasma of transgenic mice carrying MT–GH genes promoted Palmiter and his colleagues to suggest that transgenic animals could be used in this fashion (Palmiter *et al.*, 1982). However, many human biomedical proteins are themselves plasma proteins and it may be difficult to separate and purify them from the equivalent host proteins. Furthermore, high circulating levels of some foreign proteins in the blood will be detrimental to the health of the producer animal.

We have suggested that targeting expression to the mammary gland may overcome some of these problems (Lathe *et al.*, 1986; Clark *et al.*, 1987). Milk protein genes are expressed efficiently during pregnancy and lactation and their expression is restricted to the mammary gland. For example, the bovine αs_1-casein gene directs the synthesis of about 60 kg of protein per lactation. Even in sheep the levels of expression are high, e.g. up to 3 kg of αs_1-casein per sheep per lactation. If this capacity to express at high levels can be used to drive expression of biomedical proteins, then the process could prove a cost effective approach for the production of a number of proteins.

A number of milk protein genes have been shown to express in the mammary gland of transgenic mice, including the rat whey acidic protein (WAP) gene (Bayna and Rosen, 1990), rat β-casein (Lee *et al.*, 1989) and bovine α-lactalbumin (Vilotte *et al.*, 1990). Recently, R. J. Wall *et al.* (1991) have shown that the mouse WAP gene expresses efficiently in the mammary gland of transgenic pigs.

We have used the ovine β-lactoglobulin (BLG) gene to target expression to the mammary gland. BLG is the major whey protein in ruminant species and is expressed abundantly and specifically in the mammary gland during pregnancy and lactation. Despite the fact that an equivalent gene is not present in rodents, it is expressed efficiently and within the appropriate cell types in transgenic mice (Simons *et al.*, 1987; Harris *et al.*, 1990).

Sequences derived from BLG have been used to target the expression of cDNA sequences encoding two human biomedical proteins, factor IX (FIX) and α_1-antitrypsin (α_1AT), to the mammary gland. These are both plasma proteins and synthesized mainly in the liver. FIX is a serine protease, essential to the normal clotting of blood and is deficient in haemophilia B (Brownlee, 1987); α_1AT is a protease inhibitor that regulates the levels of neutrophil elastase in the lung and its deficiency predisposes to emphysema (Crystal, 1989). A variety of different construct designs have now been tested in transgenic mice. Most of these constructs have not performed particularly well and only relatively low levels of expression have been obtained (Whitelaw *et al.*, 1991).

In parallel, transgenic sheep carrying either BLG–FIX or BLG–α_1AT constructs have been produced (Simons *et al.*, 1988; Clark *et al.*, 1989). As for the transgenic mice, these constructs did not perform particularly well and only low levels of the human proteins were detected in milk. Nevertheless, the presence of biologically active FIX was demonstrated in the sheep's milk, demonstrating the feasibility of this route for producing active proteins.

In these studies the poorly expressing constructs all comprised cDNA segments encoding the protein of interest. We have also produced transgenic mice carrying a BLG fusion gene encoding α_1AT, but this time using a genomic α_1AT segment (Archibald *et al.*, 1990). In contrast to constructs derived from cDNA sequences, this construct was expressed efficiently in the transgenic animals (Fig. 4). α_1AT was present in milk at concentrations as high as 7 mg ml^{-1} (about 10% of the total milk proteins) and the protein appeared to be as biologically active as that derived from human plasma. Pharmaceutical Proteins Ltd, an Edinburgh-based company set up to exploit this technology, have introduced this construct into sheep. A number of transgenic females have been produced and results on expression in some of these animals should be available in the near future. [Note added in proof: sheep expressing human α_1AT in milk at levels as high as 35 mg ml^{-1} have been produced.]

Other workers have successfully directed the expression of foreign proteins into the milk of transgenic animals. Gordon *et al.* (1987) have described the expression of human tissue plasminogen activator in the milk of transgenic mice using regulatory elements derived from mouse WAP. Yu *et al.* (1989) have also utilized elements from the WAP gene to drive the expression of human CD4 protein in transgenic mice. In both these examples the levels of expression were relatively low. In contrast, Meade *et al.* (1990) have described the production of high levels (2 mg ml^{-1}) of human urokinase in the milk of transgenic mice using elements derived from the bovine αs_1-casein gene.

The use of transgenic rabbits for the production of human proteins has also been reported (Buhler *et al.*, 1990). Rabbits have the advantage over sheep or cows of having a much shorter generation time and considerably larger litter size. Each lactating female produces large amounts of milk proteins (~ 0.5 kg per animal per lactation) although the overall levels of protein production are still considerably below those attainable with cows (~ 240 kg per animal per lactation) or dairy sheep (~ 12 kg per animal per lactation). In the rabbit experiments, 5′ flanking sequences from rabbit β-casein were fused to the structural gene encoding human interleukin 2 but only low levels of expression were obtained.

Given that high a level of expression of some proteins has now been obtained in transgenic mice, it would seem reasonable to predict that similar levels will be attainable in producer animals such as sheep. The next phase of work will

CM   CM   TM   TM

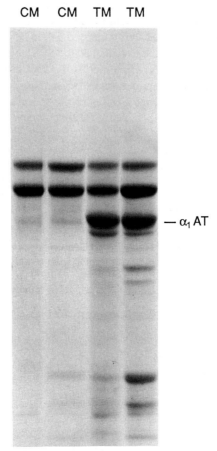

— α_1 AT

Fig. 4. Expression of human α_1-antitrypsin in the milk of transgenic mice. Whey proteins from control (CM) and transgenic (TM) mice were electrophoresed on an SDS–polyacrylamide gel and stained with Coomassie blue. In this line of transgenic mice, human α_1 AT can be seen to comprise a substantial fraction of the milk proteins. (Data from Archibald *et al.*(1990).)

be to demonstrate that active proteins can be purified from an animal source to the degree of purity required by the regulatory authorities to license for human administration.

4.3 Modification of milk composition

Since the expression of milk protein genes can be efficiently targeted to the mammary gland this raises the possibility of directly modifying milk composition for the dairy industry. The fact that about 30% of the protein in the diet of people in the West is derived from milk (Hambraeus, 1982) underlines the potential impact of this approach upon agriculture.

Most dairy processes such as cheese and butter-making utilize the caseins of cows' milk; the whey proteins, BLG and α-lactalbumin, are by-products which must be eliminated. Thus, increasing the proportion of caseins at the expense of whey could be expected to increase the value of milk. It is expected that this could be simply accomplished by introducing extra casein genes into cattle.

In the future, the development of gene-targeting techniques in domestic livestock will enable the specific deletion of whey protein genes. BLG is responsible for some of the observed allergies to cows' milk and so its elimination from milk may be advantageous. The whey protein α-lactalbumin is an essential co-factor in the synthesis of lactose, which is the major osmotic determinant of milk (Brew *et al.*, 1968). Deletion of α-lactalbumin genes would, therefore, be expected to have a profound effect on the physiology of milk production. Complete inhibition of lactose synthesis may disrupt milk production in the gland. However, partial inhibition might be expected to have some beneficial effects. Firstly, a partial reduction of milk volume, with no decrease in overall protein output, would yield a higher value milk on a per volume basis coupled with less frequent milking of the animals. Secondly, the inability to hydrolyse lactose is the cause of a widespread intolerance to milk products amongst adults; ultimately, milk with a reduced content of lactose may prove more accessible as a source of food to the ~90% of the world's population that exhibits this lactose intolerance (Mercier, 1986).

Caseins are assembled into super-macromolecular structures termed micelles which are in a colloidal suspension in milk. The three major caseins of bovine milk (αs_1, αs_2 and β) are phosphoproteins and their degree of phosphorylation determines may micelle properties, including their aggregation, stability to heat and the ability to sequester calcium. Expression of caseins with altered phosphorylation centres thus provides an approach for altering some of the basic physical and chemical properties of milk. κ-Casein is present at the surface of the micelles and also influences micelle size and stability. Changing the ratio of κ-casein to the other caseins may also provide an approach to engineering milks with altered properties.

Finally, bovine milk is not an ideal source of food for infants. In addition to the allergic reaction to β-lactoglobulin that it engenders, it contains only low levels of lactoferrin, a major constituent of human milk. As well as providing antimicrobial activity, lactoferrin may be important in the delivery of iron and other minerals to the infant by a mechanism that utilizes specific receptors in the gut. Expression of human lactoferrin in transgenic cows may produce milk better suited to human infants.

4.4. Wool production

In sheep, a major limitation to wool production is the supply of amino acids, particularly cysteine. A large amount of this amino acid is required to form the extensive disulphide bridges that link the keratins, which are the major structural proteins. Cysteine is an essential amino acid, but increasing its level in the diet is not effective because it is metabolized by the rumen bacteria. This limitation for wool production has prompted two Australian groups to attempt a gene transfer strategy to improve wool production (Ward *et al.*, 1989; Rogers, 1990).

In mammals, two additional enzymes would be required to convert the non-essential amino acid serine to cysteine, serine acetyltransferase (CysE) and *o*-acetylserine sulphydrolase (CysM). The genes encoding these enzymes have been cloned from *Escherichia coli* (Ward *et al.*, 1989) and *Salmonella typhimurium* (Rogers *et al.*, 1990), and eukaryotic promoters (MT and the Rous sarcoma virus RSV LTR) plus appropriate processing signals have been incorporated into constructs for gene transfer. The ultimate intention is to obtain expression in the mucosa of the rumen where sulphide liberated by fermentation can be utilized by acetylserine sulphydrolase to reduce *o*-acetylserine (the product of the CysE reaction) and produce cysteine. Double *cysE–cysM* constructs have been introduced into sheep and at least one transgenic animal generated (cited in Rogers, 1990) although there is, at present, no information regarding expression.

An alternative approach is to modify the protein composition of the hair fibre directly. There are four families of keratin proteins, and representative genes for each family have been cloned. The correct targeting of expression to the hair follicle has been achieved in transgenic mice using a gene encoding a sheep intermediate filament keratin protein (Powell and Rogers, 1990). In these experiments, however, overexpression of the transgene resulted in an imbalance between the filament and matrix hair proteins, disrupting the normal ordered array of these proteins and resulting in weakened fibres that were prematurely lost.

4.5. Disease resistance

An ability to produce organisms genetically tailored to resist specific diseases is, clearly, a most attractive proposition for the agricultural industry. With plants, for example, gene transfer has proved a very powerful tool for producing crops that are resistant to a variety of pathogenic organisms from viruses to insects (e.g. see Powell Abel *et al.*, 1986; Vaeck *et al.*, 1987).

Transgenic mice have been generated that respond to specific antigens by the introduction of a specific immunoglobulin (Ig) μ heavy chain (e.g. nitrophenol, Grosschedl *et al.*, 1984). This approach could be adapted, e.g. by transferring Ig genes encoding antibodies specific for viral coat proteins. Firstly, however, expression of foreign Ig genes can impair the development of the host immune system (e.g. see Herzenberg et al., 1987). Secondly, it would seem likely that pathogens such as a viruses would soon evolve and avoid the singular antigenic specificity encoded by a particular transgene.

The use of general antiviral agents, such as interferons, may be applicable. Thus, Chen et al. (1988) have introduced an MT–β-interferon construct into transgenic mice and shown that the resulting animals had an enhanced resistance to pseudorabies virus. The mouse $Mx1$ gene is known to influence resistance to influenza A and B and encodes an interferon-inducible protein. The dominant $Mx1^+$ allele has been introduced into mice and shown to confer resistance to influenza infection (Arnheiter et al., 1990). Transgenic pigs have been produced that express the $Mx1^+$ gene (Brenig et al., 1990), although these animals have not yet been appropriately challenged with influenza virus. The fact that the homologues of $Mx1^+$ exist in pigs and that, in mice, the $Mx1^-$ allele is limited to laboratory strains suggests that no great enhancement of resistance to influenza can be expected.

It may also be possible to genetically tailor resistance to bacterial pathogens such as Staphylococcus aureus, which is one of the main causes of the common mammary infection mastitis, a major problem in the dairy industry. Lysostaphin is an endopeptidase which hydrolyses the cell wall of many species of staphylococci and, when injected into the mammary glands, protects against challenge by S. aureus. The lysostaphin gene from S. simulans has been fused to regulatory elements from the sheep BLG gene and used to generate transgenic mice (C. Williamson and A. J. Bramley, Institute of Animal Health, UK, personal communication) which will be assessed for their resistance to mastitis.

In the chicken, Salter and Crittenden (1989) have described a transgenic line carrying a defective ALV genome that expresses only the envelope glycoproteins of the virus. This acts as a dominant gene for resistance since the viral glycoprotein competes with the receptor-binding sites for the virus on the cell surface. Similar approaches are being envisaged with fish because, although antibiotics and chemotherapeutics are available to control some fish diseases, there are no commercially available vaccines for fish viral diseases.

4.6. Nutrition

Lysine and threonine are essential amino acids in mammals. For non-ruminants such as pigs these amino acids limit growth, particularly when the animals are fed on plant material and, consequently, they have to be supplemented in the diet. In the future it may be possible to generate transgenic animals carrying the bacterial or fungal genes encoding the enzymes of the appropriate biosynthetic pathways (Rees *et al.*, 1990). This will be a major task since, in contrast to the biosynthesis of cysteine (see above), threonine is synthesized (from aspartic acid) by five enzymes and lysine (from 2-oxoglutarate) by seven enzymes. Nevertheless, the transfer of functional biosynthetic pathways from prokaryotic to eukaryotic cells has been demonstrated, e.g. Hartman and Mulligan (1988) have shown that *E. coli* histidine decarboxylase and tryptophan synthetase will function in animal cells to release them from a requirement for histidine and tryptophan.

Another approach being contemplated is to modify the digestive system. Monogastric animals such as pigs are unable to use cellulose as an energy source since they do not produce cellulases in the gut. Work is underway by H. Gilbert and colleagues at Newcastle University to target fungal cellulases to the gut of transgenic mice and so allow digestion of cellulose. As with the previous approach, the ultimate aim would be to produce animals better able to use poor-quality diets, particularly those comprising large amounts of plant material.

5.   Concluding remarks

The application of germ line manipulation techniques in agriculture and biotechnology is still at an early stage. Transgenic farm animals were first reported in 1985 (Hammer *et al.*, 1985a) and pronuclear injection remains the only proven route for gene transfer in these species. It is an inefficient process, although new developments in embryology can be expected to improve matters considerably. Gene transfer has also been successfully developed in poultry and fish, although progress lags behind that obtained in mammalian species.

In addition to the technical limitations, there are a number of other factors which presently restrict the exploitation of this technology. Firstly, the understanding of the physiological processes underlying conventional production traits is insufficient, and attempts to manipulate them can have unforeseen consequences. In this regard, generating animals with entirely new properties, such as producing pharmaceutical proteins, can be a distinct advantage as such animals do not have to compete with animals that have been improved by conventional genetic selection. Secondly, our knowledge of the mechanisms regulating gene expression in higher animals is limited and this restricts the precision with which transgenes can be controlled, again leading to unforeseen and, possibly, deleterious consequences. Finally, the general public will have to understand and accept this technology for it to be fully exploited, particularly if it involves genetically engineered animal material entering the human food chain.

References

Archibald, A. L., McClenaghan, M., Hornsey, V., Simons, J. P. and Clark, A. J. (1990). Efficient expression of biologically active human α_1-antitrypsin in the milk of transgenic animals *Proc. Natl Acad. Sci USA* **87**, 5178–5182.

Arnheiter, H., Skuntz, S., Noteborn, M., Chang, S. and Meier, E. (1990). Transgenic mice with intracellular immunity to influenza virus. *Cell* **62**, 51–61.

Bayna, E. M. and Rosen, J. M. (1990). Tissue-specific high level expression of the rat whey acidic protein gene in transgenic mice. *Nucl. Acids Res.* **18**, 2977–2985.

Biery, K. A., Bondioli, K. R. and DeMayo, F. J. (1988). Gene transfer by pronuclear injection. *Theriogenology* **29**, 224 (Abstr).

Bosselman, R. A., Hsu, R.-Y., Boggs, T., Hu, S., Bruszewski, J., Ou, S., Souza, L., Kozar, L., Martin, F., Nicolson, M., Rishell, W., Schultz, J., Semon, K. M. and Stewart, R. G. (1989a). Vectors of reticuloendotheliosis virus transduce exogenous genes into somatic stem cells of the unincubated chicken embryo. *J. Virol.* **63**, 2680–2689.

Bosselman, R. A., Hsu, R.-Y., Boggs, T., Hu, S., Bruszewski, J., Ou, S., Kozar, L., Martin, F., Green C., Jacobson, F., Nicolson, M., Schultz, J., Semon, K. M. and Stewart, R. G. (1989b). Germline transmission of exogenous genes in the chicken. *Science* **243**, 533–535.

Brem, G., Brenig, B., Goodman, H. M., Selden, R. C., Graf, F., Kruff, B., Springman, K., Hondele, J., Meyer, J. and Krausslich, H. (1985). Production of transgenic mice rabbits and pigs by microinjection. *Zucthygiene* **20**, 251–252.

Brem, G., Brenig, B., Muller, M., Krausslich, H. and Winnacker, E.-L. (1988). Production of transgenic pigs and possible applications to pig breeding. *Occ. Publ. Br. Soc. Anim. Prod.* **12**, 15–31.

Brenig, B., Muller, M., and Brem, G. (1990). Gene transfer in pigs. *Proc. 4th W. Cong. Genet. Appl. Livest. Prod.* **XIII**, 41–48.

Brew, K., Vanoman, T. C. and Hill, R. L. (1968). The role of α-lactalbumin and the A protein in lactose synthetase. A unique mechanism for the control of a biological reaction. *Proc. Natl Acad. Sci. USA.* **59**, 491–496.

Brinster, R. L., Chen, H. Y., Trumbauer, M. E., Yagle, M. K. and Palmiter, R. D. (1985). Factors affecting the introduction of foreign DNA into transgenic mice by microinjecting eggs. *Proc. Natl Acad. Sci. USA* **82**, 4438–4442.

Brownlee, G. G. (1987). The molecular pathology of haemophillia B. *Biochem. Soc. Trans.* **15**, 1–8.

Buhler, T. A., Bruyere, T., Went, D. F., Stranzunger, G. and Burki, K. (1990). Rabbit β-casein promoter directs secretion of human interleukin-2 into the milk of transgenic rabbits. *Bio/Technology* **8**, 140–143.

Chen, T. T. and Powers, D. A. (1990). Transgenic fish. *Trends Biotech.* **8**, 209–215.

Chen, X.-Z., Yun, Y. S. and Wagner, T. E. (1988). Enhanced viral resistance in transgenic mice expressing the human beta-1 interferon. *J. Virol.* **62**, 3883–3887.

Church, R. B., McRae, A. and McWhir, J. (1986). Embryo manipulation and gene transfer in livestock production. *Proc. 3rd W. Cong. Genet. Appl. Livest. Prod.* **XII**, 133–138.

Clark, A. J., Simons, J. P., Wilmut, I. and Lathe, R. (1987). Pharmaceuticals from transgenic livestock. *Trends Biotech.* **5**, 20–24.

Clark, A. J., Ali, S., Archibald, A. L., Brown, P., Bulfield, G., Harris, S., McClenaghan, M., Perry, M., Sang, H., Simons, J. P., Whitelaw, C. B. A. and Wilmut, I. (1988). Transgenic Biology. AFRC-IAPGR, Edinburgh Research Station Report for 1986.1987, pp. 3–9. AFRC Edinburgh.

Clark, A. J., Bessos, H., Bishop, J. O., Brown, P., Harris, S., Lathe, R., McClenaghan, M., Prowse, C., Simons, J. P., Whitelaw, C. B. A. and Wilmut, I. (1989). Expression of human anti-hemophilic factor IX in the milk of transgenic sheep. *Bio/Technology* **7**, 487–492.

Crystal, R. G. (1989). The α1-antitrypsin gene and its deficiency states. *Trends Genet.* **5**, 411–417.

Davies, P. L., Fletcher, G. L. and Hew, C. L. (1989). Fish antifreeze protein genes and their use in transgenic studies. *Oxford Surveys on Eukaryotic Genes* **6**, 85–109.

Ebert, K. M., Low, M. J., Overstom, E. W., Buonomo, F. C., Baile, C. A., Roberts, T. M., Lee, A., Mandel, G. and Goodman, R. H. (1988). A Moloney MLV–rat somatotropin fusion gene produces biologically active somatotropin in a transgenic pig. *Mol. Endocrinol.* **2**, 277–283.

Fukui, Y., Glew, A. M., Gandolfi, F. and Moor, R. M. (1988). *In vitro* culture of sheep oocytes matured and fertilised *in vitro*. *Theriogenology* **29**, 883–891.

Gordon, K., Lee, E., Vitale, J. A., Smith, A. E., Westphal, H. and Hennighausen, L. (1987). Production of human tissue plasminogen activator in transgenic mouse milk. *Bio/Technology* **5**, 1183–1187.

Grosschedl, R., Weaver, D., Baltimore, D. and Constantini, F. (1984). Introduction of a *v* immunoglobulin gene into the mouse germline: specific expression in lymphoid cells and synthesis of functional antibody. *Cell* **38**, 647–658.

Guyomard, R., Chourrout, D., Leroux, C., Houdebine, L. M. and Pourrain, F. (1989). Integration and germline transmission of foreign genes microinjected into fertilised trout eggs. *Biochimie* **71**, 857–863.

Hambraeus, L. (1982). Nutritional aspects of milk proteins. In: Fox, P. F. (ed.) *Developments in Dairy Chemistry*, vol. I, pp. 289–313. Applied Science, London.

Hammer, R. E., Pursel, V. G., Rexroad, C. E., Wall, R. J., Bolt, D. J., Ebert, K. M., Palmiter, R. D. and Brinster, R. L. (1985a). Production of transgenic rabbits, sheep and pigs by microinjection. *Nature* **315**, 680–683.

Hammer, R. E., Brinster, R. L., Rosenfeld, M. G. and Mayo, K. E. (1985b). Expression of human growth hormone releasing factor in transgenic mice results in increased somatic growth. *Nature* **315**, 413–416.

Handyside, A., Hooper, M. L., Kaufman, M. H. and Wilmut, I. (1987). Towards the isolation of embryonal stem cells in sheep. *Roux's Arch. Dev. Biol.* **196**, 185–190.

Harris, S., McClenaghan, M., Simons, J. P., Ali, S. and Clark, A. J. (1990). Gene expression in the mammary gland. *J. Reprod. Fert.* **88**, 707–715.

Hartman, S. C. and Mulligan, R. C. (1988). Two dominant acting selectable markers for gene transfer studies in mammalian cells. *Proc. Natl Acad. Sci. USA* **85**, 8047–8051.

Harvey, M. J. A., Hettle, S. J. H., Cameron, E. R., Johnston, C. S. and Onions, D. E. (1990). In: Church, R. B. (ed.) *Transgenic Models in Medicine and Agriculture*, pp. 11–19. Wiley, Liss.

Herzenberg, L. A., Stall, A. M., Braun, J., Weaver, D., Baltimore, D., Herzenberg, L. A. and Grosschedl, R. (1987). Depletion of the predominant B-cell population in immunoglobulin μ heavy chain transgenic mice. *Nature* **329**, 71-73.

Jaenisch, R. (1988). Transgenic animals. *Science* **240**, 1468-1474.

Kraft, L. A., Baker, P. K., Ricks, C. A., Lance, V. A., Murphy, W. A. and Coy, D. H. (1985). Stimulation of growth hormone release in anaesthetised and conscious pigs by synthetic human pancreatic growth hormone-releasing factor. *Dom. Anim. End.* **2**, 133-137.

Lathe, R., Clark, A. J., Archibald, A. L., Bishop, J. O., Simons, J. P. and Wilmut, I. (1986). Novel products from livestock. In: Smith, C. King, J. W. B. and McKay, J. C. (eds) *Exploiting New Technologies in Animal Breeding*, pp. 91-102. Oxford University Press, Oxford.

Lee, K. F., Atiee, S. H. and Rosen, J. M. (1989). Tissue-specific regulation of the rat β-casein gene in transgenic mice. *Nucl. Acids Res.* **16**, 1027-1041.

Lu, K. H., Gordon, I., Gallagher, M. and McGovern, H. (1987). Pregnancy established in cattle by transfer of embryos derived from *in-vitro* fertilisation of oocytes matured *in-vitro*. *Vet. Rec.* **121**, 259-260.

Lu, K. H., Gordon, I., McGovern, H. and Gallagher, M. (1988). Production of cattle embryos by *in vitro* maturation and fertilisation of follicular oocytes and their subsequent culture in sheep. *Theriogenology* **29**, 272.

McGrane, M. M., deVente, J., Yun, J. Bloom, J., Park, E., Wynshaw-Boris, A., Wagner, T. and Hanson, R. W. (1988). Tissue-specific expression and dietary regulation of a chimeric phosphoenolpyruvate carboxykinase/bovine growth hormone gene in transgenic mice. *J. Biol. Chem.* **263**, 11443-11451.

Marx, J. L. (1988). Cloning sheep and cattle embryos. *Science* **239**, 463-464.

Meade, H., Gales, L., Lacy, E. and Lonberg, N. (1990). Bovine αs_1 casein gene sequences direct high level expression of human urokinase in mouse milk. *Bio/Technology* **8**, 443-446.

Mercier, J.-C. (1986). Genetic engineering: some expectations. In: Smith, C., King, J. W. B. and McKay, J. (eds) *Exploiting New Technologies in Animal Breeding*, pp. 122-131. Oxford University Press, Oxford.

Miller, K. F., Bolt, D. J., Pursel, V. G., Hammer, R. E., Pinkert, C. A., Palmiter, R. D. and Brinster, R. L. (1989). Expression of human or bovine growth hormone gene with a mouse metallothionein-1 promoter in transgenic swine alters the secretion of porcine growth hormone and insulin-like growth factor-I. *J. Endocrinol.* **120**, 481-488.

Murray, J. D., Nancarrow, C. D., Marshall, J. T., Hazelton, I. G. and Ward, K. A. (1989). Production of transgenic merino sheep by microinjection of ovine metallothionein–ovine growth hormone fusion genes. *Rep. Fert. Dev.* **1**, 147-155.

Notarianni, E., Galli, C., Laurie, S., Moor, R. M. and Evans, M. J. (1990). Derivation of pluripotential, embryonic cell lines from porcine and ovine blastocysts. *Proc. 4th W. Cong. Genet. Appl. Livest. Prod.* **XIII**, 58-64.

Palmiter, R. D. and Brinster, R. L. (1986). Germline transformation of mice. *An. Rev. Genet.* **20**, 465-499.

Palmiter, R. D., Brinster, R. L., Hammer, R. E., Trumbauer, M. E., Rosenfeld, M. G., Birnberg, N. C. and Evans, R. M. (1982). Dramatic growth of mice that develop from eggs microinjected with metallothionein–growth hormone fusion genes. *Nature* **300**, 611–615.

Palmiter, R. D., Norstedt, G., Gelinas, R. E., Hammer, R. E. and Brinster, R. L. (1983). Metallothionein–human GH fusion genes stimulate growth of mice. *Science* **222**, 809–814.

Petters, R. M., Johnson, B. H. and Shuman, R. M. (1988). Gene transfer to swine embryos using an avian retrovirus. *Genome* **30**(Suppl. 1), 448.

Perry, M. M. (1988). A complete culture system for the chick embryo. *Nature* **331**, 70–72.

Perry, M. M. and Sang, H. M. (1990). *In vitro* culture and approaches for DNA transfer in the chick embryo. *Proc. 4th W. Cong. Genet. Appl. Livest. Prod.* **XVI**, 115–118.

Petitte, J. N., Clark, M. E., Liu, G., Verrinder- Gibbons, A. M. and Etches, R. J. (1990). Production of somatic and germline chimaeras in the chicken by transfer of early blastodermal cells. *Development* **108**, 185–189.

Pinkert, C. A., Dyer, T. J., Kooyman, D. L. and Kiehm, D. J. (1990). Characterisation of transgenic livestock production. *Dom. Anim. Endocrinol.* **7**, 1–18.

Polge, E. J. C., Barton, S. C., Surani, M. A. H., Miller, J. R., Wagner, T., Rottman, F., Camper, S. A., Elsome, K., Goode, J. A., Foxcroft, G. R. and Heap, R. B. (1989). Induced expression of a bovine growth hormone construct in transgenic pigs. In: Heap, R. B., Prosser, C. G. and Lamming, G. E. (eds) *Biotechnology in Growth Regulation*, pp. 189–199. Butterworths, London.

Powell, B. C. and Rogers, G. E. (1990). Cyclic hair loss and regrowth in transgenic mice over expressing an intermediate filament gene. *EMBO J.* **9**, 1485–1494.

Powell-Abel, P., Nelson, R. S., De. B., Hoffman, N., Rogers, S. G., Fraley, B. T. and Beachy, R. N. (1986). Delay of disease in transgenic plants that express the tobacco mosaic virus coat protein gene. *Science* **232**, 738–742.

Pursel, V. G., Rexroad, C. E., Bolt, D. J., Miller, K. F., Wall, R. J., Hammer, R. E., Pinkert, C. A., Palmiter, R. D. and Brinster, R. L. (1987). Progress on gene transfer in farm animals. *Vet. Imm. Immunopath.* **17**, 303–312.

Pursel, V. G., Pinkert, C. A., Miller, K. F., Bolt, D. J., Campbell, R. G., Palmiter, R. D., Brinster, R. L. and Hammer, R. E. (1989). Genetic engineering of livestock. *Science* **244**, 1281–1288.

Rees, W. D., Flint, H. J. and Fuller, M. F. (1990). A molecular biological approach to reducing dietary amino acid needs. *Bio/Technology* **8**, 629–633.

Rexroad, C. E. and Pursel, V. G. (1988). Status of gene transfer in domestic livestock. *Proc. 11th Int. Cong. Anim. Reprod. (Dublin)* **5**, 28–35.

Rexroad, C. E., Behringer, R. R., Bolt, D. J., Miller, K. F., Palmiter, R. D. and Brinster, R. L. (1988). Insertion and expression of a growth hormone fusion gene in sheep. *J. Anim. Sci.* **66**(Suppl. 1), 267.

Rogers, G. E. (1990). Improvement of wool production through genetic engineering techniques. *Trends Biotech.* **8**, 6–11.

Roschlau, K., Rommel, P., Anrewa, L., Zackel, M., Roschlau, D., Zackel, B., Schwerin, M., Huhn, R. and Gazarjan, K. G. (1989). Gene transfer in cattle, *J. Reprod. Fert.* Suppl. 38, 153–160.

Salter, D. W. and Crittenden, L. B. (1989). Transgenic chickens: insertion of retroviral vectors into the chicken germline. *Theor. Appl. Genet.* **77**, 457–461.

Salter, D. W., Smith, E. J., Hughes, S. H., Wright, S. E., Fadly, A. M., Witter, R. L. and Crittenden, L. B. (1986). Gene insertion onto the chicken germ line by retroviruses. *Poultry Sci.* **65**, 1445–1458.

Salter, D. W., Smith, E. J., Hughes, S. H., Wright, S. E. and Crittenden, L. B. (1987). Transgenic chickens: insertion of retroviral vectors into the chicken germ line. *Virology.* **157**, 235–240.

Sang, H. and Perry, M. M. (1989). Episomal replication of cloned DNA injected into the fertilised ovum of the hen, *Gallus domesticus*. *Mol. Reprod. Dev.* **1**, 98–106.

Simons, J. P., McClenaghan, M. and Clark, A. J. (1987). Alteration of the quality of milk by expression of sheep β-lactoglobulin in transgenic mice. *Nature* **328**, 530–532.

Simons, J. P., Wilmut, I., Clark, A. J., Archibald, A. L., Bishop, J. O. and Lathe, R. (1988). Gene transfer into sheep. *Bio/Technology* **6**, 179–183.

Smith, C., Meuwissen, T. H. E. and Gibson, J. P. (1987). On the use of transgenes in livestock improvement. *Anim. Breed. Abst.* **55**, 1–10.

Smith, L. C. and Wilmut, I. (1989). Influence of nuclear and cytoplasmic activity in the development *in vivo* of sheep embryos after nuclear transplantation. *Biol. Reprod.* **40**, 1027–1035.

Stuart, G. W., McMurray, J. V. and Westerfield, M. (1988). Replication, integration and stable germ-line transmission of foreign DNA sequences injected into early zebrafish embryos. *Development* **103**, 403–412.

Vaeck, M., Reynaerts, A., Hofte, H., Jansens S., De Beukeleer, M., Dean, C., Zabeau, M., van Montagu, M. and Leemans, J. (1987). *Nature* **328**, 33–37.

Verrinder Gibbins, A. M., Brazolot, C. L., Petitte, J. N., Liu, G. and Etches, R. J. (1990). Efficient transfection of chicken blastodermal cells and their incorporation into recipient embryos to produce chimeric chicks. *Proc. 4th W. Cong. Genet. Appl. Livest. Prod.* **XVI**, 119–122.

Vilotte, J.-L., Soulier, S., Stinnakre, M.-G., Massoud, M. and Mercier, J.-C. (1989). Efficient tissue-specific expression of bovine α-lactalbumin in transgenic mice. *Eur. J. Biochem.* **186**, 43–48.

Vize, P. D., Michalska, A. E., Ashman, R., Lloyd, B., Stone, B. A., Quinn, P., Wells, J. R. E. and Seamark, R. F. (1988). Introduction of a porcine growth hormone fusion gene into transgenic pigs promotes growth. *J. Cell Sci.* **90**, 295–300.

Wall, R. J., Pursel, V. G., Shamay, A., McKnight, R., Pittius, C. W. and Hennighausen, L. (1991). High level synthesis of a heterologous milk protein in the mammary gland of transgenic swine. *Proc. Natl Acad. Sci. USA* **88**, 1696–1700.

Ward, K. A., Murray, J. D. and Nancarrow, C. D. (1989). The insertion of foreign genes into animal cells. *Biotechnology for Livestock Production, FAO Animal Production and Health Division*, pp. 17–28. Plenum Press, New York.

Whitelaw, C. B. A., Archibald, A. L., Harris, S., McClenaghan, M., Simons, J. P. and Clark, A. J. (1991). Targeting expression to the mammary gland: intronic sequences can enhance the efficiency of gene expression in transgenic mice. *Trans. Res.* **1**, 3–13.

Wieghart, M., Hoover, J., Choe, S. H., McGrane, M. M., Rottman, F. M., Hanson, R. W. and Wagner, T. E. (1988). Genetic engineering of livestock – transgenic pigs containing a chimeric bovine (PEPCK/bGH) gene. *J. Anim. Sci.* **66**(Suppl. 1), 266.

Yu, S.-H., Deen, K. C., Lee, E., Hennighausen, L., Sweet, R. W., Rosenberg, M. and Westphal, H. (1989). Functional human CD4 protein produced in milk of transgenic mice. *Mol. Biol. Med.* **6**, 255–261.

Zhang, P., Hayat, M., Joyce, C., Gonzalez-Villasenor, L. I., Lin, C. M., Dunham, R. A., Chen, T. T. and Powers, D. A. (1990). Gene transfer, expression and inheritance. A pRSV– rainbow trout–GH cDNA in common carp, *Cyprinus carpio. Mol. Reprod. Dev.* **25**, 13–25.

Index

MMTV-*ras* transgenic animals 196
Modifier genes 103, 109–14
Moloney murine leukaemia virus (mo-MuLV) LTR 203–4
Mo-MuLV 218, 219, 221
Monoclonal antibodies 184
Monoclonal lymphoid populations 151–2
Mosaic analysis 57, 64, 69
Mosaic inactivation 118
Mosaic phenotypes 118
Mouse β_2-microglobulin 42
Mouse embryology and development 47–77
mRNA 178, 185, 202
Multiple phenotypes, gene expression 117–8
Multiple sclerosis 163
Murine leukaemia viruses (MuLV) 213–5, 221
Mutant cells 68
Mutant development 65, 69
Mycophenolic acid, aminopterin and xanthine (MAX) medium 31
Myelin basic protein (MBP) 141–2, 177

Nervous system, genetic ablation 141–2
Neural tube, regional specification 172–5
Neurobiology 168–94
  transgenic mice versus rats 170
NOD (non-obese diabetic) mouse 163
  model system 153
Non-homologous recombination (NHR)
  altering or reducing efficiencies 37–8
  definition 28
Normal development 65, 66
Nuclear transgenic markers 51
Null mutants 63, 65–6
Nutrition 263–4

O–2A progenitors 176
O-type vector 29–30
Ω-type vector 29–30, 35
Ommatidium 63
Oncogenes

breeding and maintaining mice carrying 204–5
identifying cooperating 202–4
Oncogenesis, generation of *in vitro* cell lines 197–202
Oncology, transgenic mice in 195–210
Optic nerve, growth control 175–6

P glycoprotein 225
Packaging cell lines 220–1
Pancreas, genetic ablation 137–8
Parental chromosomes 101, 102
Paternal chromosomes 101–3
Peri-implantation period 61
Peripheral tolerance 160
Pharmaceutical proteins 257–60
Phenotypes 60–3, 103
Phenotypic units 118
Phenylethenolamine *N*-methyltransferase (PNMT) gene 184
Phosphoenolypyruvate carboxykinase (PEPCK) 257
Photoreceptor differentiation 63
Pituitary, genetic ablation 138–9
Platelet-derived growth factor (PDGF) 176, 179
Pluripotent HSCs (PHSCs) 234–6
Polyethylene glycol (PEG) 229
Polymerase chain reaction (PCR) 185
Polymorphic genes 94
Polytropic viruses 221
Position effect 17
Positive–negative selection (PNS) 38–9
Postimplantation development 62
Postimplantation lethals 62
Poultry 257
  germ line manipulation 252–3
Prokaryotic sequences 82
Prolactin (PRL)-producing lactotropes 138–9
Pronuclear microinjection 249
Proteolipid protein (PLP) 141
Purkinje cells 186–8
Purkinje neurons 188